PROCESS SYNTHESIS FOR FUEL ETHANOL PRODUCTION

BIOTECHNOLOGY AND BIOPROCESSING SERIES

Series Editor
Anurag Rathore

1. Membrane Separations in Biotechnology, *edited by W. Courtney McGregor*
2. Commercial Production of Monoclonal Antibodies: A Guide for Scale-Up, *edited by Sally S. Seaver*
3. Handbook on Anaerobic Fermentations, *edited by Larry E. Erickson and Daniel Yee-Chak Fung*
4. Fermentation Process Development of Industrial Organisms, *edited by Justin O. Neway*
5. Yeast: Biotechnology and Biocatalysis, *edited by Hubert Verachtert and René De Mot*
6. Sensors in Bioprocess Control, *edited by John V. Twork and Alexander M. Yacynych*
7. Fundamentals of Protein Biotechnology, *edited by Stanley Stein*
8. Yeast Strain Selection, *edited by Chandra J. Panchal*
9. Separation Processes in Biotechnology, *edited by Juan A. Asenjo*
10. Large-Scale Mammalian Cell Culture Technology, *edited by Anthony S. Lubiniecki*
11. Extractive Bioconversions, *edited by Bo Mattiasson and Olle Holst*
12. Purification and Analysis of Recombinant Proteins, *edited by Ramnath Seetharam and Satish K. Sharma*
13. Drug Biotechnology Regulation: Scientific Basis and Practices, *edited by Yuan-yuan H. Chiu and John L. Gueriguian*
14. Protein Immobilization: Fundamentals and Applications, *edited by Richard F. Taylor*
15. Biosensor Principles and Applications, *edited by Loï''efc J. Blum and Pierre R. Coulet*
16. Industrial Application of Immobilized Biocatalysts, *edited by Atsuo Tanaka, Tetsuya Tosa, and Takeshi Kobayashi*
17. Insect Cell Culture Engineering, *edited by Mattheus F. A. Goosen, Andrew J. Daugulis, and Peter Faulkner*
18. Protein Purification Process Engineering, *edited by Roger G. Harrison*
19. Recombinant Microbes for Industrial and Agricultural Applications, *edited by Yoshikatsu Murooka and Tadayuki Imanaka*
20. Cell Adhesion: Fundamentals and Biotechnological Applications, *edited by Martin A. Hjortso and Joseph W. Roos*
21. Bioreactor System Design, *edited by Juan A. Asenjo and José C. Merchuk*
22. Gene Expression in Recombinant Microorganisms, *edited by Alan Smith*
23. Interfacial Phenomena and Bioproducts, *edited by John L. Brash and Peter W. Wojciechowski*

24. Metabolic Engineering, *edited by Sang Yup Lee and Eleftherios T. Papoutsakis*

25. Biopharmaceutical Process Validation, *edited by Gail Sofer and Dane W. Zabriskie*

26. Membrane Separations in Biotechnology: Second Edition, Revised and Expanded, *edited by William K. Wang*

27. Isolation and Purification of Proteins, *edited by Rajni Hatti-Kaul and Bo Mattiasson*

28. Biotransformation and Bioprocesses, *Mukesh Doble, Anil Kumar Kruthiventi, and Vilas Gajanan Gaikar*

29. Process Validation in Manufacturing of Biopharmaceuticals: Guidelines, Current Practices, and Industrial Case Studies, *edited by Anurag Singh Rathore and Gail Sofer*

30. Cell Culture Technology for Pharmaceutical and Cell-Based Therapies, *edited by Sadettin S. Ozturk and Wei-Shou Hu*

31. Process Scale Bioseparations for the Biopharmaceutical Industry, *edited by Abhinav A. Shukla, Mark R. Etzel, and Shishir Gadam*

32. Processs Synthesis for Fuel Ethanol Production, *C. A. Cardona, Ó. J. Sánchez, and L. F. Gutiérrez*

PROCESS SYNTHESIS FOR FUEL ETHANOL PRODUCTION

C. A. Cardona
Universidad Nacional de Colombia
Manizales, Colombia

Ó. J. Sánchez
Universidad de Caldas
Manizales, Colombia

L. F. Gutiérrez
Universidad de Caldas
Manizales, Colombia

CRC Press
Taylor & Francis Group
Boca Raton   London   New York

CRC Press is an imprint of the
Taylor & Francis Group, an **informa** business

CRC Press
Taylor & Francis Group
6000 Broken Sound Parkway NW, Suite 300
Boca Raton, FL 33487-2742

First issued in paperback 2020

© 2010 by Taylor and Francis Group, LLC
CRC Press is an imprint of Taylor & Francis Group, an Informa business

No claim to original U.S. Government works

ISBN-13: 978-0-367-57720-9 (pbk)
ISBN-13: 978-1-4398-1597-7 (hbk)

Library of Congress Cataloging-in-Publication Data

Cardona, C. A.
    Process synthesis for fuel ethanol production / C.A. Cardona, Ó.J. Sánchez, L.F. Gutiérrez.
       p. cm. -- (Biotechnology and bioprocessing series ; 32)
    Includes bibliographical references and index.
    ISBN 978-1-4398-1597-7 (hardcover : alk. paper)
    1. Ethanol as fuel. 2. Corn--Biotechnology. 3. Sugarcane--Biogtechnology. 4. Biomass energy. I. Sánchez, Ó. J. II. Gutiérrez, L. F. III. Title. IV. Series.

TP339.C37 2010
662'.6692--dc22   2009035959

Visit the Taylor & Francis Web site at
http://www.taylorandfrancis.com

and the CRC Press Web site at
http://www.crcpress.com

Contents

Preface.. xv
Acknowledgments...xvii
The Authors xix

Chapter 1   Biofuels.. 1

    1.1   Biofuels Generalities 1
           1.1.1   Solid and Gaseous Biofuels.................................... 3
           1.1.2   Liquid Biofuels... 4
                     1.1.2.1   Biodiesel 4
                     1.1.2.2   Bioethanol...................................... 5
    1.2   Gasoline Oxygenation 5
           1.2.1   Tetraethyl Lead as Antiknocking Additive 8
           1.2.2   Ethers as Gasoline Oxygenates 8
                     1.2.2.1   Methyl Tert-Butyl Ether (MTBE)............ 9
                     1.2.2.2   Ethyl Tert-Butyl Ether (ETBE)............. 10
                     1.2.2.3   Tert-Amyl Methyl Ether (TAME)
                                 and Tert-Amyl Ethyl Ether (TAEE)....... 11
                     1.2.2.4   Di-Isopropyl Ether (DIPE).................... 11
            1.2.3   Methanol.. 12
    1.3   Ethanol as a Gasoline Oxygenate....................................... 12
           1.3.1   Advantages of Fuel Ethanol 13
           1.3.2   Drawbacks of Fuel Ethanol 14
    1.4   Gasoline Oxygenation Programs with Fuel Ethanol in
          Some Countries 17
    References 22

Chapter 2   Process Design and Role of Process Synthesis 27

    2.1   Conceptual Process Design.. 27
    2.2   Knowledge-Based Process Synthesis 30
           2.2.1   Evolutionary Modification.................................... 31
           2.2.2   Hierarchical Decomposition.................................. 32
           2.2.3   Phenomena-Driven Design.................................... 33
           2.2.4   Conflict-Based Approach 33
           2.2.5   Thermodynamics-Based Process Synthesis.......... 33
           2.2.6   Analysis of the Statics 34
    2.3   Optimization-Based Process Synthesis............................. 35
           2.3.1   Mathematical Aspects... 36
           2.3.2   Superstructures.. 38

	2.3.3	Hybrid Methods	39
2.4	Final Considerations		40
References			40

Chapter 3 Feedstocks for Fuel Ethanol Production 43

3.1	Sugars		43
	3.1.1	Sugarcane	43
	3.1.2	Cane Sugar	45
	3.1.3	Sugar Beet	48
	3.1.4	Beet Sugar	48
	3.1.5	Sucrose-Containing Materials Used for Ethanol Production	50
		3.1.5.1 Cane and Beet Juices	51
		3.1.5.2 Sugarcane Molasses	51
		3.1.5.3 Beet Molasses	52
		3.1.5.4 Other Sugar-Containing Materials	52
3.2	Starchy Materials		53
	3.2.1	Starch	53
	3.2.2	Starch Sources for Ethanol Production	55
		3.2.2.1 Corn	56
		3.2.2.2 Other Grains as Starch Sources	59
		3.2.2.3 Cassava	60
3.3	Lignocellulosic Materials		61
	3.3.1	Structure of Lignocellulosic Complex	62
	3.3.2	Classification of Lignocellulosic Materials	63
		3.3.2.1 Sugarcane Bagasse	68
		3.3.2.2 Corn Stover	69
		3.3.2.3 Cereal Straws	70
		3.3.2.4 Municipal Solid Waste	70
References			71

Chapter 4 Feedstock Conditioning and Pretreatment 77

4.1	Conditioning of Sucrose-Containing Materials	77
4.2	Pretreatment of Starchy Materials	80
4.3	Pretreatment of Lignocellulosic Biomass	83
	4.3.1	Physical Methods of Pretreatment	84
	4.3.2	Physical–Chemical Methods of Pretreatment	85
	4.3.3	Chemical Methods of Pretreatment	90
	4.3.4	Biological Methods of Pretreatment	95
	4.3.5	Role of Pretreatment during Process Synthesis	95
4.4	Detoxification of Pretreated Biomass	98
	4.4.1	Physical Methods of Detoxification	100

4.4.2   Chemical Methods of Detoxification 100
4.4.3   Biological Methods of Detoxification 105
References 108

Chapter 5   Hydrolysis of Carbohydrate Polymers..115

5.1   Starch Saccharification...115
5.2   Hydrolysis of Cellulose...117
    5.2.1   Enzyme Systems for Cellulose Hydrolysis...........118
    5.2.2   Conversion of Cellulose to Glucose 122
    5.2.3   Cellulose Hydrolysis Assessment for Process
        Synthesis... 123
        5.2.3.1   Efficiency of Cellulases 123
        5.2.3.2   Modeling of Cellulose Hydrolysis 124
References 126

Chapter 6   Microorganisms for Ethanol Production......................................131

6.1   Metabolic Features of Ethanol Producing
    Microorganisms...131
6.2   Nongenetically Modified Microorganisms for Ethanol
    Production.. 134
    6.2.1   Yeasts.. 136
    6.2.2   Bacteria.. 138
6.3   Genetically Modified Microorganisms for Ethanol
    Production.. 140
    6.3.1   Mutagenesis.. 140
    6.3.2   Recombinant DNA Technology 141
        6.3.2.1   Recombinant Microorganisms for
            Starch Processing.................................143
        6.3.2.2   Recombinant Microorganisms
            for Processing of Lignocellulosic
            Biomass.. 148
References 151

Chapter 7   Ethanolic Fermentation Technologies.................................... 155

7.1   Description of Main Fermentation Technologies for
    Ethanol Production 155
    7.1.1   Features of Ethanolic Fermentation Using
        Saccharomyces cerevisiae................................ 155
    7.1.2   Fermentation of Sucrose-Based Media 156
        7.1.2.1   Batch Fermentation............................. 157
        7.1.2.2   Semicontinuous Fermentation 158
        7.1.2.3   Continuous Fermentation.....................161

7.1.2.4   Fermentation of Sugar Solutions
Using Immobilized Cells 163
7.1.3   Fermentation of Media Based on Starchy
Materials... 166
7.1.3.1   Conversion of Saccharified Corn
Starch into Ethanol 166
7.1.3.2   Very High Gravity Fermentation 167
7.1.4   Fermentation of Media Based on
Lignocellulosic Biomass.................................... 168
7.1.4.1   Fermentation of Cellulose
Hydrolyzates 168
7.1.4.2   Pentose Fermentation...........................170
7.1.4.3   Co-Fermentation of Lignocellulosic
Hydrolyzates 171
7.2   Modeling of Ethanolic Fermentation for Process
Design Purposes 173
7.2.1   Modeling of Ethanolic Fermentation from
Sugars 176
7.2.2   Modeling of Co-Fermentation of Hexoses and
Pentoses 179
7.3   Analysis of Fed-Batch Ethanolic Fermentation 181
7.4   Dynamics of Continuous Fermentation Systems 182
References 191

Chapter 8   Analysis of Ethanol Recovery and Dehydration 199

8.1   Concentration and Rectification of Ethanol Contained
in Culture Broths 199
8.2   Ethanol Dehydration.. 201
8.2.1   Pressure-Swing Distillation 202
8.2.2   Azeotropic Distillation 202
8.2.3   Extractive Distillation 206
8.2.4   Saline Extractive Distillation 209
8.2.5   Adsorption...210
8.2.6   Pervaporation..211
8.3   Evaluation of Separation and Dehydration Schemes........ 213
References 217

Chapter 9   Integrated Processes for Fuel Ethanol Production 221

9.1   Process Integration 221
9.2   Reaction–Reaction Integration for Bioethanol
Production...224
9.2.1   Process Integration by Co-Fermentation............. 224

9.2.2   Process Integration by SSF 225
         9.2.2.1   SSF of Starch 225
         9.2.2.2   SSYPF of Starchy Materials 230
         9.2.2.3   SSF of Lignocellulosic Materials 231
         9.2.2.4   Modeling of SSF of Cellulose 235
9.2.3   Process Integration by SSCF 240
9.2.4   Process Integration by Consolidated
         Bioprocessing 241
9.3   Reaction–Separation Integration for Bioethanol
       Production 244
       9.3.1   Ethanol Removal by Vacuum 246
       9.3.2   Ethanol Removal by Gas Stripping 246
       9.3.3   Ethanol Removal by Membranes 249
       9.3.4   Ethanol Removal by Liquid Extraction 263
9.4   Separation–Separation Integration for Bioethanol
       Production 275
References 276

Chapter 10 Environmental Aspects of Fuel Ethanol Production 285

10.1   Effluent Treatment during Fuel Ethanol Production 285
        10.1.1   Residues Generated in the Process of
                  Bioethanol Production 285
        10.1.2   Methods for Treatment and Utilization of
                  Stillage 286
                  10.1.2.1   Stillage Recycling 287
                  10.1.2.2   Stillage Evaporation 289
                  10.1.2.3   Solids Recovery 289
                  10.1.2.4   Stillage Incineration 289
                  10.1.2.5   Fertilization 290
                  10.1.2.6   Anaerobic Digestion 290
                  10.1.2.7   Composting 291
                  10.1.2.8   Stillage as a Culture Medium 292
                  10.1.2.9   Stillage Oxidation 292
                  10.1.2.10 Wastewater Treatment of Biomass-
                              to-Ethanol Process 293
10.2   Environmental Performance of Fuel Ethanol
        Production 298
        10.2.1   WAR Algorithm 299
        10.2.2   Life Cycle Assessment of Bioethanol
                  Production 304
        10.2.3   Other Methodologies for the Environmental
                  Analysis of Bioethanol Production 306
References 307

Chapter 11 Technological Configurations for Fuel Ethanol Production in
the Industry 311

11.1   Ethanol Production from Sucrose-Containing
Materials 311
11.2   Ethanol Production from Starchy Materials 322
11.2.1   Configuration Involving the Separate
Hydrolysis and Fermentation (SHF) of Corn
Starch 322
11.2.2   Configuration Involving the Simultaneous
Saccharification and Fermentation (SSF) of
Corn Starch 325
11.2.3   Configuration for Production of Cassava
Ethanol 333
11.3   Ethanol Production from Lignocellulosic Materials 336
11.3.1   Process Flowsheet Development for
Production of Biomass Ethanol 337
11.3.2   Optimization-Based Process Synthesis for
Ethanol Production from Biomass 348
11.4   Role of Energy Integration during Process Synthesis 351
References 353

Chapter 12 Food Security versus Fuel Ethanol Production 359

12.1   Crop Potentials for Food and Energy 359
12.1.1   Corn in the United States 360
12.1.2   Sugarcane in Colombia 361
12.1.3   Sugarcane in Brazil 362
12.1.4   Sugarcane in Tanzania 363
12.1.5   Lignocellulosics: Nonfood Alternative 364
12.2   Bioethanol and Fossil Oil Dependence 365
12.3   Bioenergy and Transgenics 366
12.4   Bioenergy and Food Market 366
12.5   Bioenergy and Food Security Project 368
12.5.1   The BEFS Project Is Developed in Basic
Phases 369
12.5.2   Purposes and Activities of Modules 369
12.5.2.1   Module 1 369
12.5.2.2   Module 2 370
12.5.2.3   Module 3 371
12.5.2.4   Module 4 371
12.5.2.5   Module 5 371
12.5.2.6   Module 6 371
12.5.3   Preliminary BEFS Results for Tanzania 372

12.6   Concluding Remarks 375
References 376

Chapter 13 Perspectives and Challenges in Fuel Ethanol Production 379

13.1   Feedstocks 379
13.2   Process Engineering 380
13.3   Food Security Impacts 384
13.4   Environmental Impacts 387

Index 391

Preface

The mitigation of climate change, energy versus food security, and equity in rural areas are some of the main global goals facing the world today. One of the real options to achieving these goals is the development of cleaner and renewable energy sources, such as biofuels. Fuel ethanol from sugarcane and corn is worldwide the most important biofuel, followed by biodiesel from rapeseed. However, speculations about the possible impacts of fuel ethanol production create confusion about its convenience.

This book stresses the need to analyze and design accurately fuel ethanol production systems based on a process engineering approach as the source of technical information for assessing the real impacts of this biofuel on energy, food, and environmental balances. The processes for producing fuel ethanol from different feedstocks are not all the same in terms of technologies, impacts, and benefits. There is diversity among scientists, engineers, governments, or decision makers who try to analyze fuel ethanol projects. The book, through the 13 chapters, describes a logic and structured strategy for further analysis and development of fuel ethanol production from different feedstocks including energy crops and lignocellulosic biomasses.

The authors aim to offer a comprehensive review as well as results from more than 15 years in process engineering and ethanol research developed by the Chemical and Biotechnological Processes Design Group at the National University of Colombia, Manizales campus. Additionally, the process intensification reached by integration of reaction and separation processes in fuel ethanol production is analyzed in detail.

Chapter 1 discusses bioenergy focusing on liquid biofuels and describes the development of biofuels production in the world. Chapter 2 discusses the role of process synthesis and design as a key strategy for rapid and high-tech analysis and design of complex biotechnological processes. Chapters 3 and 4 describe the characteristics and technological implications of using different sugary and starchy crops as well as lignocellulosic feedstocks. Chapter 5 emphasizes the hydrolysis technologies for the saccharification of carbohydrate polymers as cellulose and starch. Chapter 6 analyzes the microorganisms used in ethanol production. Chapters 7 and 8 describe in detail the fuel ethanol production technologies for different feedstocks. Chapter 9 analyzes the new technological innovations based on process integration as a way for reducing energy consumption. Chapter 10 addresses the environmental issues regarding bioethanol production. Here, environmental impacts are discussed in terms of waste reduction algorithm and the life cycle assessment. Then, these impacts are calculated as a result of overall energy and material balances obtained from a process engineering approach. Chapter 11 describes the technological configurations for fuel ethanol production in the industry. Chapter 12 discusses the possible factors that could affect food security when fuel ethanol production and consumption are encouraged in different countries. Chapter 13 summarizes the main topics discussed in the book in terms of perspectives and challenges in research and development for fuel ethanol production. Most of these chapters are supported by case studies that include calculations and discussion of results.

The authors believe that accurate analysis and precise design as well as proactive government policies in fuel ethanol production will contribute to fair and sustainable development of energy crops in the world promoting new alternatives for poor rural areas. Finally, this book is an open and dynamic work waiting for improvements and suggestions from the readers.

C. A. Cardona
Ó. J. Sánchez
L. F. Gutiérrez

Acknowledgments

The authors wish to acknowledge the following institutions for their support and collaboration:

- Pilot Plants of Biotechnology and Agro-Industry, National University of Colombia at Manizales, Colombia, for providing scientific and laboratory resources during research projects that supported part of the knowledge presented in this book.
- Research Direction of National University of Colombia at Manizales (DIMA), Colombia, for funding different research projects in fuel ethanol production.
- Colombian Institute for Science and Technology Development (Colciencias), Colombia, for granting fellowships and research projects related to the biofuels area.
- Moscow State Academy of Fine Chemical Technology M.V. Lomonosov, Moscow, Russian Federation, for providing the authors with science and technology bases in biotechnology, process design, and integration approach.
- Institute of Agricultural Biotechnology and Department of Engineering, University of Caldas, Manizales, Colombia, for financial and scientific support in fuel ethanol production projects.
- Government of the Department of Caldas, Colombia, for financial support of biofuels projects.
- Department of Chemical Engineering, University College London, United Kingdom, for scientific and computational support in process systems engineering.
- Korea Institute of Energy Research, South Korea, for scientific collaboration in biofuels research and development.

In addition, the authors wish to acknowledge Annie Cerón, M.Sc., for her assistance and support during the preparation of the manuscript.

The Authors

Carlos A. Cardona is associate professor in the Chemical Engineering Department at the National University of Colombia at Manizales since 1995. He received M.Sc. and Ph.D. degrees in chemical engineering from the Moscow State Academy of Fine Chemical Technology M.V. Lomonosov (Russian Federation). In addition, he has attended the especialization program in rheology at Moscow State University (Russian Federation) in 1994. From 1996 to 1997, he worked at the University of Caldas supporting a new program in food engineering. He has been twice awarded the research merit recognition by the National University of Colombia.

Dr. Cardona's research focuses on process system engineering for biotechnological processes, process integration, separation technologies, thermodynamics, biofuels production, fermentation technology, and agro-industry. In particular, he has worked on different research projects concerning the chemical and biochemical process design, biofuels research and development, economic and sustainable utilization of agro-industrial residues by biotechnological methods, and separation technologies development. He has authored or co-authored over 50 research papers as well as 5 research books and 14 book chapters. Additionally, he has edited two research books. He has presented over 100 works at scientific events. Dr. Cardona has been a visiting professor at several universities and research institutes abroad. Currently, he leads the research group in Chemical and Biochemical Processes Design at National University of Colombia at Manizales and coordinates the research area of biotechnology at the protechnology and agroindustry pilot plants at this university. In addition, he is a Food and Agriculture Organization (FAO) consultant in the field of technologies for sustainable bioenergy production.

Óscar J. Sánchez is associate professor of food engineering at the University of Caldas, Colombia. He received a chemical engineering degree and M.Sc. degree in Biotechnology with the highest honors from the Moscow State Academy of Fine Chemical Technology M.V. Lomonosov (Russian Federation), and a Ph.D. degree with honors in engineering from the National University of Colombia at Manizales. From 1996 to 1997, he worked in the beverage ethanol industry. He joined the faculty of engineering at the University of Caldas in 1997 where he coordinates the biotechnology course for food engineering students. In addition, he has taught biochemical engineering to chemical engineering students at the National University of Colombia at Manizales.

His research focuses on process systems engineering for biotechnological processes, biofuels production, industrial microbiology and fermentation technology, and enzyme technology. In particular, he has completed different research projects concerning the economic and sustainable utilization of agro-industrial residues by biotechnological methods. Dr. Sánchez has authored or co-authored over 15 research papers as well as 2 research books and 9 book chapters. He has presented over 20 works at scientific events. In 2005, Dr. Sánchez was awarded the Elizabeth Grose prize in the field of microbiology applied to the industry in the framework of the VII Latin American Congress of Food Microbiology. He was Honorary Research Fellow in the Department of Chemical Engineering of University College London in 2006. Currently, he co-leads the research group in foods and agro-industry at the University of Caldas and coordinates the research area of food biotechnology at the Institute of Agricultural Biotechnology of this university. In addition, he has been an FAO consultant in the field of technologies for sustainable production of fuel ethanol.

Luis F. Gutiérrez received a chemical engineering degree from the National University of Colombia at Manizales, an M.Sc. degree in chemical engineering from the University of Valle (Colombia), and a Ph.D. degree with honors in engineering from the National University of Colombia at Manizales. From 2002 to 2003, he worked in the precious metal processing industry. His research focused on chemical thermodynamics, process integration, precious metals refining, alloy production, and computer-aided process design. In particular, he has worked on different research projects for the application of the thermodynamics to chemical processes, and analysis of biodiesel production technologies as well as equipment design for chemical and biochemical processes.

He has authored or co-authored over 6 research papers as well as 3 research books and 4 book chapters. He has given over 12 presentations in scientific events.

Dr. Gutiérrez has held traineeships at Queen's University (Canada) and the National Autonomous University of Mexico. He has been a research director of ITEMSA SA, a firm manufacturing chemical equipment. Currently, he is professor in the Department of Engineering at the University of Caldas. In addition, he has been an FAO consultant in the field of technologies for sustainable production of biodiesel.

1 Biofuels

This chapter deals with the generalities of biofuels in the context of the current situation in the fossil fuels market. The importance of using alternative renewable energy sources is highlighted and the main advantages of liquid biofuels are presented. The state of the global ethanol market is analyzed and the advantages and disadvantages of fuel ethanol are described.

1.1 BIOFUELS GENERALITIES

Considering the very probable depletion of liquid fossil fuels toward 2090 and the start of declination of oil production in 2020 through 2030, humankind will face the huge challenge of maintaining its economic growth and stable technological development without compromising the welfare of the future generations (sustainable development). In addition, the quality of life for people from developing and underdeveloped countries should be improved without compromising the life level of those people in developed nations. Kosaric and Velikonja (1995) point out that the solution to this problematic situation depends on how mankind develops and implements viable technologies for the industry, transport sector, and heating based on alternative (renewable) fuels and feedstocks as well as ensuring the availability of sufficient amounts of renewable resources (energy and raw materials). Furthermore, man should develop and implement technologies for reducing the environmental pollution and CO_2 emissions. For these reasons, the renewable energies may partially or totally replace the fossil fuels, especially if humankind does not choose the dangerous pathway toward the global development of nuclear energy as a primary source of energy.

The renewable energy sources correspond to those kinds of energy that are obtained from natural sources, which are practically inexhaustible due to the huge amount of energy that they contain and to their ability to regenerate themselves by natural means. Among the energy sources of one type, solar, wind, hydraulic, geothermal, and tide energy should be highlighted. The energy sources of a second type correspond to the bioenergy or energy from biomass. One of the main features of the renewable energies is that their utilization does not imply the net generation of polluting emissions, which contribute to the greenhouse gas effect or the destruction of the ozone layer. The renewable energy sources represent 7.68% of the total energy consumed in the world, with the biomass being the most exploited resource (Energy Information Administration, 2008). Two thirds of the biomass is used for food cooking and heating in developing countries (traditional use of the biomass), e.g., through the use of firewood. The remaining third corresponds to the commercial use of the biomass for the industry (e.g., cane bagasse as energy source in sugar mills), generation of electricity (e.g., wood chips feeding small thermal plants), and transport sector (liquid biofuels production). The hydroelectric energy is the second most important renewable resource, whereas the contribution to the global energy consumption from such sources as the sun, winds, tides, and geothermal energy is marginal. A 59.2% increase in the consumption of renewable energy (corresponding to 828 million tons of oil equivalent) is expected in the period 2002 to 2030 (IEA, 2004).

One renewable solution in the search for alternative sources of energy for the world populace is the use of solar energy in the form of biomass (bioenergy). The global potential of bioenergy is represented by the energy-rich crops (mostly represented by agroenergy) and lignocellulosic biomass (including the dendroenergy from forest activities). The conversion of these feedstocks into biofuels, either for electricity generation or for their use in vehicles, is an important option for exploitation of alternative energy sources and reduction of polluting gases (Sánchez and Cardona, 2008b), mainly CO_2. The emissions generated by the combustion of biofuels are offset by the CO_2 absorption during the growth of plants and other plant materials from which these biofuels are produced. In this way, the biomass utilization releases the carbon dioxide that was fixed during its growth, compensating the emissions generated in the current scale of time. In contrast, fossil fuel usage releases into the atmosphere the carbon dioxide that was fixed by the plants million of years ago, which implies a net increase in the amount of atmospheric CO_2, provoking global warming.

Energy-rich crops comprise those crops that could be exclusively addressed to the energy production either as solid fuels for electricity generation or as liquid biofuels that can substitute for the fossil fuels (bioethanol, biodiesel). It is estimated that one hectare of energy-rich crops employed for liquid biofuel production can avoid the emissions of 0.2 to 2.0 tons of carbon into the atmosphere compared to the use of fossil fuels (Cannell, 2003). For the case of ethanol obtained from sugarcane in Brazil, its use may offset carbon dioxide emissions at a rate of 2 t C/ (Ha/year) related to the oil. This substitution is more appreciable in tropical countries, whereas the offset is more effective for European countries if the electricity is produced from biomass. Kheshgi and Prince (2005) indicate that if the CO_2 released during the alcoholic fermentation is captured and injected into the subsoil or deeply in the ocean (where it is dissolved), ethanol production may lead to a net carbon dioxide removal from the atmosphere (CO_2 sequestration) avoiding the emissions generated by the gasoline usage. Eventually, the environmental benefits can be enhanced if the feedstock employed is made up of residues or wastes.

The so-called lignocellulosic biomass includes agricultural, forestry, and municipal solid residues as well as different residues from agro-industry, the food industry, and other industries. The lignocellulosic biomass is made up of complex biopolymers that are not used for food purposes. The main polymeric components of biomass are cellulose, hemicelluloses, and lignin. For their conversion into a liquid biofuel such as ethanol, a complex pretreatment process is required in order to transform the carbohydrate polymers (cellulose and hemicellulose) into fermentable sugars. In contrast, for electricity generation, only the combustion of the biomass is needed.

It is evident that lignocellulosic biomass as feedstock for energy production is important. The lignocellulosic complex is the most abundant biopolymer on Earth and is present in such profuse materials as wood, sawdust, paper residues, straw, and grasses (Sánchez and Cardona, 2008b). It is estimated that the lignocellulosic biomass makes up about 50% of world biomass and its annual production has been estimated at 10 to 50 billion tons (Claassen et al., 1999). For instance, 35% of the material collected during the total wheat harvest in Europe corresponds to the straw, whereas 45% corresponds to the grain. In addition to the energy generation, the biomass utilization allows the economic exploitation of a wide range of residues from domestic, agricultural, and industrial activities. One of the main advantages of using lignocellulosic biomass is that this feedstock is not related to food production, which would permit the energy production without the utilization of a great number of hectares of land for cane, corn, or cassava production. Furthermore, the biomass is a resource that can be processed in different ways to produce a significant variety of products such as ethanol, synthesis gas, methanol, hydrogen, and electricity (Chum and Overend, 2001). However, some authors, such as Berndes et al. (2001), consider that the great scale implementation of biomass energy would create serious social, economic, and environmental consequences, especially if dedicated energy crops are employed. For example, these authors estimate that the labor requirements for bioenergy production at a great scale in any country should not exceed 1% of the total labor force in order to make its production feasible. Grassi (1999) points out that the development of bioenergy production technologies in the European Union (EU) can represent the creation of 200,000 direct and indirect jobs as well as the reduction of 255 million annual tons of CO_2 by 2010. Thornley (2006) summarizes the main environmental, social, and economic advantages of employing biomass as an energy source. Some of these advantages are applicable to both developed and developing countries. Among them, the reduction of greenhouse gas emissions, reduced use of agro-chemicals, diversification of rural economies, the potential for low-cost heat supply, and the potential income streams for farmers should be highlighted. On the other hand, this author presents some of the main consequences of bioenergy development: impacts on particular native species, the visual impact of crop growth and conversion plants, environmental emissions associated with thermal conversion plants, the uptake of significant amounts of water from below ground, and the requirement for policy support, among others. The need for policies that stimulate the development of technologies based on biomass through tax exemptions or subsidies should be emphasized considering the lower production and transport costs of fossil fuels.

1.1.1 SOLID AND GASEOUS BIOFUELS

In general, the biomass releases energy through its conversion into simpler compounds. This conversion can be carried out by chemical or biological methods. In this way, the biomass can be employed for producing solid, gaseous, or liquid biofuels. The biomass itself can be used as a solid biofuel for electricity production. In this case, the biomass undergoes combustion with or without coal as an auxiliary fuel (co-combustion). The biomass can also be used as a feedstock for producing gaseous fuels. For this, the biomass undergoes thermal treatment in the presence of a reduced amount of oxygen (partial oxidation) or by steam. The aim of these kinds of processes (pyrolysis, thermal gasification) is to obtain a gaseous fuel that can be mixed in a better way with the air leading to a cleaner and more complete combustion than is the case of the solid biomass. Moreover, the biomass can be converted into biogas, a mixture of CH_4 and CO_2, using anaerobic bacteria that assimilate the organic matter contained in the biomass forming more bacterial cells and releasing methane and carbon dioxide as a result of the methanogenic metabolism in absence of oxygen.

1.1.2 LIQUID BIOFUELS

The worldwide transport sector depends almost totally on fossil fuels. While the production of electricity is more diversified, vehicles require gasoline, diesel, or natural gas for the mobilization. Only in specific cases, has the total substitution of gasoline with liquid biofuels been achieved, as in the case of the ethanol obtained from energy-rich crops such as sugarcane. The advantage of liquid biofuels compared to solid ones consists in their ease of transport and in the utilization of the supply chains of fossil fuels. Biodiesel and bioethanol are the two main liquid biofuels.

1.1.2.1 Biodiesel

Biological diesel is an oxygenated fuel obtained as a result of the transesterification of vegetable oils and animal fats with an alcohol in the presence of a catalyst (Figure 1.1). In general, the employed alcohol is methanol or ethanol. Thus, the biodiesel is a mixture of methyl or ethyl esters of fatty acids from oil and fats. Usually, rapeseed oil (in central Europe), sunflower oil (in southern Europe), and palm oil (in tropical countries of Southeast Asia and South America) are employed as a source of triglycerides. Used frying oil of animal or vegetable origin is employed for their conversion into biodiesel as well. For biodiesel production, acid, basic, and biological (enzymes) catalysts are used, with KOH and NaOH being the most utilized catalysts worldwide.

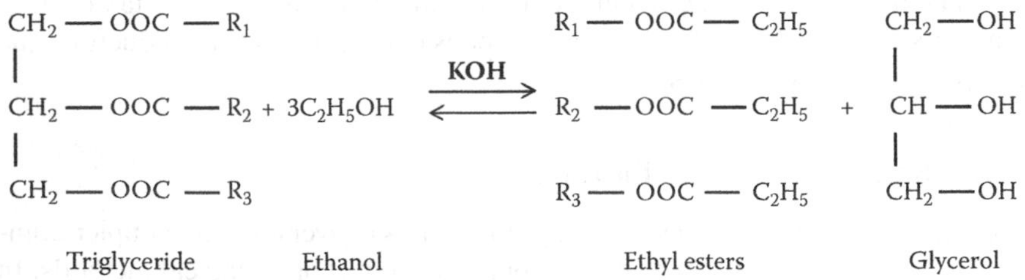

$$
\begin{array}{llll}
CH_2 - OOC - R_1 & & R_1 - OOC - C_2H_5 & CH_2 - OH \\
| & \xrightarrow{\text{KOH}} & & | \\
CH_2 - OOC - R_2 + 3C_2H_5OH & \rightleftharpoons & R_2 - OOC - C_2H_5 \quad + & CH - OH \\
| & & & | \\
CH_2 - OOC - R_3 & & R_3 - OOC - C_2H_5 & CH_2 - OH \\
\end{array}
$$

      Triglyceride   Ethanol   Ethyl esters   Glycerol

FIGURE 1.1   Transesterification reaction of triglycerides.

Biodiesel is employed in heating systems and as an oxygenating additive (oxygenate) in diesel engines. For these purposes, blends containing 10% (B10) or 20% (B20) of biodiesel and the remaining fossil diesel are being utilized. In some cases and depending on its purity, the biodiesel is directly used in internal combustion engines (Ma and Hanna, 1999). The main advantages offered by the diesel usage are the reduction of polluting gases (mostly CO) and particulate matter emissions, as well as the net equilibrium in the balance of atmospheric CO_2. Unlike the oil-derived diesel, the biodiesel does not contain sulfurs, making its combustion much cleaner.

Worldwide biodiesel production achieved about 7.70 billion liters in 2007. The production of biodiesel in the United States was about 1.82 billion liters in 2007, whereas the production in the EU was about 5.71 billion liters (European Biodiesel Board, 2008). Germany is the world leader in biodiesel utilization. In this country, pure biodiesel is used in adapted vehicles, while in France, 30% and 50% biodiesel blends with fossil diesel are employed (Demirbas and Balar, 2006). In Colombia, the agro-industry of oilseeds is being developed through the production of biodiesel to be used as an oxygenate of local diesel in B5 blends. Thus, the productive chain of the oil palm can be boosted.

1.1.2.2 Bioethanol

Bioethanol (ethyl alcohol, fuel ethanol) is the most-used liquid biofuel in the world. It is obtained from energy-rich crops, such as sugarcane and corn. Ethanol can be directly employed as a sole fuel in vehicles or as gasoline oxygenate increasing its oxygen content and allowing a better hydrocarbon oxidation that reduces the amount of aromatic compounds and carbon monoxide released into the atmosphere. For this reason, fuel grade ethanol (FGE) is the market with the most rapid growth rate in America and Europe.

The fuel ethanol can be obtained from lignocellulosic biomass as well, but its production is much more complex. Nowadays, great efforts are being made to diminish the production costs of lignocellulosic ethanol. It is expected that the evolution of biomass conversion technologies will allow the massive oxygenation of gasoline with fuel ethanol and make possible the substitution of a significant portion of fossil fuels considering the huge availability of lignocellulosic worldwide (Bull, 1994).

1.2 GASOLINE OXYGENATION

Gasoline is a mixture of hydrocarbons that correspond to the light fractions obtained during the fractionated distillation of oil, but it can also be produced from other heavier fractions through catalytic cracking. These hydrocarbons have between 5 and 12 atoms of carbon in their molecules. The typical composition of the gasoline is 30 to 50% aliphatic hydrocarbons with linear and branched chains, 20 to 30% cyclic aliphatic hydrocarbons, and 20 to 30% aromatic hydrocarbons. The content of the latter can reach 50% (by volume) corresponding mostly to benzene, toluene, ethylbenzene, and xylenes, which are highly toxic (Nadim et al., 2001). The physical and chemical properties of the gasoline depend on its formulation, which can be different for each refinery and varies according to the season (winter or summer). The comparison of the properties of gasoline related to its main oxygenating additives is presented in Table 1.1.

During gasoline combustion and in addition to CO_2, a great amount of substances resulting in the incomplete combustion of hydrocarbons contained in this fuel is released. The unburned compounds constitute the toxic emissions that contribute to the atmospheric pollution and have a direct influence on human health as in the case of aromatic hydrocarbons. Additionally, nitrogen oxides are generated (generically represented as NO_x) as a consequence of the partial oxidation of nitrogen contained in the air. As among the most important properties of the gasoline, the octane number should be highlighted. This parameter quantifies the tendency of this fuel to resist the spontaneous combustion in the case of spark ignition engines. If inside the engine cylinders, the air–gasoline mixture explodes, due to its compression and the heat, before the spark plugs generate the spark, engine knocking (detonation) occurs leading to reduced efficiency and possible damage to the motor. This preignition can become manifest when the vehicle transports high loads or accelerates, or during a prolonged climb (Tshiteya et al., 1991). To quantify the antiknocking properties of the gasoline, the 2,2,4-methylpentane, an isomer of the n-octane, also called iso-octane, was used as a comparison standard (octane number). The iso-octane is one of the gasoline components with good antiknocking properties and was assigned a value of 100. The minimum value of the octane number corresponds to the n-heptane, which indicates that the more branched the carbon chain of one hydrocarbon is, the greater its octane number. The different blends of gasoline can have lower or higher values of the octane number. The greater the octane number, the lower the possibility of engine knocking. Thus, the regular gasoline in the United States can have an octane number of 87, whereas the premium gasoline can reach octave numbers of 91, which means a higher quality gasoline.

Another important gasoline property is the Reid vapor pressure (RVP), which is a measure of the vapor pressure of a fuel at 100°F in a vessel with a 4:1 vapor-to-liquid volumetric ratio (French and Malone, 2005). The greater the RVP of the gasoline components, the greater the gasoline volatility is. Gasoline with a higher RVP has higher evaporative emissions, which are released into the atmosphere while supplying fuel to gas stations and when the vehicles are not running. These emissions are ozone precursors (smog) at ground level and are higher when the outside temperature increases. For this reason, some countries like the United States, impose a maximum limit to the gasoline RVP during the summer season. For instance, this limit is 9 psi in northern states, 8 psi in southern states, and even 7.8 psi in California, the most populated state in the United States (Thomas and Kwong, 2001). In Colombia, the limit is 65 kPa (9.43 psi).

TABLE 1.1

Some Properties of Gasoline and Main Oxygenates

Physical Properties	Gasoline	Methanol	Ethanol	MTBE	ETBE	TAME	DIPE
Molecular weight	—	32.04	46.07	88.15	102.18	102.18	102.17
Specific gravity (15.6°C/15.6°C)	0.72–0.78	0.796	0.794	0.74	0.746	0.764	0.725
Density (g/L at 15.6°C)	719.0–778.9	794.5	792.1	740.4	745	764	723
Boiling point (°C)	27–225	65	78	55.2	72	85–87	101
Reid Vapor pressure (RVP)							
Net (psi)	—	4.6	2.3	—	—	—	—
Mixed (psi)	8–15	93–98	12–27	8–10	3–5	3–5	0–2
Octane number							
Net	72–76	98	97	109	110	104	105
Mixed	—	115	111	—	—	—	—
Solubility in water (% en vol. a 21.1°C)	Insignificant	100	100	51.26	1.2	12	—
Latent heat vaporization							
kJ/L at 15.6°C	250.8	930.7	662.7	240.55	240.38		285.53
kJ/kg at 15.6°C	348.9	1,177.0	921.1	320.99	322.6		207.1
Heating value (low)							
kJ/L at 15.6°C	30,374–33,160	15,828	21,178	26,073	26,960	28,008	
kJ/kg at 15.6°C	41,868–44,194	19,934	26,749	35,101	36,315	36,392	
Stequiometric air/fuel, weight	14.7	6.45	9.00	11.7		13.44	

Sources: Data collected from the European Fuel Oxygenates Association (2009) and Tshiteya et al. (1991).

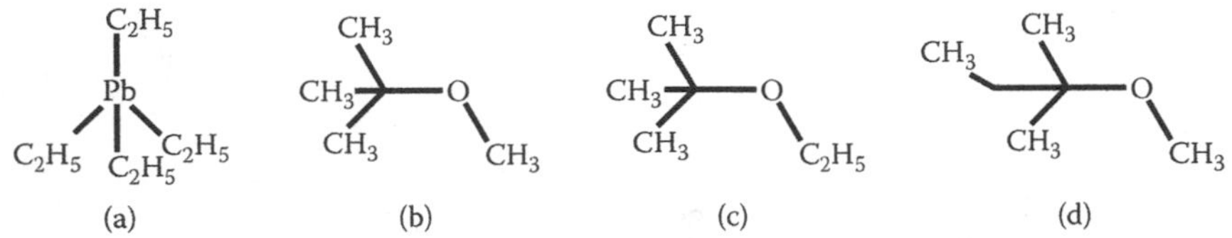

FIGURE 1.2 Structure of some compounds with antiknocking properties in gasoline: (a) TEL, (b) MTBE, (c) ETBE, (d) TAME.

1.2.1 TETRAETHYL LEAD AS ANTIKNOCKING ADDITIVE

Tetraethyl lead (TEL) is the major oxygenating additive of gasoline that has been used in the world in the recent past (Figure 1.2a). The U.S. scientist Thomas Midgley discovered the excellent properties of TEL in December 1921. One liter of TEL was enough to treat 1,150 liters of gasoline. Thus, the oil companies began the addition of this compound to the gasoline instead of ethanol. In the case of the latter, its use would have reduced the utilization of gasoline by 20 to 30% making the vehicles less oil dependent. Leaded gasoline has an average octane number of 89 (Nadim et al., 2001). However, some concerns were expressed in 1923 about the negative effect of TEL on human health. In particular, it was pointed out that each liter of consumed gasoline would emit 1 g of lead oxide. Later studies demonstrated the negative neurological effects of this compound, especially in children (Thomas and Kwong, 2001). It wasn't until January 1996 that leaded gasoline was banned in the United States. In Colombia, leaded gasoline is no longer produced. In fact, its import has been banned since 1994 (Ministerio del Medio Ambiente, 1994). Nevertheless, there exist some African and Mideast countries where leaded gasoline is still being used because of its low cost (Thomas and Kwong, 2001).

One of the ways of avoiding TEL usage is the modernization of refineries in order to elevate the production of aromatic and aliphatic hydrocarbons, but these modifications are costly. In addition, the augmentation in the concentration of aromatic compounds in the gasoline increases the risks related to the benzene exposition. Another option consists in the substitution of TEL with other less toxic compounds. Both options represent additional expenses in the refineries that explain why the phase-out of TEL utilization has been very slow in spite of having known its toxic properties for a number of years.

1.2.2 ETHERS AS GASOLINE OXYGENATES

In the past, and as a consequence of the concerns generated by the atmospheric pollution in cities, the better combustion of gasoline through the incorporation of oxygen into its composition has been pursued. Again, the United States boosted the development of new additives for the world automobile industry. In this sense, the signing of the Clean Air Act in 1970 and the Clean Air Amendments in 1990 introduced reformulated gasoline as an answer to the need for diminishing the air pollution levels in the biggest cities of the United States. This type of gasoline should have a minimum 2% (by weight) oxygenate and benzene levels less than 1% (by volume; Nadim et al., 2001). The employing of oxygenates is aimed at reducing the atmospheric contamination (smog during summer, CO in winter, and toxic emissions throughout the year) because of the better combustion toward CO_2 due to the involvement of one atom of oxygen in their molecules. In addition, these additives present significant antiknocking properties. In this way, the oxygenates play two important roles for elevating the gasoline quality. Precisely, the most employed oxygenates as antiknocking additive in gasoline and has a branched carbon chain like TEL (see Figure 1.2) that suggests the increase of the octane number in its fuel blends. In this way, the oxygenates have allowed the replacement of lead in the gasoline.

1.2.2.1 Methyl Tert-Butyl Ether (MTBE)

Methyl tert-butyl ether (MTBE) has been used since 1979 in the United States. This compound is totally miscible with the gasoline, has a similar volatility, and does not absorb water, which confers a low susceptibility to the water–gasoline phase separation. MTBE was the main oxygenating compound and was added to the gasoline up to 15% (by volume), which is equivalent to 2.7% oxygen. Its utilization causes the reduction of ozone concentrations at ground level during seasons with high smog and CO formation as well as the decrease in benzene usage for gasoline, which led to the reduction of this powerful carcinogenic substance (Braids, 2001).

The production pace of MTBE has increased remarkably since the beginning of a reformulated gasoline program in the United States from 12,000 barrels/day in 1970 to 250,000 barrels/day in 1998 (Braids, 2001). In fact, it was the chemical with the greatest growth during the 1990s with annual increments in its production of 10 to 20%. The worldwide MTBE production reached 30 million ton/year at the end of the twentieth century (Oudshoorn et al., 1999). MTBE is produced by the reaction of isobutene with methanol. Both isobutene and methanol are from fossil origins, although the latter can be produced from natural gas and even from renewable sources through biomass gasification to methane followed by its oxidation toward methanol. The isobutene is obtained from fractions with four atoms of carbon (C_4 fractions) from the catalytic cracking process in oil refineries as well as from processing of ethylene (Ancillotti and Fattore, 1998).

The solubility of MTBE in water is in the range of 50,000 ppm or 50 g/L, which makes this compound the most soluble compared to any other gasoline component. When gasoline contacts water, MTBE has a tendency to be dissolved in water but also continues to be dissolved in gasoline. In addition, MTBE is resistant to degradation by chemical or biological means. These two features suggest that MTBE presents a high mobility in natural water streams, especially in groundwater where it can migrate as a result of leakages in the storage and transport systems of gasoline. Furthermore, MTBE at low concentrations can alter the taste and odor of potable water (Nadim et al., 2001). Numerous cases of water sources contaminated with MTBE without the presence of the remaining gasoline hydrocarbons have been reported. One of the most effective methods for MTBE removal is by means of adsorption using granulated activated coal obtained from coconut peels (Braids, 2001). Due to these negative features, the use of MTBE was banned in California in 2004 and its total prohibition is projected by 2010.

1.2.2.2 Ethyl Tert-Butyl Ether (ETBE)

Although MTBE volatility is relatively low, the standards for reformulated gasoline in the developed countries are very exigent during summer in order to reduce the smog formation in the cities. This makes the RVP specifications for gasoline lower than the corresponding value for MTBE. For this reason, the employing of ethers with more branched carbon chains and lower volatilities has been explored. Precisely, the ethyl tert-butyl ether (ETBE), the second most utilized ether as an oxygenate, has an RVP less than MTBE and antiknocking properties slightly higher (see Table 1.1). ETBE is produced by the exothermic reaction between isobutene and ethanol. As in the case of MTBE production, this reaction requires ionic exchange resins as a catalyst (Ancillotti and Fattore, 1998). ETBE is produced mainly in the United States and Europe. The production of ETBE incorporates reactive distillation using acid ionic exchange resins (Thiel et al., 1997) or structured zeolites as packing materials (Oudshoorn et al., 1999). One substantial difference in ETBE production is that one of its feedstocks, ethanol, can be obtained from renewable resources like the biomass, which entails the integration of the petrochemical industry with the biotechnological sector. One example is the case of France where bioethanol has been produced from sugar beets since 1990. In this way, the ETBE obtained is partially renewable, which implies greater environmental benefits compared to MTBE whose production is totally from fossil origin. These benefits are represented in a lower emission of greenhouse gases (N_2O, CH_4, and CO_2), a lower contribution to the depletion of natural resources, and better air quality in the cities (less unburned hydrocarbons and formed formaldehyde, although more acetaldehyde produced). In fact, a 3.8% reduction in the amount of unburned hydrocarbons emitted and 17.6% decrease in benzene emissions has been noted in vehicles provided with catalyst employing gasoline blends with 15% (by volume) ETBE (Poitrat, 1999). On the other hand, Spain also employs the ETBE as an oxygenate for its gasoline. In this case, the ETBE production has reached its maximum due to the availability of isobutene in the Spanish refineries (Espinal et al., 2005).

The solubility of ETBE in water is also relatively high compared to the remaining gasoline components (12.2 g/L). Like MTBE, ETBE can also be easily transferred to groundwater from gasoline leakages from storage tanks, pipelines, or other distribution systems (Ancillotti and Fattore, 1998). The ETBE biodegradation under aerobic or anaerobic conditions has been researched to evaluate the potential for its decontamination (Kharoune et al., 2001).

1.2.2.3 Tert-Amyl Methyl Ether (TAME) and
Tert-Amyl Ethyl Ether (TAEE)

The production of tert-amyl methyl ether (TAME) began at the commercial level at the end of the 1980s both in the United States and in Europe. TAME is used as a gasoline oxygenate like MTBE, but it uses another type of feedstock: tertiary olefins with five carbon chains (tert-C_5 olefins) that exist among the components of the different kinds of gasoline produced in an oil refinery. The tert-C_5 olefins are selectively converted into tertiary ethers employing methanol. In this way, TAME production represents an expansion of the raw material basis for producing oxygenates as well as the possibility of reducing the content of tertiary olefins and increasing the oxygen content in gasoline (Huttunen et al., 1997). TAME is obtained by the liquid-phase reaction of methanol and two of the three isoamylenes (branched olefins of five atoms of carbon): 2-methyl-1-butene and 2-methyl-2-butene. The third isoamylene (3-methyl-1-butene) is not reactive. This reaction is accomplished using acid ionic exchange resins (Oost et al., 1995). TAME solubility is comparable to that of ETBE (12 g/L) and can be transferred to surface water streams and groundwater as well. TAME is not easily biodegradable and is persistent in the soil and sediments (Huttunen et al., 1997). In some cases, the refineries employ blends of MTBE, ETBE, and TAME. This results in water streams having contents of these three ethers. The possibility of aerobically degrading this kind of blend using fixed bed reactors has been demonstrated (Kharoune et al., 2001).

Alternatively, the production of the ethyl homologue of TAME, the tert-amyl ethyl ether (TAEE), has been proposed as an attractive option when the ethanol production costs decrease (Ancillotti and Fattore, 1998). TAEE can be produced from isoamylenes as well, but the use of ethanol could make it renewable. In addition, TAEE has one of the lowest RVPs compared to the main oxygenates used in the refineries (see Table 1.1).

1.2.2.4 Di-Isopropyl Ether (DIPE)

Di-isopropyl ether (DIPE) is not currently produced at an industrial level, but it represents an important option as an oxygenate due to its antiknocking properties and low volatility, among other features. However, it presents the disadvantage of the auto oxidation with the subsequent production of low solubility and explosive peroxides. To neutralize this effect, the addition of antioxidants in amounts as small as 20 ppm is enough, which does not allow the production of more peroxides than the ones formed during MTBE synthesis (Ancillotti and Fattore, 1998). DIPE is produced by the addition of water to isopropylene for producing isopropylalcohol, which is added to other molecules of isopropylene obtaining DIPE. Hydration reaction is carried out with catalysts of the type ZSM-5 or acid polysulfone resins at 700 to 24,000 kPa. The second reaction is accomplished using acid zeolites at 450 to 7,000 kPa (Harandi et al., 1992). In this case, the need for employing alcohols is eliminated because oxygen is supplied by the water. Like the ETBE, the production of DIPE by reactive distillation has been proposed (Cardona et al., 2000, 2002).

Ethers, as gasoline oxygenates, have demonstrated several advantages derived from their oxygen content, antiknocking properties, and compatibility with the gasoline, among other factors. Nevertheless, the concerns arising from the low biodegradability and high mobility of MTBE have imposed serious limitations to the production of this type of oxygenates in some countries, particularly in the United States where even the usage of ETBE and TAME has been restricted in California. A limiting factor is that the toxicological properties and the environmental impacts of these ethers are not sufficiently known. It is considered that due to the similarity in the molecular structure of the different ethers, their properties should be analogous to those of MTBE. For this reason, contamination problems in groundwater as a consequence of using oxygenates like ETBE, TAME, TAEE, or DIPE are expected (Graham et al., 2000; Nadim et al., 2001).

1.2.3 METHANOL

Alcohols are an oxygenation alternative for gasoline considering the environmental disadvantages of the ethers mentioned above. The blends of alcohols and gasoline have comparable properties related to the traditional fuels based on oil. Despite their lower combustion heat compared to gasoline, the increase in fuel consumption when alcohols are used as oxygenates is not significant in principle. Moreover, the possibility of increasing the conversion of the blend and, therefore, the engine efficiency, represents a great advantage for alcohol-containing gasoline. Furthermore, the emissions of hydrocarbons and carbon monoxide are reduced, although a considerable increase in the emission of aldehydes is presented (Rasskazchikova et al., 2004).

In the 1970s, several researches were carried out in countries like Japan, the United States, and Germany aimed at the utilization of methanol (CH_3OH) as an additive for enhancing the octane number of gasoline (see Table 1.1). Ethanol was pushed into the background due to its high comparative costs. However, despite its high octane number, methanol usage was limited and even banned in many countries due to its high toxicity, volatility (the highest RVP values of the analyzed oxygenates), and hygroscopicity, which generates a series of technical difficulties for the use of methanol–gasoline blends. Furthermore, the formaldehyde formed during methanol oxidation results in the formation of a substance considered dangerous (Rasskazchikova et al., 2004).

At the beginning of the 1980s, the mixture of methanol and tert-butyl alcohol was commercialized under the trademark Oxynol™. Nevertheless, the high volatility of the blends containing methanol, due to the formation of one azeotrope with the gasoline hydrocarbons, caused the market to refuse it (Ancillotti and Fattore, 1998). Currently, methanol is employed as feedstock for the production of MTBE and TAME.

1.3 ETHANOL AS A GASOLINE OXYGENATE

Ethanol (C_2H_5OH), also known as ethyl alcohol, fuel ethanol (when it is used as a fuel or gasoline oxygenate), or bioethanol (when it is obtained from biomass [energy-rich crops or lignocellulosic materials]), is the most widespread alcohol employed in the transport sector. Ethanol usage has many advantages and some disadvantages. An objective proof on advantages of this biofuel against its disadvantages is the rising number of countries that have chosen it as a gasoline oxygenate or even as a direct fuel. Sometimes, the disadvantages of using fuel ethanol are emphasized with arguments that have been the subject of controversial debates between supporters and detractors of biofuels. This polemic is even more exacerbated when considering the temporal character of the governments, the main promoters of gasoline oxygenation programs using fuel ethanol. This discussion goes beyond the academic field (which is desirable) and falls into the political debate with the corresponding manipulation doses that it implies.

1.3.1 ADVANTAGES OF FUEL ETHANOL

The utilization of ethanol as an oxygenate has many benefits: higher oxygen content (lesser amount of required additive), high octane number (see Table 1.1), greater reduction in carbon monoxide emissions, and nonpollution of the water sources. Compared to methanol, ethanol is less hygroscopic, has a higher combustion heat, and has less heat from evaporation, and, most important, it is much less toxic. In addition, the acetaldehyde formed during ethanol oxidation is much less dangerous than the formaldehyde formed during methanol combustion. In fact, acetaldehyde predominates in comparison to formaldehyde in exhaust gas from vehicles using ethanol–gasoline blends (Rasskazchikova et al., 2004). Ethanol–gasoline blended fuels increase the emission of formaldehyde, acetaldehyde, and acetone 5.12 to 13.8 times more than gasoline. Although the aldehyde emissions will increase when ethanol is used as a fuel, the damage to the environment by the emitted aldehydes is far less than that caused by the polynuclear aromatics emitted from burning gasoline (Yüksel and Yüksel, 2004).

From the viewpoint of combustion properties, the autoignition temperature and flash point (temperature at which the liquid generates sufficient vapor to form a flammable mixture with the air) of ethanol are higher than those of gasoline, which makes it safer to transport and store. Ethanol has a latent heat of evaporation 2.6 times greater than that of gasoline, which makes the temperature of the intake manifold lower and increases the volumetric efficiency of the engine. Nevertheless, this property causes the engine's cold start ability to be reduced because the alcohols require more heat to vaporize than does gasoline in order to form an appropriate air–fuel mixture that can be burned. Ethanol heating value is also lower than that of gasoline and, therefore, it is necessary to have 1.5 to 1.8 times more fuel ethanol to release the same amount of energy if it is used in a pure form rather than in gasoline blends. On the other hand, the stoichiometric air–fuel ratio of ethanol is about two thirds to one half that of gasoline, hence, the required amount of air for complete combustion is less for ethanol (Yüksel and Yüksel, 2004).

From a socioeconomic point of view, the utilization of fuel ethanol presents important advantages as well (Chaves, 2004; Sánchez and Cardona, 2008a). Bioethanol contributes to the decrease of imports of gasoline or oil in consuming countries through the partial substitution of these fossil fuels. Thus, this biofuel has the potential of compensating and reducing the impact of periodical rises of oil prices in the context of exhausting national reserves. Therefore, significant currency savings can be achieved that otherwise would have to be directed to fossil fuel imports. In each country, ethanol usage favors the economic utilization of raw materials and renewable resources, such as sugarcane, cassava, corn, and sorghum, as well as a great amount of lignocellulosic residues having the potential to be converted into ethyl alcohol. The use of fuel ethanol boosts the economic and productive reactivation of many rural communities through the increase in demand for agricultural production. By means of the development of productive projects for obtaining fuel ethanol, the base for creating and expanding actual agro-industrial chains where several links are integrated with the participation of private and public sectors is provided. This integration spreads benefits to different segments of the economy such as the energy, agricultural, industrial, and financial sectors. This makes possible the development of commercial relationships as well as the creation of jobs in depressed rural areas avoiding the migration of population to urban centers, especially in developing countries. In addition, the large-scale utilization of ethanol will promote the scientific and technological development of many countries in the biofuel field even when turnkey technology is acquired. Once installed, this type of technology generates new challenges, such as increase in productivity, improvement of the different crops varieties that are used as feedstocks, enhancement of process efficiency, and reduction of the environmental impact caused in production facilities, as well as many others.

From the environmental point of view, the utilization of ethanol as a gasoline oxygenate offers net reductions in the amount of greenhouse gas emissions per traveled mile of 8 to 10% in gasoline blends containing 10% ethanol by volume (known as E10 blends). For blends containing 85% ethanol (E85 blends), this reduction can reach up to 68 to 91% (Wang et al., 1999). In general, it is considered that the greater the percentage of ethanol in gasoline blends, the better the environmental benefits mostly expressed through the net reduction of greenhouse gas emissions during the entire life cycle of ethanol. Numerous studies have proved the environmental benefits of fuel ethanol usage in terms of its impact on the emission of the combustion production from ethanol–gasoline blends. According to data compiled by the Canadian Renewable Fuels Association (2000), the gasoline oxygenated with 10% ethanol reduces the levels of carbon monoxide by 25 to 30% as well as the net CO_2 emission by 10%. The main benefits of using ethanol as an oxygenate are presented in Table 1.2. One of the features of ethanol lies in the fact that it can be utilized as a feedstock for ETBE production using isobutene obtained from the petrochemical industry. This duality converts the ethanol to a very promising product in the international energy market, especially if the legislation of different countries continues to be aimed at using renewable gasoline oxygenates (Ancillotti and Fattore, 1998) as in the case of the EU.

1.3.2 DRAWBACKS OF FUEL ETHANOL

The main disadvantage of producing fuel ethanol is that its production is more expensive than the production of fossil fuels. From a technical viewpoint, gasoline

TABLE 1.2

Main Benefits of Using Ethanol as an Oxygenate in Gasoline Blends

Emission	Mix E10	Mix E85
Carbon monoxide (CO)	Reduction of 25–30%	Reduction of 25–30%
Carbon dioxide (CO_2)	Reduction of 10%	Reduction up to 100% (E100)
Nitrogen oxides (NO_x)	Increment or reduction of 5%	Reduction up to 20%
Volatile organic compounds	Reduction of 7%	Reduction of 30% or more
Sulfur dioxide (SO_2) and particulate matter	Reduction	Significant reduction
Aldehyde	Increment of 30–50% (very low with catalyst)	Insufficient data
Aromatic compounds (benzene and butadiene)	Reduction	Reduction up to 50%

Source: Canadian Renewable Fuels Association. 2000. *Environmental Benefits of Ethanol.*

blended with ethanol conducts electricity and its RVP is higher than nonblended gasoline. This implies a higher volatilization rate that can lead to ozone and smog formation (Thomas and Kwong, 2001). Although the addition of ethanol increases the evaporation rate of organic volatile compounds, many experts consider that the reduction of CO emissions effectively offsets the volume losses due to the volatility increase (Ghosh and Ghose, 2003).

One of the most troubling issues regarding the utilization of gasoline blends is the tendency of ethanol to form two liquid phases in the presence of water: one aqueous phase with an important ethanol content and one organic phase. If water contaminates the fuel, the water dissolves into the ethanol and disperses through the tank. Once it exceeds the tolerance level, the alcohol–water mixture will separate from the gasoline. Depending on individual conditions, about 40 to 80% of the ethanol will be drawn away from the gasoline by the water, forming two distinct layers. The top layer will be a gasoline that is a lower octane and perhaps out of specification, while the bottom layer is a mix of water and ethanol that will not burn (Central Illinois Manufacturing Company, 2006). To avoid the phase separation, ethanol–gasoline blends are not directly transported by pipelines. In general, ethanol is added to gasoline in bulk terminals (retail outlets, the end link of the supply chain for wholesale distribution) or in tanker trucks at the terminal immediately before delivery to the service station. In these points, reception and storage tanks are steel-made to minimize the exposure to water that can infiltrate into the distribution and storage systems for gasoline as well. This problem can be overcome by using stabilizing additives such as higher alcohols, fusel oils (mixture of higher alcohols, fatty acids, and esters), aromatic amines, ethers, and ketones. For instance, the addition of 2.5 to 3% isobutanol ensures the stability of ethanol–gasoline blends containing up to 5% water at temperatures down to –20°C. In fact, the phase stability of gasoline blends increases with high ethanol concentrations (Rasskazchikova et al., 2004). Another approach for stabilization of these gasoline blends is the modification of the carburetor that also can allow the utilization of blends with higher ethanol contents (Yüksel and Yüksel, 2004).

On the other hand, ethanol, like all the alcohols in general, is highly corrosive, depending on water content. The higher the molecular weight of the alcohol, the less corrosive it is. To neutralize this effect, corrosion inhibitors can be added. Among these inhibitors are hydroxyethylated alkylphenols, alkyl imidazolins, and different oils obtained during cyclohexane production. Ethanol can have a negative effect on rubber and plastic materials because it penetrates hoses and tight seals, which increases fuel losses due to evaporation. Nevertheless, the current level of development of the polymer industry makes it possible to select materials resistant to penetration of alcohols so that fuel losses are eliminated. These new polymers are being used in the automobile parts industry (Rasskazchikova et al., 2004).

From an economic point of view, fuel ethanol also presents some drawbacks that depend on the situation in all countries of the world. Particularly, in the case of the sugar sector, there exists the risk that sugar producers involved in ethanol production can reduce the amount of ethanol produced when sugar prices are especially high on the international market. For this reason, some governments, like Colombia's for instance, have adopted measures to avoid this situation by linking the international sugar price to the value paid to ethanol producers. However, several authors and nongovernmental organizations (NGOs) have expressed their concern regarding the fact that this price structure and related tax exemptions favor the economic groups controlling sugar markets in each country (Chaves, 2004). Another great concern, when an ethanol oxygenation program for gasoline is being implemented, is in the pressure over food prices related to feedstocks from which ethanol is produced, especially sugar and corn. In particular, the biofuels were hardly criticized when both oil and food commodities prices reached their historical peaks during 2008. Specifically, it was estimated that the "biofuels effect" could have provoked an increase in the international price of food commodities and crops of about 35% in that year. This statement was weakened when the oil price fell at the end of 2008 (corresponding to the beginning of the global financial crisis) and the price of food commodities also fell to a percentage higher than 35%. This could indicate than the elevated value of food feedstocks was more linked to the oil price than to the biofuels prices. However, it should be pointed out that the price of biofuels in the international market depends on the oil price as well. In any case, the effect of producing bioethanol and biodiesel from agricultural resources on food and feed prices cannot be neglected and should be thoroughly assessed. In this regard, the production of the so-called second-generation biofuels represents an important option for producing biofuels from sources other than those related to food and feed production. In this case, bioethanol can be produced from lignocellulosic residues or by-products, such as cane bagasse, corn stover, or wheat straw, that have no influence on food production structure.

Considering the effect of bioethanol production on the environment, some concerns have been expressed based on the fact that ethanol usage in gasoline blends increases the aldehyde level, mostly acetaldehyde, compared to the combustion of conventional gasoline. The aldehydes are formed during the incomplete combustion of ethanol and have been linked to some potentially harmful effects on human health. Nonetheless, it should be emphasized that all oxygenates form higher amounts of aldehyde emission than nonoxygenated gasoline. Furthermore, the effect on the health is negligible as proven by the Royal Society of Canada taking into account, in addition, that the catalytic convertors of new vehicles reduce these aldehyde levels to a higher degree (Canadian Renewable Fuels Association, 2000). In the case of old automobiles that do not have this kind of convertor, the level of formed aldehydes, although increased when ethanol is employed as the gasoline oxygenate, is always below the permissible limits. Actually, the emission of this type of organic compound is very low compared to other types of dangerous emissions (e.g., aromatic hydrocarbons), which are effectively reduced when fuel ethanol is used (see Table 1.2).

There exists a great debate on the environmental suitability of ethanol usage as an oxygenate, especially when blends with low ethanol contents are employed. Reviewing various literature sources, Niven (2005) points out that gasoline blends with 10% content of ethanol (E10) offer few advantages in terms of greenhouse gas emissions, energy efficiency, or environmental sustainability. In addition, this author indicates that E10 blends increase both the risk and severity of soil and groundwater contamination, although greenhouse gas benefits for 85% ethanol blends (E85) are recognized. By contrast, the Argonne National Laboratory (USA) estimates that an 8 to 10% reduction in greenhouse gas emissions per vehicle mile traveled is achieved when biomass ethanol is used in E10 blends and 68 to 91% reduction when used in E85 blends (Wang et al., 1999). This controversy has arisen in countries with an important ethanol industry, as in the case of the United States. Authors such as Pimentel (2003) have maintained for several years that the energy required for producing ethanol is greater than the energy contained in the ethanol itself, particularly when starchy materials like corn are used. This implies that important natural resources of a national economy are being squandered by maintaining an artificial biofuels program. Nevertheless, most studies have concluded that the energy invested in ethanol production is less than its energy content, which allows achieving significant environmental benefits. These studies have been accomplished by both independent research groups and governmental centers and have belied Pimentel's arguments, as shown in the works of Shapouri et al. (2003) and Wang et al. (1999).

1.4 GASOLINE OXYGENATION PROGRAMS WITH FUEL ETHANOL IN SOME COUNTRIES

Fuel ethanol production has increased remarkably because many countries look to reduce oil imports, boost rural economies, and improve air quality. The world ethyl alcohol production was about 64 billion liters in 2008 (Renewable Fuels Association, 2009), with the United States being the first producer (see Table 1.3). On average, 86% of produced ethanol worldwide corresponds to fuel ethanol, 17% to beverage ethanol, and 10% to industrial ethanol. Reported data taken from

TABLE 1.3
World Production of Ethanol (in Million Liters)

Country	2008	2007	2006[a]	2005[a]	References
1. USA	34,065.00	24,597.20	18,376.00	16,139.00	Renewable Fuel Asociation (2009)
2. Brazil	24,497.28	18,997.67	16,998.00	15,999.00	Renewable Fuel Asociation (2009)
3. China	1,899.69	1,839.51	3,849.00	3,800.00	Renewable Fuel Asociation (2009)
4. Canada	899.69	799.77	579.00	231.00	Renewable Fuel Asociation (2009)
5. Thailand	339.89	299.77	352.00	299.00	Renewable Fuel Asociation (2009)
6. Colombia	300.11	283.50	269.00[b]	27.00[b]	Renewable Fuel Asociation (2009); Londoño (2007)
7. India	249.81	199.85	1,900.07	1,699.00	Renewable Fuel Asociation (2009)
Total	64,259.48	49,024.27	44,329.07	40,199.00	Renewable Fuel Asociation (2009)

[a] Industrial and beverage alcohol are included.
[b] These data correspond to the fuel ethanol produced in new distilleries whose construction started in 2005 (Londoño, 2007); industrial and beverage alcohol are not included.

different sources indicate that Brazil and the United States account for 82% of world production of fuel ethanol, though this percentage is changing constantly due to the dynamics of this biofuel on the global market. Asia and the EU are following the large producers. China has the biggest plant for ethanol production with an annual capacity of 320 million gallons; this plant is located in the Jilin province and currently produces 240 million gallons per year (Murray, 2005). India is seeking to enhance its ethanol production to meet its auto biofuel program; it has envisioned the use of ethanol not only for gasoline blends, but also for ethanol–diesel blends and for biodiesel production (Subramanian et al., 2005).

Europe's fuel ethanol sector was a slow starter. It took over 10 years to grow production from 60 million liters in 1993 to 525 million liters in 2004. In the following two years, we saw a true explosion in production. In 2005 and 2006, there were double-digit growth levels of over 70%. However, it was not a sustainable growth; in 2007, production increased by only 11%. Total EU production in 2008 was 2.8 billion liters, up from 1.8 billion liters the previous year. This represents a significant increase of 56%. This increase was due to the growth in French production, which almost doubled to 1 billion liters in 2008 (539 million liters in 2007). This made France the biggest EU fuel ethanol producer in 2008, followed by Germany that also expanded its production to 568.5 million liters, which represented a 32.5% increase in internal Germany fuel ethanol production. The third highest producing country was Spain with 317 million liters. In

2008, fuel ethanol production capacity increased in Belgium. Finland resumed its production in 2008 and, in Austria, fuel ethanol has been produced for the first complete year (European Bioethanol Fuel Association, 2009).

Many countries have implemented or are implementing programs for addition of ethanol to gasoline (Table 1.3). Through the ProAlcool program, Brazil has been utilizing hydrous ethanol as a fuel and anhydrous ethanol as an oxygenate. This country produces ethanol from sugarcane. Although, at the beginning of this program, concerns about the disjunction between food production versus fuels (sugar versus ethanol) were expressed, it has been demonstrated that when the food shelves remained empty it was caused by the destination of food production to the export markets. In the same way, when the shelves remained full, it was because large proportions of the population lacked the purchasing power to buy food (Rosillo-Calle and Hall, 1987). However, due to the liberalization of the Brazilian fuel market, the official end of the ProAlcool program, the gradual elimination of subsidies, and the political and economic conjuncture, production and consumption paces of fuel ethanol diminished at the end of twentieth century, although a reactivation is expected in the coming years (Rosillo-Calle and Cortez, 1998; Wheals et al. 1999). Brilhante (1997) indicates that the pursuit of ethanol fuel in Brazil was not based on long-term plans with deep-set values, but has been a response to such circumstances as a depressed sugar industry, an ambitious attempt to reduce oil dependency, and more recently, "green" arguments.

Brazil is the second largest producer of fuel ethanol in the world and the first in world's exports (Renewable Fuels Association, 2007). Brazil is considered as the best economical model for ethanol production in the world because it does not implement subsidies for fuel ethanol production. However some authors consider that the successful Brazilian ethanol model is sustainable only in Brazil due to its advanced technology and its enormous amount of arable land available (Sperling and Gordon, 2009). Currently, Brazil is actively seeking export markets for its bioethanol (Thomas and Kwong, 2001), especially in Japan, which will become a net importer of ethanol (Orellana and Neto, 2006). Brazil's 30-year-old fuel ethanol program is based on the most efficient agricultural technology for sugarcane cultivation in the world, using modern equipment and cheap sugarcane as feedstock. The residual cane waste (bagasse) is used to process heat and power, which results in a very competitive price and also in a high energy balance (output energy/input energy), which varies from 8.3 for average conditions to 10.2 for best practice production (Macedo et al., 2004).

In the case of the United States, ethanol is used either as a fuel substitute or as an oxygenate. At present, both Ford and Chrysler offer standard models designed to run on either 85% ethanol (E85) or gasoline. The addition of gasoline to the E85 or E95 mixtures is done for improving cold start in engines using these kinds of fuel. However, the best perspectives for bioethanol can be found in the oxygenate market. Fuel ethanol production was boosted by the passage of the Clean Air Act in 1970, and especially by the Clean Air Amendments in 1990. Current regulations aimed at controlling CO emissions have already produced a significant demand for ethanol as an oxygenate. The tax credit gives oil companies an economic incentive to blend ethanol and gasoline. It is currently $0.51/gallon for pure ethanol, scaled to the amount of ethanol in the blend. Additionally, a $0.54/gal tariff on ethanol from Brazil was established to protect the domestic ethanol industry. The 2008 farm bill reduced the blenders credit to $0.46/gal and maintained the tariff through 2010 (Keeney, 2009). The establishment of the Renewable Fuel Standard (biofuels) provided additional incentives for ethanol use, mainly in the form of binding mandates. The 2006 act required that 4.0 billon gallons per year (BGPY) be mixed with gasoline, 6.1 BGPY by 2009, and 7.5 BGPY by 2012. The Energy Independence and Security Act (EISA) of 2007 then revised the mandates to require 15 BGPY by 2015. No doubt the industry is highly subsidized; one analysis estimates that the biofuels industry in the United States receives a $60-billion subsidy each year (Keeney, 2009).

The EU has issued different directives about the addition of renewable oxygenates to fuels. The oxygenation target for fuels estimates the addition of up to 2 wt.% by 2005 and 5.75% by 2010. However, the implementation of these directives varies too much among the different countries. Spain and France are leading in the production of bioethanol in Europe. In contrast, Germany has developed the production of methyl esters of fatty acids obtained from rapeseed as biodiesel. For this country, it is considered that the production of fuel ethanol is not economically feasible in comparison to gasoline due to the high costs of feedstocks (grains, sugar beet; Henke et al., 2005; Rosenberger et al., 2002). This situation can change dramatically if volatility and high prices in the oil market remain.

Besides Brazil, other Latin American countries are implementing fuel ethanol programs. Colombia is a country with important oil reserves. However, the country imports gasoline due to its refining limitations. In addition, the pace of new oil discoveries in the country has decreased. Colombia faces the depletion of its reserves and possible oil imports in 2015. For this reason, besides environmental considerations, the country has decided on replacing part of the fossil fuels with alternative energy sources. The production and use of liquid biofuels (bioethanol and biodiesel) is the strategy adopted by the Colombian government to diminish the oil dependency and reduce the polluting gases released by the transportation sector in the Colombian cities. Moreover, it is expected that the implementation of the biofuels programs allows the development of the rural sector that, in certain regions, is attacked by the violence associated with the use of land for illegal crops.

Colombia started the addition of 10% (v/v) ethanol to gasoline in cities with more than 500,000 inhabitants in November 2005 with the Act 693 of 2001 issued by the Congress. This disposition will be extended to the entire country in the coming years. The perspectives for the ethanol market in Colombia are bright. The amount of fuel ethanol needed in the zones where the E10 program is being implemented reaches 1,050,000 L/d. The expansion of the program to the rest of the country requires a total ethanol production of about 1,600,000 L/d. In addition, Colombia has all the conditions to become an ethanol-exporting country considering its agro-ecological conditions, its geographical location, as well as the growth of potential markets in North America, Europe, and Asia. To make the most of these conditions, the country should produce an estimated 3,800,000 L/d of fuel ethanol by 2020 to meet the national demand and export the surplus to new growing markets (Proexport Colombia, 2005).

The implementation of the E10 program depends on raw materials and technological limitations. For this reason and driven by the availability of the feedstock and the capacity of the well-structured sugar industry, the main cities of the West Southern and West Central departments (administrative regions into which Colombia is divided) started this program. The Cauca River valley is the major sugarcane production region and it is located in these departments. This region concentrates the best cultivable lands for sugarcane cropping (about 200,000 Ha) because it has the most appropriate conditions: extensive and fertile alluvial valleys located at an average of 1,000 m above sea level in tropical latitude. Currently, five ethanol production plants co-located in large sugar mills are operating. However, the synergies of the sugar sector, controlled by a few economic groups, have not been allowed to achieve a great impact on the creation of new rural jobs. According to estimations by the Colombian Ministry of Agriculture (Ministerio de Agricultura y Desarrollo Rural, 2006), the surface cropped with sugarcane in 2006 did not increase compared to data obtained in 2005 (by the Columbian sugar industry). Despite the enhancement in the production of fuel ethanol during these two years, Colombia has not projected any increase in cane plantations. This means that the generation of new rural jobs related to ethanol production is practically nil. In fact, part of the cane produced is diverted to ethanol distilleries, decreasing the volumes of sugar for export. Regulations issued by the government have included the sugar value in the international market as a variable considered in calculating the price of fuel ethanol produced in the country. Thus, future increases in the international price of sugar will not discourage the national production of fuel ethanol. This situation allows the development of new commercial projects for ethanol production in order to guarantee permanent use of ethyl alcohol for the E10 program.

The Colombian government expects that the construction of new ethanol-producing facilities in other regions of the country will lead to real improvement in the economic situation of the rural communities through new demands for agricultural raw materials. Therefore, other feedstocks for ethanol production are being analyzed considering the future growth of the fuel ethanol market. Corn, cassava, and beets have been considered as potential feedstocks. Sugarcane is also considered for ethanol production in zones other than in the Cauca River valley. This encourages the cropping of cane by specific rural communities not related to the big sugar companies. Many rural communities cultivate sugarcane for producing noncentrifugal sugar called panela (solid brown sugar), a sweetener and low-cost beverage base widely used by popular segments in Colombia. However, the quality of life of panela producers is traditionally low within the Colombian context. For this reason, the government is actively encouraging the organization of the communities linked to the panela economy for them to supply the feedstock for new projects of fuel ethanol production.

The relatively mature technology for ethanol production from corn is one of the options to be considered under Colombian conditions because corn is an important crop mainly cultivated in northern and eastern regions. The construction of ethanol plants using corn could offer the production of valuable co-products (e.g., dried distillers grains with solubles for cattle food). Furthermore, this technology may be the base for the development of ethanol production processes from other starchy materials such as cassava or potatoes, crops with a significant economic importance in Colombia. Additionally, the projected signing of a free trade agreement with the United States, the first world producer of corn, implies the search for new markets for local corn production.

Ethanol can be obtained synthetically from coal and natural gas. The major producer of this type of alcohol is South Africa (Thomas and Kwong, 2001). It can be obtained by oxidation of olefins as well. However, 95% of world ethanol is produced by fermentation from carbohydrate-containing feedstock (Berg, 2004). The main part of ethanol is produced in batches.

Substrate concentration at the beginning of fermentation is 15 to 25% (w/v) solids and the pH is adjusted to a value of 4 to 5 with the aim of reducing infection risks. The process is carried out at 30 to 35°C. Generally, yield reaches 90% of theoretical maximum, expressed in g EtOH/g substrate. The rest of the substrate is converted into biomass and other by-products such as glycerol and acetate. Formed biomass can be utilized many times. Ethanol concentration at the end of fermentation is 80 to 100 g/L (Claassen et al., 1999). In most of the distilleries, fermentation time is 24 h, employing 6 h for yeast sedimentation in the vessels (Pandey and Agarwal, 1993).

REFERENCES

Ancillotti, F., and V. Fattore. 1998. Oxygenate fuels: Market expansion and catalytic aspect of synthesis. *Fuel Processing Technology* 57:163–194.

Berg, C. 2004. *World Fuel Ethanol. Analysis and Outlook.* F.O. Licht. http://www.distill. com/World-Fuel-Ethanol-A&O-2004.html (accessed February 2005).

Berndes, G., C. Azar, T. Kåberger, and D. Abrahamson. 2001. The feasibility of large-scale lignocellulose-based bioenergy production. *Biomass and Bioenergy* 20:371–383.

Braids, O. 2001. MTBE—Panacea or problem. *Environmental Forensics* 2:189–196.

Brilhante, O.M. 1997. Brazil's alcohol programme: From an attempt to reduce oil dependence in the seventies to the green arguments of the nineties. *Journal of Environmental Planning and Management* 40 (4):435–449.

Bull, S.R. 1994. Renewable alternative fuels: Alcohol production from lignocellulosic biomass. *Renewable Energy* 5 (II):799–806.

Canadian Renewable Fuels Association. 2000. *Environmental Benefits of Ethanol.* Canadian Renewable Fuels Association. http://www.sentex.net/~crfa/emissionsimpact.html (accessed October 2006).

Cannell, M.G.R. 2003. Carbon sequestration and biomass energy offset: Theoretical, potential and achievable capacities globally, in Europe and the UK. *Biomass and Bioenergy* 24:97–116.

Cardona, C.A., L.G. Matallana, A. Gómez, L.F. Cortés, and Yu.A. Pisarenko. 2002. Phase equilibrium and residue curve analysis in DIPE production by reactive distillation. Paper presented at the IV Iberoamerican Conference on Phase Equilibria and Fluid Properties for Process Design, Foz de Iguazu, Brazil.

Cardona, C.A., Yu. A. Pisarenko, and L.A. Serafimov. 2000. ДИПЭ как альтернативная добавка к моторным топливам (DIPE as alternative additive to motor fuels, in Russian). *Science and Technology of Hydrocarbons* 4 (11):72–73.

Central Illinois Manufacturing Company. 2006. *Switching to Ethanol? Guidelines for Conversion of Retail Service Station Tanks from Gasoline to EthanolBlended Gasoline.* Central Illinois Manufacturing Company. http://www.cimtek.com/pdfs/resources_switchingtoethanol.pdf (accessed February 2009).

Chaves, M. 2004. *Etanol: Un biocombustible para el futuro* (Ethanol: A biofuel for the future, in Spanish). Paper presented at Antecedentes y Capacidad Potencial de Cogenerar Energía y Producir Etanol por Parte del Sector Azucarero Costarricense, at San José, Costa Rica.

Chum, H.L., and R.P. Overend. 2001. Biomass and renewable fuels. *Fuel Processing Technology* 171:187–195.

Claassen, P.A.M., J.B. van Lier, A.M. López Contreras, E.W.J. van Niel, L. Sijtsma, A.J.M. Stams, S.S. de Vries, and R.A. Weusthuis. 1999. Utilisation of biomass for the supply of energy carriers. *Applied Microbiology and Biotechnology* 52:741–755.

Demirbas, M.F., and M. Balar. 2006. Recent advances on the production and utilization trends of bio-fuels: A global perspective. *Energy Conversion and Management* 47:2371–2381.

Energy Information Administration. 2008. *International Energy Outlook 2008,* Energy Information Administration, Washington, D.C. www.eia.doe.gov/oiaf/ieo/index.html.

Espinal, C.F., H.J. Martínez, and E.D. González. 2005. The chain of oilseeds, fats, and oils in Colombia: An overview of its structure and dynamics 1991–2005, in Spanish. Documento de Trabajo No. 93, Obervatorio Agrocadenas Colombia, Ministerio de Agricultura y Desarrollo Rural.

European Biodiesel Board. 2008. *Statistics the EU Biodiesel Industry.* European Biodiesel Board. http://www.ebb-eu.org/stats.php (accessed February 2009).

European Bioethanol Fuel Association. 2009. *The EU Market* (accessed February 2009).

European Fuel Oxygenates Association. 2009. *Ethers improve air quality and reduce CO2 Emissions.* The European Fuel Oxygenates Association. http://www.efoa.org/ (accessed February 2009).

French, R., and P. Malone. 2005. Phase equilibria of ethanol fuel blends. *Fluid Phase Equilibria* 228–229:27–40.

Ghosh, P., and T.K. Ghose. 2003. Bioethanol in India: Recent past and emerging future. *Advanced Biochemical Engineering and Biotechnology* 85:1–27.

Graham, M., P. Pryor, and M.E. Sarna. 2000. Refining options for MTBE-free gasoline. Paper presented at the NPRA (National Petrochemical and Refiners Association) Annual Meeting, San Antonio, TX.

Grassi, G. 1999. Modern bioenergy in the European Union. *Renewable Energy* 16:985–990.

Harandi, M.N., W.O. Haag, H. Owen, and W.K. Bell. 1992. Ether production. U.S. patent 5144086.

Henke, J.M., G. Klepper, and N. Schmitz. 2005. Tax exemption for biofuels in Germany: Is bio-ethanol really an option for climate policy? *Energy Policy* 30:2617–2635.

Huttunen, H., L.E. Wyness, and P. Kalliokoski. 1997. Identification of environmental hazards of gasoline oxygenate *tert*-amyl methyl ether (TAME). *Chemosphere* 35 (6):1199–1214.

IEA. 2004. *World Energy Outlook 2004.* Paris: International Energy Agency (IEA) Publications.

Keeney, D. 2009. Ethanol USA. *Environmental Science and Technology* 43:8–11.

Kharoune, M., A. Pauss, and J.M. Lebeault. 2001. Aerobic biodegradation of an oxygenates mixture: ETBE, MTBE and TAME in an upflow fixed-bed reactor. *Water Research* 35 (7):1665–1674.

Kheshgi, H.S., and R.C Prince. 2005. Sequestration of fermentation CO_2 from ethanol production. *Energy* 30:1865–1871.

Kosaric, N., and J. Velikonja. 1995. Liquid and gaseous fuels from biotechnology: Challenge and opportunities. *FEMS Microbiology Reviews* 16:111–142.

Londoño, L.F. 2007. *Informe anual.* Annual report. Colombian sugar sector, in Spanish. Asocaña. http://www.asocana.com.co/informes/default.aspx (accessed September 2007).

Ma, F., and M. Hanna. 1999. Biodiesel production: A review. *Bioresource Technology* 70:1–15.

Macedo, I.C., L.V.L. Manoel, and J. Da silva. 2004. *Assessment of greenhouse gas emissions in the production and use of fuel ethanol in Brazil,* Government of the State of São Paulo, Brazil.

Ministerio de Agricultura y Desarrollo Rural. 2006. Perspectives of the farming sector: Second semester of 2006. Bogotá, Colombia: Ministerio de Agricultura y Desarrollo Rural, República de Colombia.

Ministerio del Medio Ambiente. 1994. Resolución No. 415 del 24 de noviembre de 1994. Ministerio del Medio Ambiente, República de Colombia.

Murray, D. 2005. *Ethanol's potential: Looking beyond corn. Eco-economy updates.* Earth Policy Institute. http://earth-policy.org/Updates/2005/Update49.htm (accessed July 2005).

Nadim, F., P. Zack, G.E. Hoag, and S. Liu. 2001. United States experience with gasoline additives. *Energy Policy* 29:1–5.

Niven, R.K. 2005. Ethanol in gasoline: Environmental impacts and sustainability review article. *Renewable and Sustainable Energy Reviews* 9:535–555.

Oost, C., K. Sundmacher, and U. Hoffmann. 1995. Synthesis of tertiary amyl methyl ether (TAME): Equilibrium of the multiple reactions. *Chemical Engineering & Technology* 18:110–117.

Orellana, C., and R.B. Neto. 2006. Brazil and Japan give fuel to ethanol market. *Nature Biotechnology* 24 (3):232.

Oudshoorn, O.L., M. Janissen, W.E.J. van Kooten, J.C. Jansen, H. van Bekkum, C.M. van den Bleek, and H.P.A. Calis. 1999. A novel structured catalyst packing for catalytic distillation of ETBE. *Chemical Engineering Science* 54:1413–1418.

Pandey, K., and P.K. Agarwal. 1993. Effect of EDTA, potassium ferrocyanide, and sodium potassium tartarate on the production of ethanol from molasses by *Saccharomyces cerevisiae. Enzyme and Microbial Technology* 15:887–898.

Pimentel, D. 2003. Ethanol fuels: Energy balance, economics, and environmental impacts are negative. *Natural Resources Research* 12 (2):127–134.

Poitrat, E. 1999. The potential of liquid biofuels in France. *Renewable Energy* 16:1084–1089.

Proexport Colombia. 2005. *Colombia. Sectorial Profile Agro-Industry.* Bogotá, Colombia: Proexport Colombia.

Rasskazchikova, T.V., V.M. Kapustin, and S.A. Karpov. 2004. Ethanol as high-octane additive to automotive gasolines. Production and use in Russia and abroad. *Chemistry and Technology of Fuels and Oils* 40 (4):203–210.

Renewable Fuel Association. 2009. *Statistics.* Renewable Fuel Asociation. http://www. ethanolrfa.org/industry/statistics/#E (accessed February 2009).

Renewable Fuels Association. 2007. *Industry statistics.* Renewable Fuels Association. http://www.ethanolrfa.org/industry/statistics (accessed September 2007).

Rosenberger, A., H.-P. Kaul, T. Senn, and W. Aufhammer. 2002. Costs of bioethanol production from winter cereals: The effect of growing conditions and crop production intensity levels. *Industrial Crops and Products* 15:91–102.

Rosillo-Calle, F., and L. Cortez. 1998. Towards Proalcool II—A review of the Brazilian ethanol programme. *Biomass and Bioenergy* 14 (2):115–124.

Rosillo-Calle, F., and D.O. Hall. 1987. Brazilian alcohol: Food versus fuel? *Biomass* 12:97–128.

Sánchez, Ó.J., and C.A. Cardona. 2008a. *Producción de Alcohol Carburante: Una Alternativa para el Desarrollo Agroindustrial (Fuel ethanol production: An alternative for agro-industrial development*, in Spanish). Manizales: Universidad Nacional de Colombia.

Sánchez, Ó.J., and C.A. Cardona. 2008b. Trends in biotechnological production of fuel ethanol from different feedstocks. *Bioresource Technology* 99:5270–5295.

Shapouri, H., J.A. Duffield, and M. Wang. 2003. The energy balance of corn ethanol revisited. *Transactions of the ASAE* 46 (4):959–968.

Sperling, D., and D. Gordon. 2009. *Two billions cars: Driving toward sustainability*. New York: Oxford University Press.

Subramanian, K.A., S.K. Singal, M. Saxena, and S. Singhal. 2005. Utilization of liquid biofuels in automotive diesel engines: An Indian perspective. *Biomass and Bioenergy* 29:65–72.

Thiel, C., K. Sundmacher, and U. Hoffmann. 1997. Synthesis of ETBE: Residue curve maps for the heterogeneously catalysed reactive distillation process. *Chemical Engineering Journal* 66:181–191.

Thomas, V., and A. Kwong. 2001. Ethanol as a lead replacement: Phasing out leaded gasoline in Africa. *Energy Policy* 29:1133–1143.

Thornley, P. 2006. Increasing biomass based power generation in the UK. *Energy Policy* 34:2087–2099.

Tshiteya, R.M., E.N. Vermiglio, and S. Tice. 1991. Basic chemistry of alcohol fuels. In *Properties of Alcohol Transportation Fuels*. Alexandria, VA: Meridian Corporation.

Wang, M., C. Saricks, and D. Santini. 1999. *Effects of fuel ethanol use on fuel-cycle energy and greenhouse gas emissions*. Report no. ANL/ESD-38, Argonne National Laboratory, Center for Transportation Research. www.transportation.anl.gov/pdfs/TA/58.pdf

Wheals, A.E., L.C. Basso, D.M.G. Alves, and H.V. Amorim. 1999. Fuel ethanol after 25 years. *TIBTECH* 17:482–487.

Yüksel, F., and B. Yüksel. 2004. The use of ethanol–gasoline blend as a fuel in an SI engine. *Renewable Energy* 29:1181–1191.

2 Process Design and Role of Process Synthesis

In this chapter, the importance of conceptual process design during the development of innovative process configurations is analyzed. The utilization of some useful tools for process design is highlighted, such as mathematical modeling and simulation. The key role played by process synthesis methodologies is emphasized. The two major strategies for accomplishing process synthesis are discussed. Main trends in knowledge-based process synthesis are briefly presented as well as the main approach to carrying out optimization-based process synthesis.

2.1 CONCEPTUAL PROCESS DESIGN

Process systems engineering deals with the development of procedures, techniques, and tools to undertake the generic problems of design, operation, and control of productive processes related to the different sector of process and chemical industry (Perkins, 2002). As part of process systems engineering, process design plays a fundamental role during the development of efficient technologies, especially in technoeconomic and environmental terms, in order to produce a wide range of industrial products. In this way, its main objective consists of the selection and definition of a process configuration that makes possible the conversion of feedstocks into the end product. This should be done in such a way that the products meet given specifications and that the configuration performance is superior to existing ones, or nonexisting ones in the case of a new product introduced into the market. In general, process design can be accomplished from the perspective of sequential engineering, concurrent engineering, reverse engineering, or reengineering.

From the viewpoint of the life cycle of an industrial process, the sequential engineering in its classical version involves all the steps presented in Table 2.1 in a sequential manner. Reverse engineering is based on obtaining technical information of a product in order to determine what it is made of, what makes it work, and how it was constructed. This approach is particularly useful in the case of pharmaceuticals production, especially during the production of generic drugs. For ethanol, this approach is not applicable because its chemical structure and ways of production are relatively well known. Reengineering looks for the radical change of process design through its fundamental revision to achieve decisive improvements in terms of quality, costs, celerity, flexibility, customer satisfaction, etc., all of them simultaneously. Reengineering does not look for incremental improvements, but drastic changes that allow reaching the predefined targets. On the other hand, concurrent engineering proposes the creation of a design environment where all the actors involved in the development of a product participate,

TABLE 2.1

Steps Involved in the Development of an Industrial Process during Its Life Cycle

Stages	Features
Analysis of the chemical reaction	Study of chemical synthesis
	Experimental development
	Selection of the best stages for the synthesis
Conceptual process design	Analysis of integration possibilities
	Integration of function
	Heuristic design
	Superstructure design
	Superstructure optimization
Process development	Experimental data obtained
	Test reaction and separation process
	Obtain kinetic and physical data
	Pilot plant
Process assessment	Evaluation of cost of equipment
	Definition of control schemes
Detailed engineering	Plant layout
	Definition and construction of equipment
	Pipeline design
	Energy networks definition
	Utilities design
Plant operation	Scheduling
	Stocks handling
	Preventive maintenance and repave
End cycle of life	Second use possibilities
	Dismantle and recycle parts

not only the designers. In this way, processing features and market demands will be taken into account during early stages of design, when the changes are easier and less expensive to implement. Therefore, the problems, conflicts, and change needs can be detected in time to carry out the necessary modifications with substantially less effort than by means of sequential engineering. In this framework, concurrent engineering applied to design implies the integration of all life cycle steps of a process in the early stages, attaining the achievements of several goals. Thus, the research and development activities and conceptual design have to consider not only the building of a plant, its operation, control, and maintenance, but also the achievement of technoeconomic, market, environmental, and even social objectives. Practically, and for the case of commodities, the three first stages of the process life cycle are accomplished in a concurrent (simultaneous) way. If the synthesis pathways for a given product are already known, as in the case of fuel ethanol, process design procedures are focused on the second and third stages in Table 2.1.

In this book, these two steps are analyzed for fuel ethanol production emphasizing the related integrated processes. In fact, concurrent engineering elements are considered when taking into account as evaluation criteria not only technical indexes (yield, productivity, energy consumption), but also financial and environmental indicators in the framework of process intensification. Financial and environmental criteria correspond to the macro and mega scale levels of analysis (plant and unit integration and interaction between market conditions and environmental impact, respectively), as reported by Li and Kraslawski (2004). Just these two levels of analysis have been developed with more intensity in the past 15 years as a result of the globalization of the economy and worsening environment. On the other hand, process and apparatus integration corresponds to the micro scale level of analysis. At this level, process intensification through integrated and hybrid processes with higher efficiency and less size has become the most important development trend. This forced the change of the old paradigm of a chemical process made up of a series of unit operations where the processes and apparatus are coupled (meso scale). Finally, the nano scale (molecular design and new materials) has become crucial for designing processes to obtain very high value-added products.

The task of defining an appropriate process configuration requires the generation and evaluation of many technological schemes (process flowsheets) in order to find those exhibiting better performance indicators. This task is called *process synthesis*. In a process synthesis problem, system inputs (type, composition, conditions, and flowrates of raw materials) and outputs (product flowrate and specifications, effluent streams constraints) are given and the task consists of defining the configuration of the process flowsheet or, in other words, the topology of the technological scheme, which allows the synthesis of the product from the feedstocks entering the process. For this, at least one comparison criterion should be established with the aim of evaluating different alternative process flowsheets proposed in order to choose that with the better performance.

The configuration comprises the type and amount of unit processes and operations required by the overall process as well as their interconnection (intermediate, recycle, and purge streams) and the parameters of that configuration (mostly those ones related to operating conditions: flowrates, temperatures, pressure, compositions). Process synthesis procedures can be applied not only to the conceptual design of new processes, but also to the retrofitting of existing ones. Some approaches for process synthesis involve and apply fundamental concepts of thermodynamics as the starting point for generating new alternative process configurations. Thus, energy consumption (calculated by enthalpy balances) of different flowsheets can be helpful for selection of the best alternatives. In a similar way, the concept of useful energy or exergy (widely employed in mechanical engineering) can also be employed as a criterion for selection of alternatives. Recently, more global concepts from the ecology field, such as emergy, have been used for choosing the best configuration of a process (see Section 2.2.5).

During the next step of the life cycle of an industrial process, *process analysis*, the structure of the selected technological scheme is established in order to improve the global process through its more comprehensive insight. The type of problems undertaken by process analysis is summarized as follows: given the process inputs and once determined the technological configuration of the process that includes each one of the unit operations and processes involved, as well as their parameters, find the system outputs. The analysis is aimed at predicting how the given process behaves. It involves the process decomposition into its constituent elements for the individual study of each unit performance. The detailed features of the process (flowrates, pressures, temperatures, compositions) are predicted by using mathematical models, empiric correlations, and computer-aided process simulation. Moreover, experimental methods are employed to study the system behavior as well as validate the theoretical approach used for its description. It could be considered that the conceptual design stage corresponds to process synthesis activities and the development stage to process analysis activities, although there is no clear boundary between these two activities of process system engineering.

Though process synthesis is carried out prior to process analysis, both tasks interact with each other in order to achieve their goals. Thus, for evaluating the performance of an alternative technological configuration during process synthesis, mathematical models are required in order to predict the behavior of process units. This involves a distinctive task of process analysis. Usually, aggregated models are employed in process synthesis. Such models simplify the synthesis problem in a considerable way through the representation of one aspect or objective that tends to dominate the problem. Furthermore, short-cut models are utilized as well. In this kind of model, the description of the units involved in each proposed design is done through relatively simple nonlinear models with the aim of reducing the computational costs or exploiting the algebraic structure of the equations. During process analysis, more rigorous and complex models are involved to predict the performance of the different units, which make up part of a technological scheme (Grossmann et al., 2000). On the other hand, *process optimization* plays a very important role during process design. Once the structure with the best performance (this can, in turn, imply the employment of optimization tools) is defined, and knowing the structure of the system components that allow the prediction of its behavior, the optimization of the technological scheme can be accomplished in order to find the optimal operating parameters making possible the maximization or minimization of an objective function that evaluates the performance of the overall system based on one or more criteria (technical, economic, environmental).

2.2   KNOWLEDGE-BASED PROCESS SYNTHESIS

The generation of technological schemes includes the determination of the optimal (or near optimal) process flowsheet plus the alternative configurations. This implies that the task of process synthesis is a complex activity that can be divided into several subproblems (Gani and Kraslawski, 2000). In this book, these subproblems can comprise the following jobs applied to the production of fuel ethanol:

- Generation of alternative processing pathways
- Identification of required unit operations
- Analysis of many sequences of unit operations to assemble an optimal or at least a feasible configuration
- Configuration of distillation columns for separation of different mixtures (including azeotropic ones)
- Configuration of reaction–reaction and reaction–separation integration processes
- Other additional issues as heat integration

To undertake this task, a series of solving strategies has been proposed. They can be classified into two large groups: knowledge-based process synthesis and optimization-based process synthesis. The second group of strategies is oriented to the formulation of a synthesis problem in the form of an optimization problem. In turn, the first group of strategies is concentrated on the representation of the design problem as well as the organization of the knowledge required by this problem (Li and Kraslawski, 2004), i.e., it is oriented to the development of a representation that is rich enough to allow all the alternatives to be included, and "smart" enough to automatically ignore illogical options. In general, the procedure of generating multiple alternative process flowsheets is carried out through different strategies implying, for instance, the combination of heuristic rules with evolutionary strategies for process design. Heuristic methods are particularly useful when very large and complex problems are dealt with. Usually, these methods are often based on the observation of how many other problems of the same type are solved. Knowledge-based methods imply an evolution in the learning of the research subject as the proposed problem is being solved. This, in turn, leads to the generation of new and better alternatives. Nevertheless, it is impossible to ensure that the optimal structure can be achieved when heuristic methods are employed (Westerberg, 2004). Some approaches for knowledge-based process synthesis mostly used in process systems engineering, as well as the most innovative and perspective ones, are briefly described below.

2.2.1 Evolutionary Modification

Evolutionary modification is the conventional approach for process synthesis based on the experience of engineers and researchers. This approach condenses this experience into a programmed set of heuristic rules intended to making decisions on the process structure. Considering that this method corresponds to a trial-and-error approximation, its main limitation consists in the difficulty of obtaining relevant information in a way suitable for computational calculations.

2.2.2 HIERARCHICAL DECOMPOSITION

The hierarchical heuristic method is an extension of the purely heuristic approach that entails the evolutionary modification and combines the heuristic rules with an evolutionary strategy for process design (Li and Kraslawski, 2004). Douglas (1988) proposed a method by which any process can be decomposed into five levels of analysis for its design. This strategy has a hierarchical sequential character considering that in each level of analysis different decisions are made based on heuristic rules. This allows generating different alternatives, which are evaluated from an economic point of view using short-cut models. As the method is applied in each one of the five levels, more information becomes available and the technological scheme of the process evolves until its completion. According to this author, hierarchical decomposition comprises the analysis of the process in the following levels:

- Batch versus continuous
- Input–output structure of the flowsheet
- Recycle structure of the flowsheet
- Separation system synthesis
- Heat recovery network

A slightly different hierarchical decomposition scheme corresponds to the onion diagram (Smith, 2005), which starts from the reaction step toward effluent treatment:

- Reactor
- Separation and recycle system
- Heat recovery system
- Heating and cooling utilities
- Wastewater and effluent treatment

The design begins with the reactor selection to move toward the outer surface of the diagram by adding other layers, such as the separation and recycle system. These heuristic methods emphasize the decomposition strategy and screening of alternatives, which allow the fast identification of technological configurations often located near optimal solutions (Li and Kraslawski, 2004). However, the main drawback of these methods is the impossibility of handling the interactions among the different levels or layers due to their sequential character. In spite of this, the nature of this approach allows rapidly discarding many alternative configurations not leading to "good" designs. In addition, the analysis by design levels permits the utilization of process simulators with which the process flowsheets are completed in an evolutionary way. This methodology has been mostly applied to chemical and petro-chemical processes. Nevertheless, the utilization of these procedures and design schemes is less frequent in the processes of the microbiologic industry.

2.2.3 Phenomena-Driven Design

The phenomena-driven approach for process synthesis considers as a starting point for the design, not the unit operations, but the phenomena occurring in them at a lower level of aggregation (Gavrila and Iedema, 1996). Reactions, phase changes, heat and mass transfer, and mixing are considered among the phenomena included in this method. The design problem is divided (decomposed) into three tasks: role assignment, phenomena grouping, and operating condition analysis. The goals of these tasks can be formulated through the following questions:

- What should occur in the process in order to achieve the global design target?
- Where should it occur?
- When and how should it occur?

In the second task, the alternative designs are proposed by grouping the phenomena in units and the continuous variables are described as Boolean variables (for instance, is the rate of a phenomenon equal to zero or greater than zero?). In the third task, the favorable conditions in the units are defined employing ordinal relations among quantities ($dR/dt > dS/dt$; Gavrila and Iedema, 1996). This method is oriented to explore innovative units and processes in order to support the creativity during the design process. However, the method is based on an opportunistic identification and integration of the tasks as pointed out by Li and Kraslawski (2004). This approach has been applied to very particular cases as in the production of methyl tert-butyl ether (MTBE) by reactive distillation (Tanskanen et al., 1995).

2.2.4 Conflict-Based Approach

In the framework of the conflict-based approach, process synthesis is defined as the process of decision making for identifying and handling the design conflicts in order to satisfy multiobjective requirements (Li, 2004). Under this concept, a design problem is decomposed into subproblems instead of applying a hierarchical design. This allows for overcoming the drawback related to the interactions among several hierarchical levels of analysis. The design problem is represented by the conflicts among the interrelated design objectives or by the features of the technological scheme (Li and Kraslawski, 2004). Undoubtedly, this approach represents a change of paradigm in process design, although some aspects should be developed, such as the quantification of the conflicts and heuristic rules in function of their contribution to the conflicts as well as the development of the solving algorithms.

2.2.5 Thermodynamics-Based Process Synthesis

Although the thermodynamics imposes constraints to different chemical processes, it also sets itself up as a powerful source of design guidance for generating alternative technological configurations with better performance. For instance, the consideration of the second law of thermodynamics for analyzing the energetic efficiency of technological schemes through the exergy balance allows evaluating and choosing the best alternative configuration for a given process. This balance takes into account that the energy always will be degraded, which implies a lost of work by the mass and energy flows entering or leaving the system, according to the second principle of thermodynamics (Bastianoni and Marchettini, 1997). Sorin et al. (2000) proposed the application of some indexes based on the energy balance, such as the utilizable exergy coefficient and local contribution of exergy for each operation. They used them as criteria for reducing a superstructure of technological schemes in the case of benzene production from cyclohexane. The process synthesis methodology developed allowed obtaining a technological configuration with better performance with respect to the feedstock consumption, amount of emissions, and utilizable exergy coefficient compared with process synthesis procedures based on hierarchical decomposition or mathematical programming for the same benzene production process. In fact, the analysis of the second law of thermodynamics is the base for exploration and generation of technological alternatives in the framework of the algorithmic methods for process design (Seider et al., 1999).

For synthesizing sustainable and environmentally friendly processes, the employment of the emergy concept has been proposed. This approach exploits the fact that industrial systems also obey the laws of nature as ecological systems. Since all the materials and services are transformed and stored forms of solar energy, the amount of solar energy directly or indirectly used to make any product can be employed as a measure of the ecological input or investment in that product or service. In this way, the incorporated solar energy (emergy) is used as a comparison standard for process synthesis (Bakshi, 2000).

2.2.6 ANALYSIS OF THE STATICS

The analysis of the statics is another of the thermodynamics-based approaches that has allowed the synthesis of integrated processes of the reaction–separation type (Pisarenko et al., 2001b). This type of analysis is based on the principles of the topologic thermodynamics and has been widely used during the design of reactive distillation processes (Pisarenko et al., 1999; Pisarenko et al., 2001a), although it is also applicable to the synthesis of distillation trains. Thus, the analysis of the statics provides the fundamentals and tools needed for the preliminary design of distillation, reactive distillation, and more recently, reactive extraction and extractive fermentation processes through the development of short-cut methods based on a graphic representation that allows the visualization of the process trajectory. With the help of these methods, it is possible to specify the operating conditions and regimes corresponding to stable steady states. This information is used later during the rigorous modeling of the processes or for their simulation using commercial packages in the subsequent design steps. For this, the information on static (not varying with time) properties, not only of phase equilibrium (separation) but also of chemical equilibrium (reaction), is required. While the statics of distillation processes under infinite separability regime have been sufficiently represented through the topologic thermodynamics (Serafimov et al., 1971, 1973a, 1973b, 1973c), the statics of the chemical transformations are less developed. Without doubt, this approach is very useful for process design when applying the principle of integration, although its application has been mainly oriented to the basic organic chemical and petro-chemical industries (Cardona et al., 2000, 2002). The application of this approach to biological processes has been very limited, though it is difficult to undervalue the potential of integration in the development of innovative biotechnological processes with a high performance.

Besides the above-mentioned approaches, other types of knowledge-based process synthesis strategies are being developed as case-based reasoning, axiomatic design, and mean-end analysis (Li and Kraslawski, 2004). Case-based reasoning is supported in very specific data of prior situations and reuses previous results and experience to adjust them to the solution of new design problems. Axiomatic design is based on the principle that a good design maintains the independence of the functional requirements. This approach also applies the axiom that the information content of a good design is minimized. Finally, mean-end analysis considers that the purpose of a chemical process consists in the application of several operations in such a sequence that all the differences between the properties of the feedstocks and products are eliminated. The birth of new paradigms is expected for generating designs boosted, for instance, by the development of the artificial intelligence (Barnicki and Siirola, 2004).

2.3   OPTIMIZATION-BASED PROCESS SYNTHESIS

Optimization-based process synthesis makes use of optimization tools to identify the best configuration of a process flowsheet. For this, the definition of a superstructure that considers a significant amount of variations in the topology of the technological configurations for a given process is required. In general, the evaluation and definition of the best process flowsheet is carried out solving a mixed-integer nonlinear programming (MINLP) problem. In this way, process synthesis is accomplished in an automatic way excluding so far as possible the formulation of heuristic rules.

The great advantage of this approach lies in the generation of a generic framework to solve a large variety of process synthesis problems carrying out a very rigorous analysis of the global process structure. In particular, all the equations corresponding to the models of each process unit should be specified. This allows the definition of the accuracy level during the description of unit processes and the operations involved. This implies solving an optimization problem while simultaneously taking into consideration all the models of the units (equation-oriented approach). In contrast, most of the commercial process simulators sometimes used for knowledge-based strategy are based on the modular–sequential approach in which the calculation scheme involves the models for each unit to which the user does not have direct access. Thus, the simulation is solved taking into account the strict order of the units (from feedstocks to end products) making up the process flowsheet. Main drawbacks of the optimization-based strategy are related to the fact that the optimal configuration can only be found within the alternatives considered in the formulated superstructure (Li and Kraslawski, 2004). Furthermore, this approach has the additional disadvantages of having a significant mathematical complexity as well as the difficulties arising during the definition of the superstructure of technological configurations, i.e., the difficulty to ensure that the initial superstructure contains the "best" solution (Barnicki and Siirola, 2004). In this sense only, it is possible to formulate the design problem as a mathematical programming (optimization) problem when all the alternatives to be considered can be enumerated and evaluated quantitatively (Westerberg, 2004). In addition, this approach presents a number of difficulties when dealing with the optimization of under-defined design problems and uncertainties that result from the multi-objective requirements of the design problem (Li and Kraslawski. 2004).

2.3.1 MATHEMATICAL ASPECTS

The mathematical complexity inherent to this synthesis strategy is related to the nondifferentiable, discontinuous, and nonconvex nature of the resulting MINLP problems. However, new optimization techniques have been developed in the last decade that, along with the great advances in computational resources, have become more feasible in the utilization of this approach for process synthesis (Barnicki and Siirola, 2004), although the difficulty in the case of very large and complex synthesis problems persists.

One of the concepts related to the solution of MINLP-type problems for process synthesis is shown in Figure 2.1. For instance, this strategy has been used by Floudas (1995). After defining the superstructure containing all the possible process units for transformation of a given feedstock into a specific end product (and that includes all the possible connections among these units), an optimization loop that employs mixed-integer linear programming (MILP) tools is started. During the calculation of this optimization loop, the streams connecting the process units of the first configuration to be evaluated are chosen. This is done by specifying the value of the integer (discrete) variables (for example, 1 if the stream connects two units, 0 if this stream does not exist). The selected technological configuration represents the model of the process, which is involved in a nonlinear programming (NLP) loop. In this loop, the optimal values of the continuous process variables (temperatures, pressures, flowrates, compositions, etc.) are determined completing the optimization of the evaluated process model. Once completing the optimization of this first configuration by NLP, a second configuration of the technological scheme is defined by varying the values of the discrete variables, i.e., by choosing new connection streams among the process units, and the procedure is repeated. In this way, the different configurations of the process flowsheets are selected in an outer loop represented by the MILP

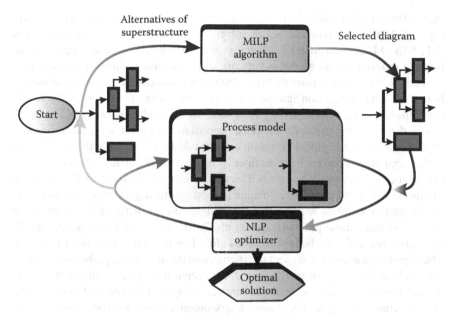

FIGURE 2.1 Conceptual diagram of one of the algorithms most used for optimization-based process synthesis.

algorithm, whereas the optimal values of the operating parameters are found in the inner loop based on NLP. The best technological configuration of the process that maximizes or minimizes the employed objective function is identified by repeating and executing these two loops. For this configuration, the optimal values of its operating parameters are defined as well.

The solution of NLP problems is usually based either on successive quadratic programming (SQP) or on reduced gradient methods. Among the main solution methods of MINLP problems are the branch and bound method (Gupta and Ravindran, 1985), generalized Benders decomposition (Geoffrion 1972), and outer approximation (Duran and Grossmann, 1986). A new trend for solving MINLP problems is the generalized disjunctive programming whose formulation includes the condition that one of a set of three types of constraints should be exactly satisfied (Lee and Grossmann, 2005; Raman and Grossmann, 1991). The three types of constraints comprise global independent inequalities concerning discrete decisions, disjunctions that are conditional constraints involving the operator *OR*, and logic pure constraints involving Boolean variables (Grossmann et al., 2000).

The formulation and solution of the main types of mathematical programming problems can be accomplished in an effective way using specialized computer programs, such as GAMS (Generic Algebraic Modeling System; GAMS Development Corporation, 2007). This system requires that the models and the formulation of the optimization problem be explicitly introduced in algebraic form. It automatically creates an interface with the solution codes for various types of problems (solvers). This offers a great advantage because it makes its main efforts to formulate the problem itself and not to develop methods for solving it. GAMS has a powerful set of solvers for different optimization problems (LP, NLP, MILP, and MINLP, among others). Moreover, it makes possible the input of indexed equations that is very useful in the case of large-sized models.

The NLP methods ensure that the global optimum can be found if and only if both the objective function and the constraints are convex (Floudas, 1995). If this is not the case, the location of the global solution cannot be guaranteed. The rigorous or deterministic methods for global optimization ensure an arbitrarily close approximation to the global optimum and, in addition, carry out the verification if this approximation has been attained. These methods include the branch and bound method, methods based on interval arithmetic (Byrne and Bogle, 1996; Stadherr, 1997) and generalized disjunctive programming (Lee and Grossmann, 2005), and procedures with multiple starting points in which a local optimizer is invoked from these points (Edgar et al., 2001). In the past few years, significant advances in the methods of rigorous global optimization have been achieved. These methods assume that special structures in these types of problems are present, such as bilinear, linear, fractioned, and separable concave functions. It has been demonstrated that the algebraic models always can be reduced to these simpler structures if they do not involve trigonometric functions (Grossmann et al., 2000). Floudas (2005) points out that the most important advances in deterministic global optimization belong to the following categories: convex envelopes and convex underestimators, twice continuously differentiable constrained nonlinear optimization problems, mixed-integer nonlinear optimization problems, bilevel nonlinear optimization problems, optimization problems with differential-algebraic equations, grey-box and factorable models, and enclosure of all solutions.

Other types of techniques for global optimization correspond to nonrigorous or heuristic search methods. These methods can find the global optima, but do not guarantee and generally are not able to prove that the global solution is found even if they do so. However, these procedures often find good solutions and can be successfully applied to MINLP problems. In such cases, the heuristic method starts with a starting solution and explores all the solutions in a certain vicinity to that starting point, looking for a better solution. The method repeats the procedure every time a better solution is found. The metaheuristic methods direct and improve the search with a heuristic algorithm. Tabu search, sparse search, simulated annealing, and genetic algorithms belong to this category (Edgar et al., 2001). In particular, stochastic methods, such as simulated annealing (Kirkpatrick et al., 1983) and the genetic algorithms (Goldberg, 1989), have gained more and more popularity, do not make any assumptions on the form of the functions, and require some type of discretization, and the violation of the constraints are tackled through penalization functions (Grossmann et al., 2000).

2.3.2 SUPERSTRUCTURES

One of the main challenges in the optimization-based synthesis of technological schemes is the definition of the superstructure of alternatives that contains the best solution. Most of the works reported have been based on the hand representation of the superstructure for each particular problem without following general rules. To generalize this procedure, the definition of the representation type for the superstructure is needed. Among the main types of representation proposed are the state–task network and the state–equipment network. In the first type, two classes of nodes (states and tasks) are used for the representation; the assignment of equipment is dealt implicitly through the model. In the second type of representation, the states and the equipments are employed as the nodes; the tasks in this case are treated implicitly through the model (Grossmann et al., 2000).

Other crucial aspect during optimization of superstructures consists in how to generate in a systematic way the superstructure so that all the alternatives of interest are included. One of the trends in this field is the automatic generation of superstructures, which has been developed for the case of lineal process networks employing an algorithm based on graphs and ensuring a search space large enough to include the optimal solution (Friedler et al., 1993). Fraga (1998) and Fraga et al. (2000) have developed the Jacaranda system that is able to solve a process synthesis problem through the automatic generation of the superstructure at the same time that it executes the search procedure. The Jacaranda system is based on the use of an implicit enumeration procedure that generates and analyzes a directed graph representing the synthesis problem.

The level of detail of the optimization model also plays an important role for the optimization-based approach. In general, the models can be classified into three main classes: aggregated models, short-cut models, and rigorous models (Grossmann et al., 2000). The first class of models employs high-level representations in which the synthesis problem is simplified by one aspect or objective that tends to dominate the problem as mentioned above. Models for predicting the minimum utilities and the minimum amount of units in a heat exchange and mass exchange networks belong in this category. Short-cut models are employed in superstructures with a high level of detail and involve the optimization of capital and operating costs, but in which the performance of individual units is predicted with relatively simple nonlinear models in order to reduce the computational effort. Among the examples of this type, distillation sequences and technological schemes can be highlighted. Finally, rigorous models are also applied in detailed superstructures for predicting the behavior of the units as in the case of the synthesis of distillation trains for separation of ideal and nonideal mixtures.

2.3.3 HYBRID METHODS

Hybrid methods for process synthesis make use of mathematical programming techniques based on optimization in combination with the knowledge-based approach. For instance, the synthesis of heat exchange networks has been improved by optimization employing a heuristic approach based on physical phenomena (Gundersen and Grossmann, 1990). A method for process synthesis exploiting the advantages of mathematical programming that uses MINLP techniques and hierarchical decomposition maintaining the consistency of its fundamental principles has been developed. The method consists of solving the whole technological scheme in each decomposition level. Aggregated models ("black box" models) are utilized for the subsystems that are in the lower hierarchical levels, evaluating in this way their interaction with the detailed MINLP problem of the current subsystem that is being analyzed. The subsystems are: reaction, separation, and heat integration (Daichendt and Grossmann, 1997). In this way, the solution of a large-scale MINLP problem is avoided.

2.4 FINAL CONSIDERATIONS

The absolute majority of methods for conceptual process design have been developed for processes of basic chemical and petro-chemical industries. Most of these approaches are especially oriented to the design of separation schemes, particularly distillation trains. For this reason, there is a paramount interest in the application of these methodologies to biotechnological processes, which are of a complex and highly multicomponent nature. None of the described approaches can be applied in a generic way to any type of synthesis problem, even more so in the case of biological processes. Considering this, the application of a minimum of two strategies or synthesis procedures is required to undertake the design of processes for fuel ethanol production with a high technoeconomic and environmental performance. In this way, a higher amount of possibilities can be covered allowing the creativity during the design process that is the source of innovation. Therefore, the employment of synthesis strategies that systematically compile and utilize the accumulated knowledge around a research subject, such as fuel ethanol, directly contributes to screening and filtering the best alternative process configurations.

Further, some of the approaches mentioned above were applied to the generation of different process flowsheets with a high technical, economic, and environmental performance. Thus, both the hierarchical decomposition method and optimization-based strategies were used for the main process steps needed for ethanol production. For specific process steps, such as the synthesis of separation and ethanol dehydration scheme, the analysis of the statics and the exergy balance were also employed. These issues will be discussed in following chapters.

REFERENCES

Bakshi, B.R. 2000. A thermodynamic framework for ecologically conscious process systems engineering. *Computers and Chemical Engineering* 24:1767–1773.

Barnicki, S.D., and J.J. Siirola. 2004. Process synthesis prospective. *Computers and Chemical Engineering* 28:441–446.

Bastianoni, S., and N. Marchettini. 1997. Emergy/exergy ratio as a measure of the level of organization of systems. *Ecological Modelling* 99:33–40.

Byrne, R.P., and I.D.L. Bogle. 1996. An accelerated interval method for global optimisation. *Computers and Chemical Engineering* 20 (Suppl.):S49–S54.

Cardona, C.A., L.G. Matallana, A. Gómez, L.F. Cortés, and Yu.A. Pisarenko. 2002. Phase equilibrium and residue curve analysis in DIPE production by reactive distillation. Paper presented at the IV Iberoamerican Conference on Phase Equilibria and Fluid Properties for Process Desing, Foz de Iguazu, Brazil.

Cardona, C.A., Yu.A. Pisarenko, and L.A. Serafimov. 2000. ДИПЭ как альтернативная добавка к моторным топливам (DIPE as alternative additive to motor fuels, in Russian). *Science and Technology of Hydrocarbons* 4 (11):72–73.

Daichendt, M.M., and I.E. Grossmann. 1997. Integration of hierarchical decomposition and mathematical programming for the synthesis of process flowsheets. *Computers and Chemical Engineering* 22 (1–2):147–175.

Douglas, J.M. 1988. *Conceptual Design of Chemical Processes*. New York: McGraw-Hill.

Duran, M.A., and I.E. Grossmann. 1986. An outer approximation algorithm for a class of mixed-integer nonlinear programs. *Math Programming* 36 (3):307–339.

Edgar, T.F., D.M. Himmelbleau, and L.S. Lasdon. 2001. *Optimization of Chemical Processes*. New York: McGraw-Hill.

Floudas, C.A. 1995. *Nonlinear and Mixed-Integer Optimization*. New York: Oxford University Press.

Floudas, C.A. 2005. Research challenges, opportunities and synergism in systems engineering and computational biology. *AICHE Journal* 51 (7):1872–1884.

Fraga, E.S. 1998. The generation and use of partial solutions in process synthesis. *Transactions of the Institution of Chemical Engineers* 76 (A1):45–54.

Fraga, E.S., M.A. Steffens, I.D.L. Bogle, and A.K. Hind. 2000. An object-oriented framework for process synthesis and simulation. In *Foundations of Computer-Aided Process Design*, ed. M.F. Malone, J.A. Trainham, and B. Carnahan, Vol. 323 of AlChE Symposium Series.

Friedler, F., K. Tarjan, Y.W. Huang, and L.T. Fan. 1993. Graph-theoretic approach to process synthesis: Polynomial algorithm for maximum structure generation. *Computers and Chemical Engineering* 17:929–942.

GAMS Development Corporation. 2007. *GAMS—A User's Guide*. Washington, D.C.: GAMS Development Corporation.

Gani, R., and A. Kraslawski. 2000. *Whole Process Synthesis & Integration—State of the Art*. http://capenet.chemeng.ucl.ac.uk (accessed June 2004).

Gavrila, I.S., and P. Iedema. 1996. Phenomena-driven process design, a knowledge-based approach. *Computers and Chemical Engineering* 20 (Suppl.):S103–S108.

Geoffrion, A.M. 1972. Generalized Benders decomposition. *Journal of Optimization Theory and Applications* 10 (4):237–260.

Goldberg, D.E. 1989. *Genetic Algorithms in Search, Optimization and Machine Learning*. Reading, MA: Addison-Wesley.

Grossmann, I.E., J.A. Caballero, and H. Yeomans. 2000. Advances in mathematical programming for the synthesis of process systems. *Latin American Applied Research* 30:263–284.

Gundersen, T., and I.E. Grossmann. 1990. Improved optimization strategies for automated heat exchanger synthesis through physical insights. *Computers and Chemical Engineering* 14 (9):925–944.

Gupta, O.K., and V. Ravindran. 1985. Branch and bound experiments in convex nonlinear integer programming. *Management Science* 31 (12):1533–1546.

Kirkpatrick, S., C.D. Gelatt, and M.P. Vechi. 1983. Optimization by simulated annealing. *Science* 220 (4598):671–680.

Lee, S., and I.E. Grossmann. 2005. Logic-based modeling and solution of nonlinear discrete/continuous optimization problems. *Annals of Operations Research* 139:267–288.

Li, X. 2004. Conflict-based method for conceptual process synthesis. Ph.D. Thesis, Lappeenranta University of Technology, Finland.

Li, X., and A. Kraslawski. 2004. Conceptual process synthesis: Past and current trends. *Chemical Engineering and Processing* 43:589–600.

Perkins, J. 2002. Education in process systems engineering: Past, present and future. *Computers and Chemical Engineering* 26:283–293.

Pisarenko, Yu.A., C.A. Cardona, R.Yu. Danilov, and L.A. Serafimov. 1999. Design method of optimal technological flowsheets for reactive distillation processes. *Récents Progrès en Génie des Procédés* 13:325–334.

Pisarenko, Yu.A., C.A. Cardona, and L.A. Serafimov. 2001a. *Reaktsionno-rektifikatsionnye protsessy: Dostizhenya v ovlasti issledovanya i prakticheskogo ispol'zovanya (Reactive distillation processes: Advances in research and practical application,* in Russian). Moscow: Luch.

Pisarenko, Yu.A., L.A. Serafimov, C.A. Cardona, D.L. Efremov, and A.S. Shuwalov. 2001b. Reactive distillation design: Analysis of the process statics. *Reviews in Chemical Engineering* 17 (4):253–325.

Raman, R., and I.E. Grossmann. 1991. Relation between MILP modelling and logical inference for chemical process synthesis. *Computers and Chemical Engineering* 15:73–84.

Seider, W.D., J.D. Seader, and D.R. Lewin. 1999. *Process design principles. Synthesis, analysis and evaluation.* New York: John Wiley & Sons.

Serafimov, L.A., V.T. Zharov, and V.S. Timofeev. 1971. Rectification of multicomponent mixtures, I. Topological analysis of liquid-vapour phase equilibrium diagrams. *Acta chimica (Hungarian Academy of Sciences)* 69 (4):383–396.

Serafimov, L.A., V.T. Zharov, and V.S. Timofeev. 1973a. Rectification of multicomponent mixtures, III. Local characteristics of the trajectories of continuous rectification processes at finite reflux ratios. *Acta chimica (Hungarian Academy of Sciences)* 75 (3):235–254.

Serafimov, L.A., V.T. Zharov, and V.S. Timofeev. 1973b. Rectification of multicomponent mixtures, IV. Non-local characteristics of continuous rectification trajectories for ternary mixtures at finite reflux ratios. *Acta chimica (Hungarian Academy of Sciences)* 75 (3):255–270.

Serafimov, L.A., V.T. Zharov, V.S. Timofeev, and M.I. Balashov. 1973c. Rectification of multicomponent mixtures, II. Local and general characteristics of the trajectories of rectification processes at infinite reflux ratio. *Acta chimica (Hungarian Academy of Sciences)* 75 (2):193–211.

Smith, R.M. 2005. The nature of chemical process design and integration. In *Chemical process: Design and integration.* Hoboken, NJ: John Wiley & Sons.

Sorin, M., A. Hammache, and O. Diallo. 2000. Exergy-based approach for process synthesis. *Energy* 25:105–129.

Stadherr, M.A. 1997. Large-scale process simulation and optimization in a high performance computing environment. Paper presented at the Aspen World 97, in Boston, MA.

Tanskanen, J., V.J. Pohjola, and K.M. Lien. 1995. Phenomenon driven process design methodology: Focus on reactive distillation. *Computers and Chemical Engineering* 19 (Suppl.):S77–S82.

Westerberg, A.W. 2004. A retrospective on design and process synthesis. *Computers and Chemical Engineering* 28:447–458.

3 Feedstocks for Fuel Ethanol Production

Currently, worldwide production of fuel ethanol is carried out by using sugar-containing or starch-containing materials. In this chapter, main feedstocks for ethanol production (sugarcane, sugar beet, corn, wheat, cassava) are discussed in reference to their production, advantages, and drawbacks. Lignocellulosic biomass as feedstock for bioethanol production is also analyzed due to the recent demand for second-generation biofuels. Finally, some evaluation data related to these feedstocks are highlighted from the viewpoint of process systems engineering.

3.1 SUGARS

Bioethanol can be produced from raw materials containing fermentable sugars, especially sucrose-containing feedstocks, such as sugarcane or sugar beet. Sugarcane is the main feedstock for ethanol production in tropical countries like Brazil, India, and Colombia. This feedstock can be used either in the form of cane juice or cane molasses. About 79% of ethanol produced in Brazil comes from fresh sugarcane juice and the remaining fraction corresponds to cane molasses (Wilkie et al., 2000). Sugarcane molasses is the main feedstock for ethanol production in India (Ghosh and Ghose, 2003). Cane molasses and different sugar-containing streams like syrup and B-type cane juice are used for producing bioethanol in Colombia. Beet molasses is the main feedstock for ethanol production in France along with wheat (Decloux et al., 2002).

3.1.1 SUGARCANE

Sugarcane (*Saccharum officinarum* L.) is a perennial grass cropped in tropical and subtropical zones from Spain to South Africa. The origin of sugarcane is in the South Pacific islands and New Guinea. The main feature of sugarcane is that a sucrose-enriched juice is formed and accumulated in its stalk. This juice is extracted and used for sugar production. For its growth, sugarcane requires a moist and warm climate alternating with dry seasons. It grows better in plain or slightly sloping lands with alluvial or clay soil with abundant luminosity. Nevertheless, sugarcane grows in any soil of good quality provided there is appropriate humidity. This crop is grown at 16 to 29.9°C and with a pH of 4.3 to 8.4 in soils with annual precipitations of 47 to 429 mm. Cane tolerates flooding (Duke, 1998). Sugarcane is cut every 12 months on average, although the range goes from 6 to 24 months. One plantation can last up to five years.

TABLE 3.1

World Production of Sugarcane (2007)

No.	Country	Production/Ton	Yield/Ton/ha
1	Brazil	514,079,729	76.59
2	China	355,520,000	86.05
3	India	106,316,000	72.56
4	Thailand	64,365,682	63.71
5	Pakistan	54,752,000	53.21
6	Mexico	50,680,000	74.53
7	Colombia	40,000,000	88.89
8	Australia	36,000,000	85.71
9	USA	27,750,600	77.62
10	Philippines	25,300,000	63.25
11	Indonesia	25,200,000	72.00
12	South Africa	20,500,000	48.81
13	Argentina	19,200,000	66.21
14	Guatemala	18,800,000	83.56
15	Egypt	16,200,000	119.56
16	Vietnam	16,000,000	56.14
17	Cuba	11,100,000	27.75
18	Venezuela	9,300,000	74.40
19	Peru	8,246,406	121.75
20	Iran	5,700,000	87.69
	World	1,557,664,978	

Source: FAO. 2008. *FAOSTAT.* Food and Agriculture Organization of the United Nations (FAO). http:// faostat.fao.org (accessed February 2009)

Sugarcane is one of the most important crops in the world and plays a crucial role in the economies of many developing and emerging countries. Brazil is the major sugarcane producer followed by China and India (FAO, 2008; Table 3.1). Brazil fielded about 5.14 million ha in 2007, while China fielded 3.55 million. Among main cane producers, Colombia has the highest yields achieving on average 88.89 ton/ha (FAO, 2008). If considering only the cane intended for sugar and ethanol production, the Colombian cane yield reaches more than 122 ton/ha (Asocaña, 2006; Espinal et al., 2005b). In some developing countries, sugarcane is cultivated by many rural communities for producing noncentrifugal sugar (solid brown sugar), a low-cost sweetener with significant content of minerals and traces of vitamins, widely used by the populace in those countries. This product is known as *gur* in India or *panela* in Colombia. In particular, the Colombian government is encouraging the use of the cane varieties normally employed for *panela* production (generally, low-yield varieties) in order to utilize them for fuel ethanol production. In this way, these communities can improve their socioeconomic conditions.

TABLE 3.2

Average Sugarcane Composition

Components	Percentage (by Weight)
Cellulose	6.48
Fructose	0.60
Glucose	0.90
Fat	0.30
Hemicellulose	5.40
Lignin	1.42
Protein	0.40
Sucrose	13.50
Water	69.70
Nonfermentable sugar	0.35
Other reduced compounds	0.35
Organics acids	0.10
Ash	0.50

The content of sucrose, glucose, and fructose in sugarcane is significant. Process microorganisms to synthesize ethanol assimilate these sugars. Besides these carbohydrates, cane contains fiber (mainly cellulose, hemicellulose, and lignin), proteins, fats, ash, and small amounts of other substances, such as other nonfermentable sugars (e.g., raffinose), organic acids, and other reducing compounds. The composition of sugarcane depends on the conditions under which it was cultivated. In particular, there are great variations in the content of moisture, sugars, and ash. The composition of feedstocks is very important during the simulation of ethanol production processes. As process simulation is a fundamental tool during process synthesis, the suitable specification of cane components directly influences the quality of obtained results. As an example of such specification, average percentages (by mass) of the cane constituents are shown in Table 3.2. The data were taken from information corresponding to cane varieties from Brazil, Venezuela, Colombia, and Cuba (Andrade et al., 2004; González and González, 2004; Sánchez and Cardona, 2008a; Suárez and Morín, 2005).

3.1.2 Cane Sugar

Sugarcane is the raw material for producing sugar. The sugar is one of the most important commodities in the world market. Besides cane, sugar can be obtained from the sugar beet. The sugar is extracted from these two feedstocks in the form of sucrose crystals (Figure 3.1) with high purity. The annual worldwide production of sugar was 163.3 million ton (raw value) in 2007, according to the International Sugar Organization (ISO, 2008). The world sugar consumption, in turn, was 157.7 million ton (raw value) in the same year. Main producers of sugar are Brazil, India, the European Union (EU), and China (Table 3.3). Sugar is mostly produced

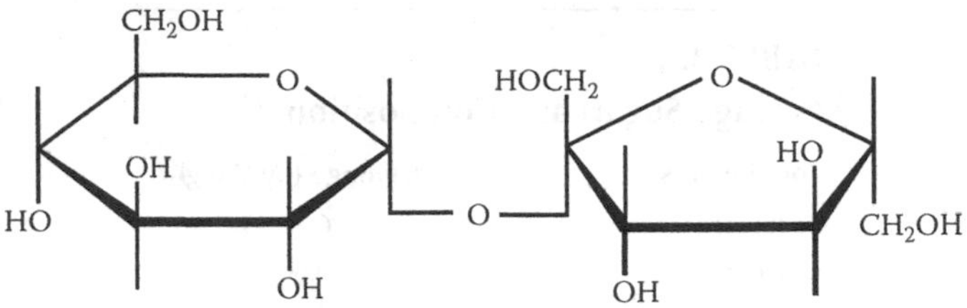

FIGURE 3.1   Chemical structure of sucrose.

TABLE 3.3
Main Sugar Producers (2007)

No.	Country	Production/Ton	Participation/%
1	Brazil	33,200,000	17.63%
2	India	29,090,000	14.78%
3	China	13,900,000	7.38%
4	EUA	7,680,000	5.42%
5	Thailand	7,150,000	5.12%
6	Australia	4,630,000	3.65%
7	Mexico	5,420,740	3.63%
8	European Union	18,450,000	2.90%
10	Pakistan	4,360,000	2.67%
15	Russia	3,400,000	1.40%
	World	149,752,383	

from sugar beets in the case of the EU countries. Brazil is the main actor in the world sugar market (first producer, third consumer, and first exporter), thus, any change in the production in this country, due to either changes in climate or the balance of ethanol production related to sugar, provokes changes in the international sugar price. This situation will continue in the future considering that Brazil is expanding the area dedicated to sugarcane plantations. It is expected that about 40 and 50 new sugar mills will be put into operation by 2010 (Decision News Media SAS, 2006).

Mature cane is collected manually or mechanically. Hand cutting is the most usual method. The cane is transported to the sugar mills where the sucrose is extracted and crystallized. The first step in the production of cane sugar in mills consists in the cane juice extraction (Figure 3.2). For this, the cane is sampled, washed, weighed, and prepared using rotating knives to shred the stalks into pieces. Then, the shredded cane is ground in heavy-duty roller mills, which extract the raw juice while hot water is sprayed onto the cane bed to dissolve the sugars. The sucrose extraction yield reaches 90 to 95% in modern sugar mills. The fiber fraction of the cane, called bagasse, is generated in this step. This material is pre-dried and utilized in the same sugar mill as an energy source burning it in special

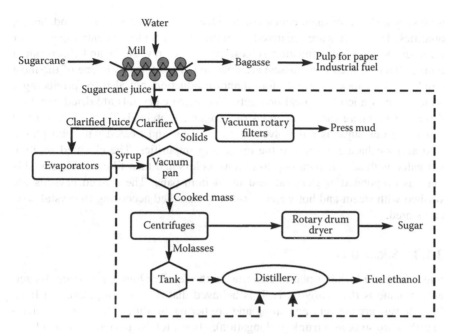

FIGURE 3.2 Simplified scheme of the process for sugar production from sugarcane.

boilers to produce steam that generates electricity through turbo generators. The raw juice is clarified by adding sulfur dioxide (SO_2) to oxidize color substances, destroy microorganisms, and favor the agglutination of colloids. Then, lime is added to neutralize the juice, thus avoiding, in this way, the sucrose hydrolysis into glucose and fructose (sucrose inversion). The limed juice is heated to 105 to 110°C in the clarifiers, to which flocculants also are added. Solid separation is favored because the calcium sulfite formed is insoluble and precipitates removing the suspended particles and other impurities. This precipitate (known as mud) is mixed with flocculants, lime, and small amounts of the light fraction of bagasse, separated from the juice, and sent to vacuum rotary filters where the filter cake is recovered. The filter cake (press mud) is a fibrous material that can be used for animal feed or composted to produce manure. Part of the clarified juice can be used for ethanol production. In the evaporation step, the clarified juice is concentrated by removing up to 75% of the water in multiple-effect evaporators. The removed water is used in other process steps either as steam or as condensate. The concentrated juice or syrup contains 58 to 62% solids (about 60° Brix).

The obtained syrup is treated with direct steam to reduce its viscosity. Then, the syrup is clarified by adding SO_2, phosphoric acid, and lime. After this treatment, the syrup is sent to a flotation tank where it is clarified. This clarified syrup can be used for ethanol production. In the next step or pan stage, the sucrose contained in the syrup is crystallized by removing excess water with vacuum evaporators called pans. In order to increase the speed of this process, the pan stage operates in a manner that utilizes seed crystals and a combination of products with varying levels of sugar content to produce a range of crystal sizes and, hence, qualities. The crystals are separated from the mother liquor in centrifuges. After three evaporation/centrifugation stages, crystals of raw sugar and the residual mother liquor (called C-molasses) are obtained. The molasses is one of the most employed feedstocks, not only for ethanol production, but also for producing a wide range of microbiological products. The sucrose crystals are dried in rotary drum driers where the commercial raw sugar is obtained. To produce refined sugar, the raw sugar is redissolved in clean water and treated with phosphoric acid and saccharate to remove the remaining impurities. The clarified solution is treated with activated carbon to remove colors and then filtered. Finally, this mass is evaporated and crystallized in vacuum pans. The refined crystals are washed with steam and hot water, air-dried, classified according to crystal size, and stored.

3.1.3 Sugar Beet

Sugar beet is a biennial plant belonging to the family *Quenopodiaceae*. Its scientific name is *Beta vulgaris* L. It is believed that the beet originated in Italy. Beet flowers are not very appealing and are hermaphroditic. The roots are pivoting (they are sunk as a trunk prolongation), almost totally buried, with a yellow-greenish rough peel. The root is the organ where most sugar is accumulated in the plant. The seeds are adhered to the calyx. The different beet varieties are used for human food (table or red beet), animal feed (fodder beet), and sugar production (sugar beet). The latter variety (*B. vulgaris* var. *altissima*) is the most employed in temperate zones, especially in Europe and North America. To be cultivated, sugar beet requires a temperate, sunny, moist climate and deep soils with neutral pH, high water retention, and good aeration. Clay, sandy, calcareous, and dry soils are not adequate for this crop (Infoagro, 2002).

Sugar beet is an important crop in Europe, North America, and Asia. France is the major producer of sugar beet followed by Germany and the United States (FAO, 2006) as shown in Table 3.4. France had a cropped area with sugar beet of about 379,000 ha in 2005, while Russia had about 780,000 ha cropped with sugar beet. These data give an idea on the yield differences between these two countries.

3.1.4 Beet Sugar

Among the world's foremost sugar producers (presented in Table 3.4), the United States, France, Germany, and Russia produce their sugar mostly from sugar beet. In the case of the United States, 53% of its sugar production comes from sugar beet and the balance comes from sugarcane. China produces 7% of its sugar from sugar beet (700,000 ton on average). The remaining percentage corresponds to the cane mostly cropped in southeastern provinces of the country (10,000,000 ton on average; FAS, 2005).

Sugar beet contains from 12 to 15% sucrose. The beets arrive at the production plant without crown and are loaded in silos by mechanical means through

TABLE 3.4
World Production of Sugar Beet (2005)

No.	Country	Production/Ton
1	France	29,303,000
2	Germany	25,427,000
3	USA	24,724,410
4	Russia	21,520,000
5	Ukraine	15,620,600
6	Turkey	13,500,000
7	Italy	12,000,000
8	Poland	10,972,030
9	China	7,910,000
10	United Kingdom	7,500,000
11	Spain	6,676,900
12	Holland	5,750,000
13	Belgium	5,606,025
14	Iran	4,850,000
15	Morocco	4,560,000
16	Japan	4,200,000
17	Egypt	3,429,535
18	Czech Republic	3,189,740
19	Hungary	3,108,150
20	Belarus	3,070,000

Source: FAO. 2006. *Major food and agricultural commodities and producers.* Food and Agricultural Organization of the United Nations (FAO). http://www.fao.org/es/ess/top/country.html (accessed November 2006).

channels with circulating water. The beets are washed and passed through systems retaining diverse solid materials, such as stones, leaves, and small roots. Once washed, the beets are transported to the choppers where they are shredded into very thin slices (3 mm), called *cossettes*, and passed to a machine called a diffuser to extract the sugar content into a water solution. The diffusers are large rotary drums where the cossettes are put in contact with a hot water stream flowing in the opposite direction. The sucrose is extracted from the vacuoles of the beet cells into the flowing water generating the raw juice or diffusion juice. The spent cossettes are called pulp and leave the diffuser with about a 95% moisture content, but with low sucrose content. To recover part of the sucrose contained in the pulp, it is pressed in screw presses reducing the moisture to 75%. After drying in rotary drums, the pressed pulp is sold as animal feed due to its high fiber content, while the resulting liquid is added to the diffusers. The diffusion juice is an impure sucrose solution also containing gums, pectin, amino acids, mineral salts, nitrogenous compounds, etc. This juice has about 16° Brix and 85% purity. It can be clarified by a process very similar to that of cane sugar. Thus, a clear juice with 15° Brix is obtained. From this point, the process is practically the same as for cane sugar.

3.1.5 Sucrose-Containing Materials Used for Ethanol Production

Fuel ethanol production is directly linked to the production of sugar. In fact, generally speaking, ethanol can be considered as a co-product in the sugar industry, but it also can be the main end product obtained from sugarcane. In the former case, cane molasses formed during sugar processing is employed for ethanol production. However, the requirements of fuel ethanol in cane sugar-producing countries with gasoline oxygenation programs cannot be covered with the sole molasses. Therefore, autonomous (stand-alone) distilleries not co-located near sugar mills have been put into operation in countries like Brazil in order to meet the required amounts of fuel ethanol. These distilleries use the sugarcane to produce ethanol exclusively. Besides molasses, some sugar mills use the cane juice for producing ethanol in their co-located distilleries. This means a reduced amount of produced sugar compared to when the mill does not produce ethanol. This is the case in Colombia whose sugar production has been decreased due to the implementation of the gasoline oxygenation program using E10 blends. The rise in the domestic price of sugar during 2006 through 2008 has been explained not only by the increase in the international oil price, but also by the onset of this program (Sánchez and Cardona, 2008a).

The technology for fuel ethanol production from sugar-rich materials (mostly sucrose) offers multiple alternatives regarding the use of feedstocks generated within the sugar mills. For instance, and excluding the cane juice and molasses, Colombian distilleries use part of the clarified syrup. In this way, a great flexibility for production of both sugar and ethanol is attained, exploiting the integration of the different sugar-rich streams during sugar processing. For sugar mills, this flexibility allows them to respond in a suitable way to the changes and needs of both ethanol and sugar markets.

In the case of ethanol production facilities using sugar beet, both the diffusion juice and beet molasses can be employed for producing bioethanol. The production of fuel ethanol directly from the beet juice is a nonviable technological option for most countries producing sugar from sugar beet due to the higher costs of the juice produced and to the need of covering their domestic sugar demands as in the case of European countries and North America. For these countries, ethanol is mostly produced from corn, wheat, and other grains. In fact and according to Decloux et al. (2002), there exist no stand-alone distilleries producing ethanol from beet juice in France, the first ethanol producer from sugar beet. Instead of that, the distilleries are co-located next to sugar mills. Some important issues concerning the main sucrose-containing feedstocks for ethanol production are highlighted below.

TABLE 3.5

Composition of the Raw Sugarcane Juice

Components	Percentage (by Weight)
Fructose	0.58
Glucose	0.87
Fat	0.39
Protein	0.37
Sucrose	13.37
Water	82.10
Nonfermentable sugar	0.43
Other reduced compounds	0.60
Organics acids	0.19
Ash	1.10

3.1.5.1 Cane and Beet Juices

During cane milling, the soluble fraction is separated from the rest of the plant achieving up to 97% efficiencies for industrial milling and 50% efficiencies for traditional milling (González and González, 2004). Cane juice is a viscous, greenish liquid. For South American cane varieties, the juice contains 15 to 16% soluble solids, from which 85% is sucrose yielding an approximate content of this sugar in the fresh juice of 12 to 13% (Lindeman and Rocchiccioli, 1979). The average composition of raw cane juice (before its clarification) is presented in Table 3.5. As in the case of the whole sugarcane, an example of the specification corresponding to the average composition of cane juice is shown in this table using data from India, Venezuela, Colombia, and South Africa (Bhattacharya et al., 2001; González and González, 2004; Sánchez and Cardona, 2008a; Seebaluck et al., 2008). This average composition was used during most simulations conducted for process synthesis purposes.

The clear beet juice obtained in sugar production plants can be used for ethanol production as well. In general, the beet juice is used during the harvest season when there is a high availability of beets. Out of season, the distilleries produce ethanol mostly from beet molasses or beet syrup (Decloux et al., 2002). The composition of beet juice does not differ substantially from the composition of cane juice, as reported by Ogbonna et al. (2001).

3.1.5.2 Sugarcane Molasses

Sugarcane molasses is a by-product of sugar processing. It is generated during sugar crystallization by evaporation and represents the mother liquor from which the crystals are formed. Most cane molasses used for ethanol production corresponds to the C-molasses. Due to the repeated evaporation–centrifugation process, the molasses obtained is a viscous, thick brown liquid with a higher concentration of impurities and mineral salts compared to cane juice. An example of the average composition of cane molasses is presented in Table 3.6. The data in this table correspond to cane molasses from the United States, Colombia, and Costa Rica (Curtin, 1983; Fajardo and Sarmiento, 2007; Vega-Baudrit et al., 2007). This composition is quite variable because it depends on factors such as the type of soil, temperature, humidity, cane production season, cane variety, type of cane processing in the sugar mill, and storage conditions (Curtin, 1983). This implies great variations in the content of sugar and remaining nutrients in molasses as well as its color, viscosity, and flavor.

The molasses is commercialized based on its degrees Brix, which is an indicator of the specific gravity and represents an approximation to its total solid content. The degrees Brix were originally defined for solutions of pure sucrose indicating the percentage (by weight) of this sugar in the solution. However, and besides sucrose, molasses contains other sugars, such as glucose, fructose, and raffinose, as well as a variety of nonsugar organic compounds. In this way, the value of degrees Brix for molasses does not correspond to the content of sugars or total solids. Nevertheless, the degrees Brix are still being used for trading molasses. Cane molasses contains some vitamins that are crucial if considering the subsequent fermentation. For this reason, in general, it is not necessary to add vitamins to the culture medium based on cane molasses for fermentation with the yeasts of the *Saccharomyces cerevisiae* species. Biotin, pantothenic acid (in the form of pantothenate), and inositol are probably the three vitamins essential for the normal yeast growth. To this regard, the biotin is in excess in cane molasses, while the pantothenate is found in minimal amounts required for fermentation. The inositol is present in adequate amounts in cane molasses (Murtagh, 1995).

3.1.5.3   Beet Molasses

Beet molasses is a by-product of sugar production from the sugar beet. As in the case of cane molasses, this material is obtained during the crystallization and evaporation steps of the cooked mass from vacuum pans. The composition of beet molasses has small variation related to cane molasses. An example of average composition of beet molasses that can be used for simulation purposes is shown in Table 3.6 using some data from Curtin (1983). In general, beet molasses is deficient in biotin. For this reason, the mixing of 20% cane molasses (that has an excess of biotin) and 80% beet molasses has been proposed. In contrast, beet molasses has an excess of pantothenate. Inositol content in beet molasses is adequate for fermentation purposes (Murtagh, 1995).

3.1.5.4   Other Sugar-Containing Materials

As noted above, sugar mills also employ the clarified syrup for ethanol production. The equivalent of this syrup in the sugar production plants using sugar beet, the dense juice, is also used for bioethanol production. On the other hand, a special type of syrup called high test molasses is employed for ethanol production especially in the distilleries of the southern United States. These types of molasses are usually produced when the sugar prices are very depressed in the world market, thus the sugar mills simply concentrate the cane juice to obtain syrup, to

TABLE 3.6

Average Composition of Sugarcane and Beet Molasses

Components	Sugarcane Molasses/% (by Weight)	Beet Molasses/% (by Weight)
Water	18.82	18.01
Sucrose	36.70	37.50
Fructose	8.82	9.12
Glucose	6.86	7.26
Nonfermentable sugar	4.79	2.00
Other reduced compounds	4.00	3.00
Fat	0.40	0.40
Protein	3.50	6.00
Organics acids	5.00	5.00
SO_2	0.10	0.10
$Ca(OH)_2$	1.01	1.01
Ash	10.00	10.60

which acid is added in order to partially hydrolyze the sucrose and avoid its crystallization (Murtagh, 1995). In this way, the sugar mills commercialize the high test molasses as a way of exploiting the sugarcane under low sugar price conditions. High test molasses contains about 70% sugars and 80° Brix that indicates high sugar content compared to molasses. Therefore, these types of materials, as well as the syrups, require the addition of higher amounts of nitrogen and phosphates when they are used to prepare the cultivation media for ethanolic fermentation using yeasts. Moreover, the pantothenate content of these molasses is low (Murtagh, 1995).

3.2 STARCHY MATERIALS

3.2.1 STARCH

Starch is a polymeric carbohydrate made up of glucose units linked by glycosidic bonds. The starch is the most abundant carbohydrate in nature after the lignocellulosic complex and is present in high amounts in very important crops for human food, such as corn, wheat, potato, cassava, rye, oats, rice, sorghum, and barley. About 54 million tons per year of starch are produced for industrial purposes, from which 55% comes from the United States. From the total produced starch, 44.1 million tons come from corn, 2.6 million from cassava and rice, and 2.8 million from potato (Messias de Bragança and Fowler, 2004)

Starch is composed of two types of α-glucans, i.e., polymers based on glucose monomers linked by α-type glycosidic bonds, amylose and amylopectin, which represent 98 to 99% of the dry weight of starch kernels. The properties of these two polysaccharides along with the ratio between them in the starch kernels

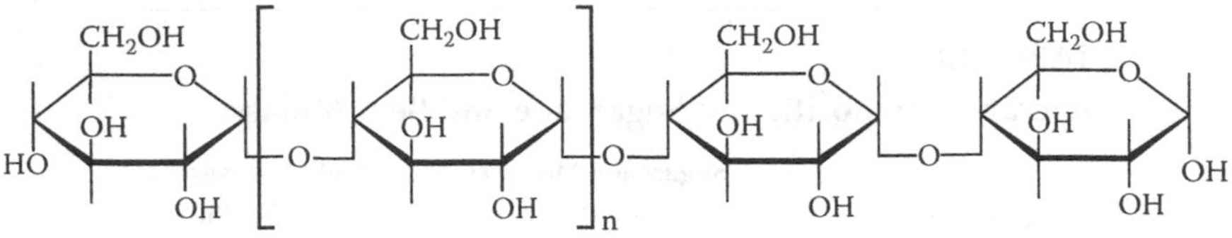

FIGURE 3.3   Structure of amylose: n ≈ 1,000.

determine the properties of the starch obtained from different plant sources. This ratio ranges from less than 15% amylose in waxy starch and 20 to 35% amylose in normal starch to greater than 40% amylose in amylo-starch. In addition, the water content of the starch kernels in equilibrium with the air goes from 10 to 12% in grains to 14 to 18% in tubers (Tester et al., 2004).

The amylose is a linear polysaccharide made up of D-glucopyranose units linked by $\alpha(1,4)$ glycosidic bonds, although it has been established that some molecules present several branching points due to the presence of $\alpha(1,6)$ (Buleón et al., 1998). These branching points are present in an amount of 9 to 20 per molecule (Sajilata and Singhal, 2005). The molecular weight of amylose is in the range $1\times10^5-1\times10^6$ Da with a polymerization degree of 324 to 4,920 (1,000 on average). These variations depend on the starch origin. The basic structure of amylose is depicted in Figure 3.3. Two ends can be distinguished: one end with intact glycosidic hydroxyl (corresponding to the glucose residue of the right-hand side of Figure 3.3) called the reducing end, and the end with the glucose residue whose glycosidic hydroxyl is participating in the glycosidic bond with the following glucose residue (the left-hand side of Figure 3.3) called the nonreducing end. These ends are very important during the process of enzymatic hydrolysis of starch.

The amylopectin is a highly branched polymer made up of chains of D-glucopyranose units linked by $\alpha(1,4)$ bonds, but with 5 to 6% $\alpha(1,6)$ bonds leading to branching points (Buleón et al., 1998). These branching points occur every 15 to 30 glucose units on average. The amylopectin has a greater molecular weight than amylose ($1\times10^7-1\times10^9$ Da) with a higher polymerization degree of 9,600–15,900 (Tester et al., 2004). Because of the branches, the amylopectin presents an elevated amount of chains that are differentiated by their inner or outer character. As in the case of amylose, the nonreducing ends of amylopectin can be distinguished (see the left-hand side of Figure 3.4a). The single reducing end of the amylopectin molecule can be observed on the right-hand side of Figure 3.4b.

The amylose and amylopectin are located in the starch kernel in a radial way in the form of concentric layers where there exist crystalline and amorphous zones of these two polysaccharides. This implies that kernel degradation, the first stage in the solubilization of starch in water, is very difficult at low temperatures. To achieve this goal, it is necessary to break the starch kernels through a hydrothermal treatment involving the adsorption of hot water by the kernels and their swelling and destruction with the corresponding release of amylose and amylopectin in a soluble form. This process is known as starch gelatinization and has a crucial

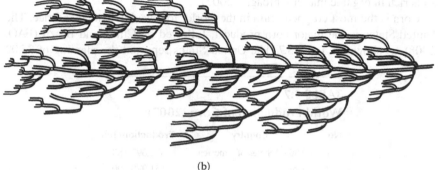

(a)

(b)

FIGURE 3.4 Amylopectin: (a) Structure of amylopectin, for outer chains a ≈ 12-23 and for inner chains b ≈ 20-30; (b) simplified diagram of an amylopectin structure.

importance during the industrial processing of this carbohydrate, especially during the production of starch hydrolyzates with sweetening properties (starch syrups), glucose, and ethanol. It is worth emphasizing that the solubilization of starch is required for the enzymatic attack of starch for producing glucose.

3.2.2 STARCH SOURCES FOR ETHANOL PRODUCTION

As the starch is a polymer exclusively composed of glucose residues, it has become a very important feedstock for ethanol production. For this, the breakdown of starch into glucose with the subsequent conversion of glucose into ethanol using appropriate fermenting microorganisms is required. The technologies for starch hydrolysis will be discussed in Chapter 5. The grains are the feedstock most used for ethanol production from starchy materials, especially corn and wheat, though bioethanol production from rye, barley, triticale (Wang et al., 1997), and sorghum (Zhan et al., 2003) has been reported. Besides the grains, some tubers, such as the cassava and potatoes, offer significant starch contents. In a similar way, ethanol from sweet potatoes, eddoes, yam (Hosein and Mellowes, 1989; Nigam and Singh, 1995), plantain, and bore (Aya et al., 2005) has been produced.

3.2.2.1 Corn

Corn (*Zea mays*) is the most employed feedstock in the world to produce starch either for the food industry or for ethanol production. Corn is a crop whose origin is in the Americas. Its stalks can reach 4 m in height and does not have branches. Its seeds are the structure with the highest value of starch in this plant and are the base for human food in many communities in Central and South America. The starch accumulates in the endosperm of corn seeds. Corn requires a temperature of 25 to 30°C and an important degree of sunlight for its growth. Corn can tolerate minimum temperatures down to 8°C. At 30°C, the adsorption of nutrients and water can be difficult. This crop is exigent in water, but adapts itself very well to all kind of soils preferring those with a pH of 6 to 7. Similarly, corn requires deep soils rich in organic matter (Infoagro, 2007).

Corn is the most cropped grain in the world, followed by wheat and rice. The United States is the major corn producer, followed by China and Brazil (FAO, 2008), as shown in Table 3.7. The United States reported the planting of 35.02

TABLE 3.7
World Production of Corn (2007)

No.	Country	Production/Ton
1	United States of America	332,092,180
2	China	151,970,000
3	Brazil	51,589,721
4	Mexico	22,500,000
5	Argentina	21,755,364
6	India	16,780,000
7	France	13,107,000
8	Indonesia	12,381,561
9	Canada	10,554,500
10	Italy	9,891,362
11	Hungary	8,400,000
12	Nigeria	7,800,000
13	South Africa	7,338,738
14	Egypt	7,045,000
15	Philippines	6,730,000
16	Ukraine	6,700,000
17	Romania	3,686,502
18	Spain	3,647,900
19	Thailand	3,619,021
	World	784,786,580

Source: FAO. 2008. *FAOSTAT.* Food and Agriculture Organization of the United Nations (FAO). http://faostat.fao.org (accessed February 2009)

TABLE 3.8
Composition of the Main Starchy Crops
Used for Ethanol Production

Component	Corn	Wheat	Cassava
Humidity	15.50	13.70	70.00
Starch	60.59	57.25	26.50
Protein	8.70	14.53	0.80
Lipids	3.64	3.50	0.30
Fiber	8.31	7.79	0.60

Source: Cardona et al., 2005.

million ha in 2007; China, 28.07 million ha; and Brazil, 13.82 million ha. The U.S. yield is very high (about 9.48 ton/ha), although Kuwait has the maximum yield (21.0 ton/ha), but it has a very small annual production: 1,050 ton in 2007 (FAO, 2008). In the United States, both private and public sectors have been systematically investing in research and development of corn cropping oriented to the production of sweeteners for the food sector and ethanol for transportation sector. In addition, the U.S. government has granted certain tax exemptions to farmers in order to support corn production, which is considered a strategic crop. It should be noted that the powerful lobby of corn producers has confronted the big oil industry supporting the implementation of renewable fuels from corn. In fact, most ethanol-producing plants in the United States employ corn as a feedstock. In 2007, 22.5% of U.S. corn production was dedicated to ethanol production (FAO, 2008; Renewable Fuel Association, 2009; Sánchez and Cardona, 2008b). China also has been increasing its ethanol production from grains including corn. This fact, along with the expected increase in ethanol production in the United States, has been pushing corn prices upward. This has had direct effects on the costs of corn-containing foods in such importing countries, such as Mexico.

Starch is the corn component that is directly used for ethanol production. Likewise, starch is corn's main component, as seen in Table 3.8 from data compiled by Cardona et al. (2005). There exist two technologies for using corn starch as a feedstock for fuel ethanol production. One technology employs the separation of starch from all the other components of the corn kernel and this is called wet milling. In this way, during the wet milling process, the corn kernel is separated into its components: starch, fiber, gluten, germ, and oil. The other technology does not involve the separation of starch. By contrast, the whole milled grain enters into the ethanol production process directly (dry milling).

The corn wet milling process allows the production of a series of value-added co-products that offset to a certain degree the fuel ethanol production costs. These co-products are directly related to the structure of corn grain. Starch represents 60% of the grain and is located in the endosperm. Grain proteins are concentrated in the gluten. Corn oil is located in the germ and represents 4% of the grain

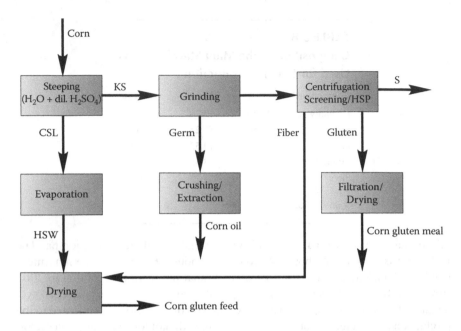

FIGURE 3.5 Simplified block diagram of corn wet milling. HSP: hydroclonic separation. Streams: CSL—corn steep liquor, HSW—heavy steep water, KS—corn slurry, S—starch.

(Gulati et al., 1996; U.S. Grains Council, 2007). The overall wet milling process is depicted in Figure 3.5. The first step in the corn wet milling process is the steeping of the grains in large tanks that can process from 1,500 to 6,000 bushels of corn, according to the typical volumes of starch industry in the United States (U.S. Grains Council, 2007). Steeping is carried out during 30 to 50 h at 49 to 54°C in water containing 0.1 to 0.2% sulfur dioxide. The sulfurous acid that is formed contributes to the separation of the starch and insoluble proteins, breaking the protein matrix of the endosperm through the destruction of disulfur bonds in gluten proteins. During this step, about 6% of dry matter is dissolved in the liquid containing the corn steep liquor. These dissolved components provide the nutritional value to the corn extractives that are condensed and fermented after the partial dehydration of steeping in water.

After grain steeping, the swelled corn grain contains about 45% water. This soft grain is ground and the germ is removed by flotation. The germ is cracked and its oil is extracted using hexane as a solvent. The corn oil is refined while the solid residue of the germ is dried to prepare the corn germ meal. Once the germ is removed, the resulting material undergoes milling that crushes starch particles and releases the fiber. This fiber is separated from the starch and gluten proteins by using a screen. The thick mixture of starch and gluten is pumped to a rotary disk column where these two components are separated by applying centrifugal force. In this centrifugal separator, the product that is obtained contains approximately 60% protein. This product is concentrated, filtered, and dried to obtain the corn gluten meal. The starch is separated again to reduce its protein content down to 0.3% (U.S. Grains Council, 2007). The fiber in the first part of milling is added to the evaporated corn steep liquor. The stream from this process evaporates in order to produce the corn gluten feed, as illustrated in Figure 3.5.

The co-products obtained are mostly intended for animal feed. The corn gluten meal has a high protein and energy content and is a good source of the amino acid methionine. This co-product is used for cattle feed as a protein supplement (about 60% protein) as well as for feeding poultry and pigs. Corn gluten feed contains about 20% protein and a reduced amount of oil and fats. This material is also used as a protein supplement, although it has a lower nutritive value in comparison to corn gluten meal. It can be used for poultry and pig feed and even for ruminants due to its significant content of digestible fiber. Corn oil is utilized for human food. In general, from each 100 kg of corn processed by this technology, 2.87 kg corn oil, 4.65 kg corn gluten meal, and 24.09 kg corn gluten feed can be obtained. According to the U.S. Department of Agriculture, the average price per tonne of these products is as follows: US$357 for corn gluten meal, US$88 for corn gluten feed, and US$662 for corn oil (Bothast and Schlicher, 2005).

3.2.2.2 Other Grains as Starch Sources

After corn, wheat (*Triticum* spp.) is the most employed grain for fuel ethanol production, especially in Europe and North America, due to its high starch content (Table 3.8). Other than the sugar beet, wheat is the main feedstock for ethanol production in France (Poitrat, 1999). In fact, France was the fifth world producer of wheat in 2007 (33.21 million tons) after China (109.86 million tons), India (74.89 million tons), United States (53.60 million tons), and Russian Federation (49.38 million tons; FAO, 2007b). The wheat yield in France is 6.25 ton/ha, while in Ireland, it reaches up to 8.1 ton/ha (FAO, 2007a).

Sorghum (*Sorghum bicolor*) is another grain proposed for ethanol production. The United States is the world's leading producer of sorghum with a volume of 12.82 million tons in 2007 (FAO, 2007a, 2007b). Sorghum is employed in many countries, such as Colombia, as a component of animal feed, though corn is replacing it. One of the features of sorghum as feedstock for ethanol production is the presence of significant amounts of tannins. These tannins provoke the decrease in the ethanol production rate during the fermentation process, although they do not affect either the ethanol yield or the enzymatic hydrolysis of starch contained in sorghum (Mullins and NeSmith, 1987).

One of the most promising crops for fuel ethanol production is sweet sorghum, which produces grains with high starch content, stalks with high sucrose content, and leaves and bagasse with high lignocellulosic content (Sánchez and Cardona, 2008a). In addition, this crop can be cultivated in both temperate and tropical countries, it requires one third of the water needed for the sugar cane harvest and half of the water needed by corn, and it is tolerant to drought, flooding, and saline alkalinity (du Preez et al., 1985; Winner Network, 2002). Grassi (1999) reports that from some varieties of sweet sorghum the following productivities can be obtained: 5 ton/ha grains, 8 ton/ha sugar, and 17 ton dry matter/ha lignocellulosics.

3.2.2.3 Cassava

The cassava (*Manihot esculenta*) is a perennial bush achieving 2 m by height and is native to South America. The main feature of this plant is its edible roots, thus the plant is uprooted after one year of growth in order to obtain its better conditions for its consumption. The cassava root is cylindrical and oblong and can reach up to 100 cm in length and 10 cm of diameter. Its pulp is firm and presents high starch content. Cassava tubers are consumed in a cooked form and represent a crucial component in the food of more than 500 million people in America, Asia, and Africa. The cassava is not a very exigent crop, but it should be grown no higher than 1,500 m above sea level. For cassava cropping, the soils should be porous because the root requires sufficient oxygen levels to grow; it also requires good drainage. The cropping temperature should be in the range 25 to 30°C (Agronet, 2007). For this reason, cassava is one of the most important tropical crops in the world. Once planted, cassava roots can be harvested after seven months and stay in the soil for three years (Alarcón and Dufour, 1998).

The world's largest cassava producer is Nigeria, followed by Brazil, Indonesia, and Thailand (FAO, 2007b). As can be observed in Table 3.9, major cassava producers are located in Africa, Southeast Asia, and South America. The most

TABLE 3.9
World Production of Cassava (2007)

No.	Country	Production/Ton
1	Nigeria	45,750,000
2	Brazil	27,312,946
3	Thailand	26,411,233
4	Indonesia	19,610,071
5	Democratic Republic of Congo	15,000,000
6	Ghana	9,650,000
7	Vietnam	8,900,000
8	Angola	8,800,000
9	India	7,600,000
10	Mozambique	7,350,000
11	United Republic of Tanzania	6,600,000
12	Paraguay	5,100,000
13	Uganda	4,456,000
14	China	4,370,000
15	Benin	2,525,000
16	Madagascar	2,400,000
17	Malawi	2,150,000
18	Côte d'Ivoire	2,110,000
19	Colombia	2,100,000
20	Cameroon	2,076,000
	World	222,138,068

elevated cassava yields among the main producers are from Thailand (20.28 ton/ha) and Brazil (13.63 ton/ha; FAO, 2007a). In general, more than 90% of cassava production is directed to human food, while the balance is used for producing starches and snacks. The substitution of corn with cassava flour has been proposed for animal feed production taking into account its high energy content (Espinal et al., 2005a). The importance of cassava cropping from the viewpoint of its agro-industrial applications lies in its high potential for energy production in the form of starch (see Table 3.8). Espinal et al. (2005a) report that the use of improved seeds and fertilizers, and a suitable weed control allows production of 20 to 30 ton/ha of fresh roots and 10 to 12 ton/ha of dried cassava in zones where other starch-producing crops like corn, sorghum, or rice do not reach yields above 4 or 5 ton/ha.

There exist two main methods for industrial production of native cassava starch: the traditional method employed in India and some Latin American countries, and the modern method of the type used by the company Alfa Laval for large-scale production. In the traditional process, fresh roots are washed and debarked before crushing in a rotary rasper. Starch is separated from the crushed pulp before passing through a series of reciprocating nylon screens of decreasing mesh size (50250 mesh). The resultant starch milk is settled over a period of four to eight hours using a shallow settling table or a series of inclined channels laid out in a zigzag pattern. Settled starch is sun-dried on large cement drying floors for approximately eight hours. During this period, the moisture content reduces from 45 to 50% down to 10 to 12%. To achieve efficient drying, sunny conditions are required with ambient temperatures of more than 30°C and relative humidity of 20 to 30%. Dried starch is ground to a fine powder and packaged for sale (FAO and IFAD, 2004). In the modern Alfa Laval–type process, roots are washed and debarked, sliced and then crushed in a rotary rasper. Starch pulp is passed through two conical rotary extractors to separate starch granules from fibrous materials, and then fed via a protective safety screen and hydro cyclone unit to a continuous centrifuge for washing and concentration. The concentrated starch milk is passed through a rotary vacuum filter to reduce water content to 40 to 45% and then flash dried. The flash drying reduces moisture content to 10 to 12% in a few seconds, so starch granules do not heat up and suffer thermal degradation.

3.3 LIGNOCELLULOSIC MATERIALS

The lignocellulosic complex represents the most abundant biopolymer on Earth and constitutes the main component of a great variety of wastes and residues from domestic and industrial activities of man. Moreover, it is present in profuse biological materials, such as wood, grass, straw, and forage. Due to its immediate origin in a biological process, this biopolymer complex has received the name of lignocellulosic biomass. Being the most plentiful material in the biosphere, the use of the lignocellulosic biomass will allow the production of a valuable biofuel like bioethanol as well as the economic exploitation of a wide range of potential feedstocks resulting from domestic, agricultural, and industrial activities.

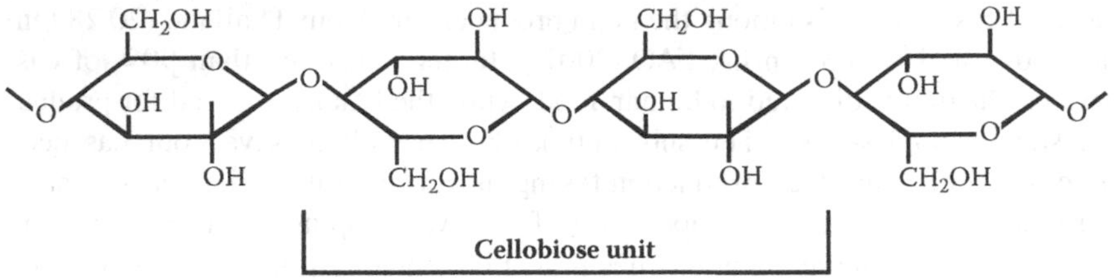

FIGURE 3.6 Structure of the cellulose chain.

3.3.1 STRUCTURE OF LIGNOCELLULOSIC COMPLEX

Lignocellulosic biomass is made up of very complex biopolymers that are not used in human food. The main components of lignocellulosic biomass are cellulose, hemicellulose, and lignin in addition to a small amount of extractives, acids, and minerals. For its conversion into ethanol, a complex process of pretreatment and hydrolysis is done in order to transform the carbohydrate polymers (cellulose and hemicellulose) into fermentable sugars.

Cellulose is a β-glucan, i.e., a polymer composed of glucose molecules linked by β(1,4) bonds. It can be considered that the cellulose is a linear polymer made up of cellobiose monomers as shown in Figure 3.6. The polymerization degree of cellulose is about 7,000 to 15,000. Due to its linear nature and to the interactions by hydrogen bonds between the OH groups of a same chain or of different chains, cellulose molecules are oriented by length leading to the formation of very stable crystalline structures. These structures allow the bundles of cellulose chains to form rigid, difficult to break microfibers. For this reason, the main function of cellulose in plants is structural, which explains its majority presence in the cell wall. In general, the cellulose composes 40 to 60% of dry matter of lignocellulosic biomass (Hamelinck et al., 2003).

Hemicellulose composes 20 to 40% of lignocellulosic biomass and consists of short, very branched chains of sugars (200 sugars on average). Among these sugars are, in their order, xylose and arabinose (both 5-carbon sugars or pentoses), and galactose, glucose, and mannose (these latter sugars are hexoses). Other carbohydrate-related compounds like glucuronic, methyl glucuronic, and galacturonic acids are also present in hemicellulose structure. Furthermore, hemicellulose contains, in a lower proportion, acetyl groups esterified to some OH groups of its different sugars. Due to the predominance of xylose, hemicellulose can be considered as a xylan. For lignocellulosic materials derived from hardwood, the xylan backbone is composed of xylose units linked by β(1,4) bonds that branch through α(1,2) bonds with the methyl glucuronic acid (Figure 3.7a). In the case of xylan from softwood, the acetyl groups are less frequent, but there exist more branches due to the presence of α(1,3) between the xylose backbone and arabinofuranose units (Figure 3.7b). Considering its branched structure, hemicellulose does not form crystalline structures, but amorphous ones. Thus, this biopolymer is more soluble in water and has a higher susceptibility to the hydrolysis (Hamelinck et al., 2003).

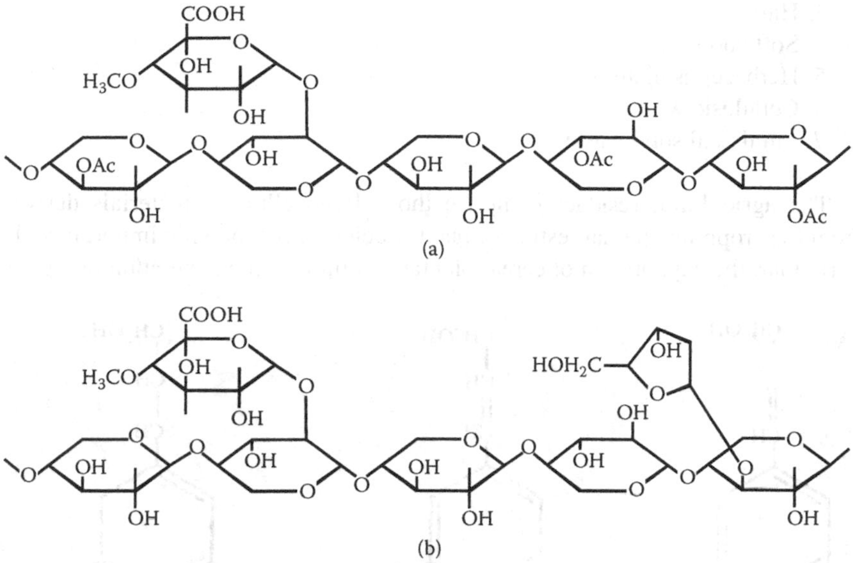

FIGURE 3.7 Partial structure of xylan chains corresponding to lignocellulosic materials obtained from hardwood (a) and softwood (b).

The lignin comprises from 10 to 25% lignocellulosic biomass. This component is a very complex phenolic polymer composed of phenyl propane units linked by C-C and C-O-C bonds forming a three-dimensional amorphous structure (Lee, 1997). The structural units of lignin are the cinnamyl alcohols, which are differentiated by the various substitutions that the aromatic ring presents (Oliva, 2003). Thus, p-hydroxyphenyl units are derived from the p-coumaryl alcohol, the guaiacyl units are derived from the coniferilyc alcohol, and the syringyl units are derived from the sinapyl alcohol (Figure 3.8). The lignin has hydrophobic character and its main function is as incrustive material of a cell wall, i.e., as a sort of cement between the cells.

The interaction and combination between the hemicellulose and lignin provide a covering shell to the cellulose making its degradation more difficult (Figure 3.9). Precisely, the main aim of biomass pretreatment is to break the lignin seal and significantly reduce the proportion of crystalline cellulose in such a way that the enzymes hydrolyzing the cellulose (cellulases) can have greater access to this polysaccharide and convert it into fermentable sugars.

3.3.2 CLASSIFICATION OF LIGNOCELLULOSIC MATERIALS

Depending on their origin, the most promising lignocellulosic materials regarding their conversion into fuel ethanol can be classified into seven big groups (see Table 3.10):

1. Agricultural residues
2. Agro-industrial residues

3. Hardwood
4. Softwood
5. Herbaceous biomass
6. Cellulosic wastes
7. Municipal solid waste

The agricultural residues comprise those lignocellulosic materials derived from the cropping and harvesting of plant species with economic importance. In particular, the exploitation of cereal plantations implies the generation of a great

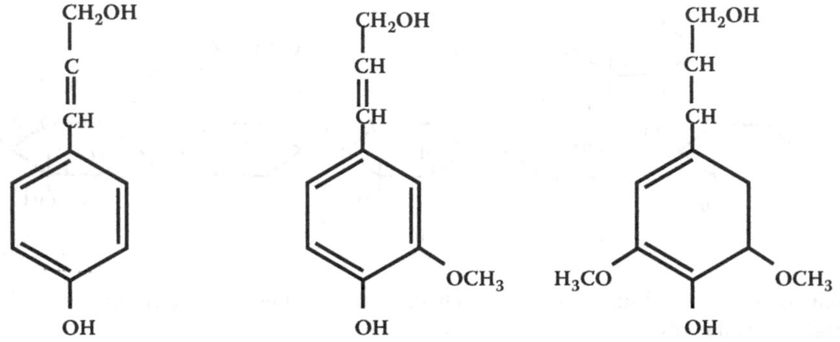

Coumaryl alcohol
p-hydroxyphenyl unit

Coniferilyc alcohol
Guaiacyl unit

Sinapyl alcohol
Syringyl unit

FIGURE 3.8 Cinnamyl alcohols that are precursors of lignin.

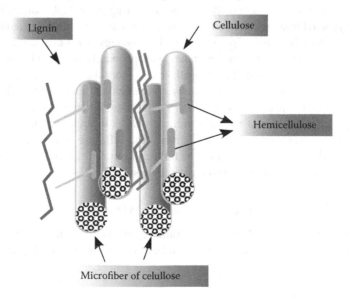

FIGURE 3.9 Schematic diagram of the structure of a lignocellulosic complex.

TABLE 3.10

Classification of Some Materials with High Content of Lignocellulosic Biomass

Agricultural Residues	Agro-Industrial Residues	Hardwood	Softwood	Cellulosic Wastes	Herbaceous Biomass	Municipal Solid Waste
Wheat straw, rice straw, and barley straw	Sugarcane bagasse, fibrous of sugarcane or extraction juice, sweet sorghum bagasse, corn stover, and olive stone and pulp	Eucalyptus, aspen, oak, maple, birch, rosewood, and mahogany	Pine, fir, spruce, larch, and cedar	Newspaper, waste office paper, paper sludge	Switchgrass, alfalfa hay, reed canary grass, coastal Bermuda grass, and timothy grass	Wasted paper, cardboard, fruit and vegetable peels, garden residues, and wood items

volume of residues, among which the straw should be highlighted. The straw is the dried, crushed or not crushed material coming from plants of the family *Gramineae* once it is separated from the grain. The straws most evaluated for ethanol production purposes are wheat straw, rice straw, and barley straw. The agro-industrial residues refer to the by-products and wastes generated during the commercial transformation of agricultural crops. The bagasse should be listed among these residues. The most studied agro-industrial residue is the sugarcane bagasse, the fibrous residue obtained after juice extraction during the milling step of sugarcane. Other promising residue is the sweet sorghum bagasse. Other materials belonging to this category are corn stover and olive stone and pulp.

The materials having origin in hardwood constitute a separate group of lignocellulosic feedstocks. These materials comprise not only the wood itself, but also its derivatives like sawdust, shavings, and the collected biomass resulting from forestry activities, such as branches, stalks, and trunk pieces. The wood obtained from trees of the angiosperm species—poplar, eucalyptus, aspen, oak, maple, birch, rosewood, and mahogany—belongs to hardwood. The softwood, in turn, comprises the wood of conifers. The wood from trees of the gymnosperm species, such as pine, fir, spruce, larch, and cedar, can be included in this group. Softwood has higher lignin content than hardwood.

The herbaceous biomass refers to the materials coming from herbaceous plants, i.e., those plants not generating wood. The grasses are plants that present neither woody stems nor woody roots. In general, their stems are green. The most studied herbaceous biomass for ethanol production purposes comprises different types of grasses like switchgrass, used in North America for hay production, alfalfa hay, reed canary grass, coastal Bermuda grass, and timothy grass that covers almost two-thirds of the livestock meadows in the United States. These herbaceous plants have a great importance because they grow very fast and have reduced nutritional requirements. Thus, these plants are excellent candidates for their exploitation as crops dedicated to bioenergy production.

Among the cellulosic wastes are residues resulting from industrial activities, mostly related to paper processing, which have an elevated content of cellulose compared to other types of lignocellulosic biomass. As examples of this group, newspaper, waste office paper, and paper sludge, one of the effluents of plants for paper recycling, should be highlighted. Finally, the organic fraction of municipal solid wastes are composed of materials with high lignocellulosic content like wasted paper, cardboard, fruit and vegetable peels, garden residues, and wood items, among others.

The composition of lignocellulosic biomass, expressed as the proportion of cellulose, hemicellulose, and lignin, depends on its origin, though some similarities can be observed independent on the group to which all material belongs. The composition of several representative lignocellulosic materials is presented in Table 3.11. As can be observed, hardwood presents high cellulose content making it very promising regarding the production of second generation bioethanol. In this sense, the hydrolysis of cellulose allows the formation of glucose, which in turn can be transformed into ethanol by fermentation. Likewise, the cellulosic

TABLE 3.11
Percentage Composition (Dry Basis) of Some Lignocellulosic Materials

Materials	Cellulose/% (by weight)	Hemicelluloses/% (by weight)	Lignin/% (by weight)	References
Sugarcane bagasse	50	25	25	Pandy et al (2000)
Hardwood	40–55	24–40	18–25	Sun and Cheng (2002)
Softwood	45–50	25–35	25–35	Sun and Cheng (2002)
Nut shells	25–30	25–30	30–40	Sun and Cheng (2002)
Corn cobs	45	35	15	Sun and Cheng (2002)
Grasses	25–40	35–50	10–30	Sun and Cheng (2002)
Paper	85–99	0	0–15	Sun and Cheng (2002)
Wheat straw	30	50	15	Sun and Cheng (2002)
Sorted refuse	60	20	20	Sun and Cheng (2002)
Leaves	15–20	80–85	0	Sun and Cheng (2002)
Cotton seed hairs	80–95	5–20	0	Sun and Cheng (2002)
Newspaper	40–55	25–40	18–30	Sun and Cheng (2002)
Waste paper from chemical pulp	60–70	10–20	5–10	Sun and Cheng (2002)
Primary wastewater solids	8–15	—	24–29	Sun and Cheng (2002)
Coastal Bermuda grass	25	35.7	6.4	Sun and Cheng (2002)
Switch grass	45	31.4	12.0	Sun and Cheng (2002)
Empty fruit bunches of oil palm	15.47	11.73	7.14	Abdul Aziz et al. (2002a, 2002b); Gutiérrez et al. (2009); Wan Zahari and Mokhtar (2004)
Palm press fiber	24.00	14.40	12.60	Abdul Aziz et al. (2002a, 2002b); Gutiérrez et al. (2009); Wan Zahari and Mokhtar (2004)

wastes have high cellulose contents. On the other hand and due to its high lignin content, softwood has a more difficult transformation process increasing the complexity of fuel ethanol production.

From the large variety of lignocellulosic materials that have been proved for fuel ethanol production, those with higher availability and production volume have been defined as the most promising. Logically, this selection is made based on the context of each region and country. Thus, the corn stover along with poplar wood has become the most potential feedstocks for bioethanol production in the United States. In Europe, the straw of different cereals is the most promising material. In fact, wheat straw and barley straw are the feedstocks defined for the first commercial plant for production of cellulosic ethanol whose construction started in 2005 and will open soon in Salamanca (Spain; Abengoa, 2008). In the case of South American countries like Brazil and Colombia, the most promising feedstock for second generation (lignocellulosic) ethanol is cane bagasse because it has great availability and production volumes. In addition, the participation of the sugar sector could provide the necessary financial resources needed for the implementation of these technologies at industrial level.

3.3.2.1 Sugarcane Bagasse

The sugarcane bagasse is one of the most produced lignocellulosic materials in the world and represents the residue of cane stems after crushing and juice extraction. It is a by-product of the sugar industry and is almost exclusively utilized in sugar mills as a fuel for steam generation. In the past few years, the research on the economic utilization of bagasse to produce electricity, paper and paper pulp, and fermentation products has intensified. For instance, the Colombian sugar sector already produces paper from bagasse at an industrial level and commercializes 15 MW of electricity obtained from this material in the national interconnected electric network (Asocaña, 2006). The technologies for production value-added products by fermentation employing bagasse are currently under development. In general, these technologies are oriented to the application of solid-state fermentation to produce animal feed, enzymes, amino acids, organic acids, and pharmaceuticals, among others (Pandey et al., 2000). The pretreated bagasse can be used as a substrate in submerged fermentations for production of xylitol, flavors, single-cell protein, cellulases, ligninases, and xylanases (Aguilar et al., 2002; Cardona and Sánchez, 2007; Pandey et al., 2000). In addition, sugarcane bagasse can be the feedstock for producing activated carbon with an exceptionally high adsorptive capacity (Cardona and Sánchez, 2007; Lutz et al., 1998).

The percentage of the main biopolymers in cane bagasse is quite similar to that of hardwood (see Table 3.11). In chemical terms, cane bagasse contains about 50% α-cellulose, 30% pentosans, and 2.4% ash. Due to its low ash content, the bagasse presents a better performance during fermentation than other residues, such as rice straw and wheat straw, which contain 17.5% and 11.0% ash, respectively. Moreover, the bagasse can be considered as a valuable means for accumulating solar energy due to its high yield (80 ton/ha) compared to wheat (1 ton/ha), grasses (2 ton/ha), and trees (20 ton/ha; Pandey et al., 2000). In general, great amounts of energy can be obtained from cane bagasse. By burning the bagasse in co-generation systems, the total amounts of process steam and electricity required to adequately cover the energy needs of sugar mills can be obtained, with an energy surplus that can be sold to the grid. Nevertheless, different energy and economic analyses indicate that more benefits can be achieved if the bagasse were used for ethanol production (Cardona and Sánchez, 2006; Moreira, 2000). To this regard, it is necessary to take into account that the electricity can be obtained from a great number of primary fuels, while the liquid fossil fuels for transportation can be substituted only by a reduced amount of renewable fuels (bioethanol and biodiesel; Moreira, 2000).

The global production of cane bagasse can be estimated at 373 to 416 million tonne per year considering data of 2004 and yields from 280 kg bagasse per tonne of cane (Moreira, 2000) up to 312 kg/ton (Kim and Dale, 2004). The ethanol yields of bagasse depend on the conversion technology employed, which will be discussed in the following chapters. In a preliminary way, a yield of 140 L EtOH/ ton bagasse can be assumed based on the efficiency reported by different trials carried out in the National Renewable Energy Laboratory in the United States (Golden, CO), which has been used by Kim and Dale (2004). Considering the mentioned yield, the global potential for producing ethanol from cane bagasse reaches 58.2 million L/year, an amount greater than all the ethanol produced in the world in 2007.

3.3.2.2 Corn Stover

Corn stover is composed of stems, leaves, and cobs resulting from the corn harvest. This lignocellulosic material is considered as one of the most promising feedstocks for ethanol production in the United States since it is the most abundant agricultural residue in that country. For instance, the corn was the second most cultivated crop in the United States in 2005 (29.8 million ha) after soybeans (29.9 million ha) and followed by wheat (20.2 million ha) and cotton (5.3 million ha; FAO, 2007a). However, in terms of product harvested, corn widely outpaces the soybean (299.9 million ton of corn versus 85.0 million ton of soybeans). In addition, the generation of residues in the case of soybeans is modest compared to corn. Corn stover availabilities have been estimated at 153 million for per year (dry basis) in the United States relative to other agricultural residues (about 58 million ton) and demonstrates the huge volumes generated (Kadam and McMillan, 2003).

A factor to be considered in the case of agricultural residues is that, unlike cane bagasse, their total utilization can lead to soil erosion and reduction of its organic matter content. For this, the sustainable fraction of residues to be collected should be defined. This fraction depends on the climate conditions, crop rotation, soil fertility, land scope, and employed agronomic practices. For example, it is estimated that it is necessary to cover the soil surface after harvesting with more than 30% of harvest residues to avoid the erosion due to water runoff. For this reason and to evaluate the availability of feedstocks for ethanol production, higher percentages are used to consider the uncertainty generated by different local conditions (Kim and Dale, 2004). Taking into account this factor, the amount of corn stover that can be collected in a sustainable way in the United States has been estimated at 80 to 100 million ton (dry basis) per year. Only a small portion of this amount (about 20 million ton) has the potential to be used for purposes other than ethanol production, such as, for instance, the production of agglomerates, pulp, or furfural. In this way, the production of 11 million L of fuel ethanol will imply the utilization of only 40% of collectable corn stover (Kadam and McMillan, 2003). The global potential for ethanol production from this material reaches 72.45 million L. Undoubtedly, the implementation of a program for fuel ethanol production from corn stover would reduce the pressure over corn prices.

3.3.2.3 Cereal Straws

Cereal straws are lignocellulosic materials with a high content of hemicellulose compared to cellulose (see Table 3.11). The straws comprise the dry stalks of a cereal plant, crushed or not, after the grain or seed has been removed. The wheat straw presents a great availability due to the vast cultivated volume of this grain in the world (about 813 million ton per year of straw). In the United States, about 33 million ton of wheat straw is available, assuming that 40% of total straw is collectable in a sustainable way (Kadam and McMillan, 2003). This amount of straw is equivalent to 9.6 million L of ethanol if considering an ethanol yield from wheat straw of 261.3 L/ton. Rice straw is another source of lignocellulosic biomass with a high availability in the world (779 million ton). In particular, the increase in environmental controls is phasing out the practice of burning rice straw in the open air. These regulations contribute to the search for new applications for these types of residues.

3.3.2.4 Municipal Solid Waste

Approximately 1.6 million ton per year of municipal solid waste (MSW) is produced in the world. This waste incurs serious problems due to the progressive deterioration of the environment and also to the costs associated with its recollection, transport, and disposal. It is estimated that the services for disposal, treatment, and economic utilization of MSW accounts for an annual market of $100 million worldwide, from which $42 million corresponds to North America, $42 million to the European Union, and only $6 million to South America, having the corresponding MSW volumes of 250, 200, and 150 million ton per year, respectively (Skinner, 2000). The organic fraction of MSW (that can reach more than 65%) is an extraordinarily heterogeneous mixture of materials in which materials with a lignocellulosic nature predominate: paper, cardboard, fruit and vegetable peels, garden residues, wood items, etc. The production of ethanol from MSW has already been patented (Titmas, 1999). MSW presents high amounts of inhibitors of its conversion process, which not only originate during the processing, but also come from the wastes themselves. Despite this, the possibility of producing ethyl alcohol from this waste has been demonstrated (Green et al. 1988; Green and Shelef, 1989). Many of these residues have significant starch contents especially in the case of wastes from marketplaces. In this case, the presence of discarded fruits with different maturity degrees provides the starch, which can be hydrolyzed using enzymes that degrade this polysaccharide and the cellulose coming from the lignocellulosic biomass present in these residues as well (Cardona et al., 2004).

REFERENCES

Abdul Aziz, A., K. Das, M. Husin, and A. Mokhtar. 2002a. Effects of physical and chemical pre-treatments on xylose and glucose production from oil palm press fibre. *Journal of Oil Palm Research* 14 (2):10-17.

Abdul Aziz, A., M. Husin, and A. Mokhtar. 2002b. Preparation of cellulose from oil palm empty fruit bunches via ethanol digestion: Effect of acid and alkali catalysts. *Journal of Oil Palm Research* 14 (1):9–14.

Abengoa. 2008. *Biocarburantes de Castilla y León informs.* Abengoa. http://www.abengoa.com/sites/abengoa/en/noticias_y_publicaciones/noticias/historico/noticias/2008/07_julio/20080715_noticias.html (accessed February 2009).

Agronet. 2007. *Propuesta técnica cultivo de yuca (Technical proposal cassava crop, in Spanish).* Ministerio de Agricultura y Desarrollo Rural, República de Colombia. http://www.agronet.gov.co/www/docs_si2/2006112717345_Propuesta%20tecnica%20cultivo%20de%20yuca.pdf (accessed February 2007).

Aguilar, R., J. Ramírez, G. Garrote, and M. Vázquez. 2002. Kinetic study of the acid hydrolysis of sugar cane bagasse. *Journal of Food Engineering* 55:309–318.

Alarcón, F., and D. Dufour. 1998. *Almidón Agrio de Yuca en Colombia. Tomo 1: Producción y Recomendaciones (Sour cassava starch in Colombia. Volume 1: Production and recommendations, in Spanish).* Cali, Colombia: Centro Internacional de Agricultura Tropical (CIAT).

Andrade, J.B., de, E. Ferrari, Jr., R.A. Possenti, I. Pozar, L. Zimback, and M.G. Landell, de A. 2004. Composição química de genótipos de cana-de-açúcar em duas idades, para fins de nutrição animal (Chemical composition of sugarcane genotypes in two ages, for animal nutrition, in Portuguese). *Bragantia* 63 (3):341–349.

Asocaña. 2006. *Aspectos generales del sector azucarero.* (General aspects of sugar sector, in Spanish.) Informe Anual 2005-2006, Asociación de Cultivadores de Caña de Azúcar de Colombia (Asocaña).

Aya, W.A., J.C. Pineda, Ó.J. Sánchez, and C.A. Cardona. 2005. Análisis comparativo de diferentes materias primas amiláceas para la obtención de alcohol carburante (Comparative analysis of different starchy feedstocks for fuel ethanol production, in Spanish). Paper presented at XXIII Congreso Colombiano de Ingeniería Química, at Manizales, Colombia.

Bhattacharya, P.K., S. Agarwal, S. De, and U.V.S. Rama Gopal. 2001. Ultrafiltration of sugar cane juice for recovery of sugar: Analysis of flux and retention. *Separation and Purification Technology* 21:247–259.

Bothast, R.J., and M.A. Schlicher. 2005. Biotechnological processes for conversion of corn into ethanol. *Applied Microbiology and Biotechnology* 67:19–25.

Buleón, A., P. Colonna, V. Planchot, and S. Ball. 1998. Starch granules: Structure and biosynthesis. *International Journal of Biological Macromolecules* 23:85–112.

Cardona, C.A., and Ó.J. Sánchez. 2006. Energy consumption analysis of integrated flowsheets for production of fuel ethanol from lignocellulosic biomass. *Energy* 31:2447–2459.

Cardona, C.A., and Ó.J. Sánchez. 2007. Fuel ethanol production: Process design trends and integration opportunities. *Bioresource Technology* 98:2415–2457.

Cardona, C.A., Ó.J. Sánchez, M.I. Montoya, and J.A. Quintero. 2005. Analysis of fuel ethanol production processes using lignocellulosic biomass and starch as feedstocks. Paper presented at the 7th World Congress of Chemical Engineering, July 10–14, Glasgow, Scotland.

Cardona, C.A., Ó.J. Sánchez, J.A Ramírez, and L.E. Alzate. 2004. Biodegradación de residuos orgánicos de plazas de mercado (Biodegradation of organic residues from market places, in Spanish). *Revista Colombiana de Biotecnología* 6 (2):78–89.

Curtin, L.V. 1983. Molasses—General considerations. In *Molasses in Animal Nutrition*. West Des Moines, IA: National Feed Ingredients Association.

Decision News Media SAS. 2006. *World sugar production increases but prices remain high, says USDA*. Decision News Media SAS. http://www.confectionerynews.com/news/ng.asp?n=65583-usda-sugar-eu-sugar-reform-brazil (accessed October 2006).

Decloux, M., A. Bories, R. Lewandowski, C. Fargues, A. Mersad, M.L. Lameloise, F. Bonnet, B. Dherbecourt, and L. Nieto Osuna. 2002. Interest of electrodialysis to reduce potassium level in vinasses. Preliminary experiments. *Desalination* 146:393–398.

Duke, J.A. 1998. *Saccharum officinarum L.* Purdue University. http://www.hort.purdue.edu/newcrop/duke_energy/Saccharum_officinarum.html (accessed October 2006).

du Preez, J.C., F. de Jong, P.J. Botes, and P.M. Lategan. 1985. Fermentation alcohol from grain sorghum starch. *Biomass* 8:101–117.

Espinal, C.F., H.J. Martínez, and X. Acevedo. 2005a. *La cadena de cereales, alimentos balanceados para animales, avicultura y porcicultura en Colombia. Una mirada global de su estructura y dinámica 1991–2005.* (The chain of grains, balanced animal feed, poultry farming and pig farming in Colombia. An overview of its structure and dynamics 1991–2005, in Spanish). Documento de Trabajo No. 87, Observatorio Agrocadenas Colombia, Ministerio de Agricultura y Desarrollo Rural.

Espinal, C.F., H.J. Martínez, L. Ortiz, and L.S. Beltrán. 2005b. *La cadena de azúcar en Colombia. Una mirada global de su estructura y dinámica 1991–2005.* (The sugar chain in Colombia. An overview of its structure and dynamics 1991–2005, in Spanish). Documento de Trabajo No. 88, Observatorio Agrocadenas Colombia, Ministerio de Agricultura y Desarrollo Rural.

Fajardo, E.E., and S.C. Sarmiento. 2007. Evaluación de melaza de caña como sustrato para la producción de *Saccharomyces cerevisiae*. (Evaluation of sugarcane molasses as substrate for *Saccharomyces cerevisiae* production, in Spanish). B.Sc. thesis, Pontificia Universidad Javeriana, Bogotá, Colombia.

FAO. 2006. *Major food and agricultural commodities and producers*. Food and Agricultural Organization of the United Nations (FAO). http://www.fao.org/es/ess/top/country.html (accessed November 2006).

FAO. 2007a. *FAOSTAT*. Food and Agriculture Organization of the United Nations (FAO). http://faostat.fao.org (accessed February 2007).

FAO. 2007b. *Major food and agricultural commodities and producers*. Food and Agriculture Organization of the United Nations (FAO). http://www.fao.org/es/ess/top/country.html (accessed February 2007).

FAO. 2008. *FAOSTAT*. Food and Agriculture Organization of the United Nations (FAO). http://faostat.fao.org (accessed February 2009).

FAO and IFAD. 2004. Cassava starch. Paper presented at the Proceedings of the Validation Forum on the Global Cassava Development Strategy, in Rome, Italy.

FAS. 2005. *World sugar situation—December 2005*. Foreign Agricultural Service (FAS), United States Department of Agriculture. http://www.fas.usda.gov/htp/sugar/2006/World%20Sugar%20Situation%20.pdf (accessed November 2006).

Ghosh, P., and T.K. Ghose. 2003. Bioethanol in India: Recent past and emerging future. *Advanced Biochemical Engineering and Biotechnology* 85:1–27.

González, D.A., and C. González. 2004. Jugo de caña y follajes arbóreos en la alimentación no convencional del cerdo. (Sugar cane juice and tree foliages in nonconventional feeding pigs, in Spanish). *Revista Computadorizada de Producción Porcina* 11 (3):2004.

Grassi, G. 1999. Modern bioenergy in the European Union. *Renewable Energy* 16:985–990.

Green, M. , and G. Shelef. 1989. Ethanol fermentation of acid hydrolysate of municipal solid waste. *The Chemical Engineering Journal* 40:B25–B28.

Green, M., S. Kimchie, A.I. Malester, B. Rugg, and G. Shelef. 1988. Utilization of municipal solid wastes (MSW) for alcohol production. *Biological Wastes* 26:285–295.

Gulati, M., K. Kohlman, M.R. Ladish, R. Hespell, and R.J. Bothast. 1996. Assessment of ethanol production options for corn products. *Bioresource Technology* 5:253–264.

Gutiérrez, L.F., Ó.J. Sánchez, and C.A. Cardona. 2009. Process integration possibilities for biodiesel production from palm oil using ethanol obtained from lignocellulosic residues of oil palm industry. *Bioresource Technology* 100 (3):1227–1237.

Hamelinck, C.N., G.V. Hooijdonk, and A.P.C. Faaij. 2003. *Prospects for ethanol from lignocellulosic biomass: techno-economic performance as development progresses*. Report NWS-E-2003-55, Utrecht University, The Netherlands.

Hosein, R., and W.A. Mellowes. 1989. Malt hydrolysis of sweet potatoes and eddoes for ethanol production. *Biological Wastes* 29:263–270.

Infoagro. 2002. *El cultivo de la remolacha azucarera*. (*The cropping of sugar beet*, in Spanish). Infoagro. http://www.infoagro.com/herbaceos/industriales/remolacha_azucarera.htm (accessed November 2006).

Infoagro. 2007. *El cultivo del maíz*. Infoagro. http://www.infoagro.com/herbaceos/cereales/maiz.asp (accessed February 2007).

ISO. 2008. *Sugar Year Book 2008*. London, U.K.: International Sugar Organization (ISO).

Kadam, K.L., and J.D. McMillan. 2003. Availability of corn stover as a sustainable feedstock for bioethanol production. *Bioresource Technology* 88:17–25.

Kim, S., and B.E. Dale. 2004. Global potential bioethanol production from wasted crops and crop residues. *Biomass and Bioenergy* 26:361–375.

Lee, J. 1997. Biological conversion of lignocellulosic biomass to ethanol. *Journal of Biotechnology* 56:1–24.

Lindeman, L.R., and C. Rocchiccioli. 1979. Ethanol in Brazil: Brief summary of the state of the industry in 1977. *Biotechnology and Bioengineering* 21:1107–1119.

Lutz, H., K. Esuoso, M. Kutubuddin, and E. Bayer. 1998. Low-temperature conversion of sugar cane by-products. *Biomass and Bioenergy* 15 (2):155–162.

Messias de Bragança, R., and P. Fowler. 2004. *Industrial markets for starch*, The BioComposites Centre, University of Wales, Bangor. www.bc.bangor.ac.uk/_includes/docs/pdf/indsutrial%20markets%20for%20starch.pdf

Moreira, J.S. 2000. Sugarcane for energy: Recent results and progress in Brazil. *Energy for Sustainable Development* 4 (3):43–54.

Mullins, J.T., and C. NeSmith. 1987. Acceleration of the rate of ethanol fermentation by addition of nitrogen in high tannin grain sorghum. *Biotechnology and Bioengineering* 30:1073–1076.

Murtagh, J.E. 1995. Molasses as feedstock for alcohol production. In *The Alcohol Textbook*, ed. T. P. Lyons, D. R. Kelsall, and J. E. Murtagh. Nottingham, U.K.: University Press.

Nigam, P., and D. Singh. 1995. Enzyme and microbial systems involved in starch processing. *Enzyme and Microbial Technology* 17:770–778.

Ogbonna, J.C., H. Mashima, and H. Tanaka. 2001. Scale up of fuel ethanol production from sugar beet juice using loofa sponge immobilized bioreactor. *Bioresource Technology* 76:1–8.

Oliva, J.M. 2003. Efecto de los productos de degradación originados en la explosión por vapor de biomasa de chopo sobre *Kluyveromyces marxianus*, (Effect of the degradation products generated during the steam explosion of poplar biomass on *Kluyveromyces marxianus*, in Spanish), Departamento de Microbiología III, Ph.D. Thesis, Universidad Complutense de Madrid, Madrid, España.

Pandey, A., C. Soccol, C. Nigam, and V. Soccol. 2000. Biotechnological potential of agro-industrial residues. I: Sugarcane bagasse. *Bioresource Technology* 74:69–80.

Poitrat, E. 1999. The potential of liquid biofuels in France. *Renewable Energy* 16:1084–1089.

Renewable Fuel Association. 2009. *Statistics*. Renewable Fuel Asociation. http://www. ethanolrfa.org/industry/statistics/#E (accessed February 2009).

Sajilata, M.G., and Rekha S. Singhal. 2005. Specialty starches for snack foods. *Carbohydrate Polymers* 59:131–151.

Sánchez, Ó.J., and C.A. Cardona. 2008a. *Producción de Alcohol Carburante: Una Alternativa para el Desarrollo Agroindustrial (Fuel ethanol production: An alternative for agro-industrial development*, in Spanish). Manizales: Universidad Nacional de Colombia.

Sánchez, Ó.J., and C.A. Cardona. 2008b. Trends in biotechnological production of fuel ethanol from different feedstocks. *Bioresource Technology* 99:5270–5295.

Seebaluck, V., R. Mohee, P.R.K. Sobhanbabu, F. Rosillo-Calle, M.R.L.V. Leal, and F.X. Johnson. 2008. *Bioenergy for sustainable development and global competitiveness: The case of sugar cane in southern Africa*. CARENSA/SEI 2008-02, Cane Resources Network for Southern Africa (CARENSA).

Skinner, J. H. 2000. *Worldwide MSW market reaches $100 billion: A report from the ISWA World Congress 2000*. Swana CEO Report, September 2000, SWANA.

Suárez, R., and R. Morín. 2005. Caña de azúcar y sostenibilidad: Enfoques y experiencias cubanas (Sugarcane and sustainability: Approaches and Cuban experiences, in Spanish). *Transformando el Campo Cubano*, http://www.desal.org.mx/article.php3?id_article=26.

Sun, Y., and J. Cheng. 2002. Hydrolysis of lignocellulosic materials for ethanol production: A review. *Bioresource Technology* 83:1–11.

Tester, R.F., J. Karkalas, and X. Qi. 2004. Starch—Composition, fine structure and architecture. *Journal of Cereal Science* 39:151–165.

Titmas, J.A. 1999. Apparatus for hydrolyzing cellulosic material. United States Patent No. 5879637.

U.S. Grains Council. 2007. *Los coproductos alimenticios derivados del proceso de la molienda húmeda del maíz. El uso de la harina de gluten de maíz.* (Feed co-products derived from corn wet-milling process. Use of the corn gluten meal, in Spanish.) U.S. Grains Council, http://www.grains.org accessed February 2007).

Vega-Baudrit, J., K. Delgado-Montero, M. Sibaja-Ballestero, and P. Alvarado-Aguilar. 2007. Uso alternativo de la melaza de la caña de azúcar residual para la síntesis de espuma rígidas de poliuretano (ERP) de uso industrial. (Alternative use of residual sugarcane molasses for synthesis of polyurethane rigid foams (PRF) with industial use, in Spanish). *Tecnología, Ciencia, Educación (IMIQ)* 22 (2):101–107.

Wan Zahari, M., and A.R. Alimon. 2004. Use of palm kernel cake and oil palm by-products in compound feed. *Palm Oil Developments* 40:5–9.

Wang, S., K. Sosulski, F. Sosulski, and M. Ingledew. 1997. Effect of sequential abrasion on starch composition of five cereals for ethanol fermentation. *Food Research International* 30 (8):603–609.

Wilkie, A.C., K.J. Riedesel, and J.M. Owens. 2000. Stillage characterization and anaerobic treatment of ethanol stillage from conventional and cellulosic feedstocks. *Biomass and Bioenergy* 19:63–102.

Winner Network. 2002. *Village level bioenergy system based on sweet sorghum.* http://www. w3c.org/TR/1999/REC-html401-19991224/loose.dtd/ (accessed October 2004).

Zhan, X., D. Wang, M.R. Tuinstra, S. Bean, P.A. Seib, and X.S. Sun. 2003. Ethanol and lactic acid production as affected by sorghum genotype and location. *Industrial Crops and Products* 18:245–255.

4 Feedstock Conditioning and Pretreatment

In this chapter, the first processing steps for fuel ethanol production from different feedstocks, especially starchy and lignocellulosic materials, are analyzed from the viewpoint of process synthesis. The importance of conditioning and pretreatment as decisive process steps for conversion of feedstocks into ethanol is highlighted. Main methods for conditioning molasses are presented as well as the enzymatic procedures for starch hydrolysis. The need of pretreatment of lignocellulosic biomass is analyzed considering the complexity of this type of raw material. Several methods for pretreatment and detoxification of biomass are briefly described.

4.1 CONDITIONING OF SUCROSE-CONTAINING MATERIALS

Sucrose-containing feedstocks for ethanol production contain concentrated solutions of sugars. In this sense, these raw materials are nearing a point for microorganisms to convert them into ethanol. However, some substances contained in these solutions can have an inhibitory effect on fermenting microorganisms because the used cultivation media are complex, i.e., their composition is not completely determined unlike the synthetic media for which the concentration of each one of their components is exactly defined. Moreover, all the components are known in synthetic media. In complex media, the composition of such components, such as cane molasses, varies due to factors as the agronomic techniques for cane cropping, climate conditions, luminosity, type of employed fertilizers, water availability, and cane-cutting procedures. In this way, it is impossible to predict the exact concentration of fermentable sugars in molasses before the corresponding analyses. Although, the presence of potential inhibitors is difficult to handle, but they should be removed or their concentrations should be reduced in order to improve the subsequent fermentation performance. In the particular case of molasses, they should be diluted and conditioned for it to be used as the main component of the fermentation media.

The molasses, obtained either from sugarcane or sugar beet, has a solids content of about 80° Brix. This high concentration makes the molasses unviable for direct fermentation using yeasts. Therefore, it is necessary to dilute them to below 25° Brix (Murtagh, 1995). Yeasts start to ferment quickly at this solids concentration. This limit is related to the osmotic pressure that molasses exert on the yeast cells because of their high content of sugar and mineral salts. Murtagh (1995) emphasizes that the calculation required to dilute the molasses should be accomplished in units of percentage by weight because the degrees Brix are measured by weight and not by volume. Considering that the specific gravity of molasses is 1.416, a liter of molasses would have 1,413.95 grams. In addition, the sugar content of molasses to be diluted should be taken into account due to the variations in compositions during each processing batch. Thus, the final concentration of sugars after the dilution of molasses down to 25° Brix is not always the same. For instance, the cane molasses in Colombian sugar mills contains 48 to 55% total sugars on average, whereas the North American molasses is poorer in sugars (46%). In general, diluted molasses containing sugar concentrations up to 18% can be used. If more concentrated molasses is employed, the ethanol formed during fermentation will make the microorganisms have a lower growth rate that leads to more prolonged fermentation times. Alternatively, a first portion of molasses diluted down to 18° Brix can be used to favor a faster yeast growth and, once the cultivation medium reaches 12° Brix, the second portion of molasses diluted down to 35° Brix is added to this medium (Murtagh, 1995).

The ash content of molasses is another important factor to be considered. If this content is greater than 10%, incrustation problems can arise in pipelines and distillation towers in the subsequent process steps. The incrustation represents the accumulation of certain salts, mostly calcium sulfate, which are formed as a consequence of using sulfuric acid for conditioning the molasses. The addition of sulfuric acid is intended to separate the sucrose hydrolysis into its two constituent sugars: glucose and fructose. Although the yeasts synthesize the enzyme required for such hydrolysis, the invertase, the previous breakdown of this disaccharide allows a faster fermentation. Furthermore, the acid addition allows adjusting the pH of the cultivation medium based on molasses in such a way that the yeast growth is increased and the development of other undesired microorganisms, mainly bacteria, is avoided. The calcium ions come from the clarification process of both cane juice and syrup. The incrustation can be prevented through sedimentation systems in the dilution tanks of molasses, fermenters, storage tanks for wine before distillation, and stillage storage tanks before its concentration. Special chelating agents can be employed in order to remove the solids causing incrustations from the molasses (Murtagh, 1995).

Different studies to neutralize the effect the osmotic pressure has on process microorganisms have been done. One approach is the development of special yeast strains with higher resistance to the salts contained in molasses, i.e., strains more tolerant to elevated osmotic pressures. Nevertheless, the most common approach to offset the negative effect of salts is the conditioning of molasses by adding different organic and inorganic compounds. Some of the substances employed for conditioning sugarcane molasses (that can be applied to beet molasses as well) are presented in Table 4.1.

The presence of metal traces in beet molasses also affects the ethanolic fermentation using yeasts. The employed agents for cane molasses, for example, EDTA (ethylene diamine tetraacetic acid), ferrocyanide, and zeolites have demonstrated their usefulness during the conditioning of beet molasses (Ergun et al., 1997). In the particular case of zeolites, it has been suggested that they can act as a pH regulator, which has been verified in fermentations with high glucose

TABLE 4.1
Some Ways for Conditioning and Supplementation of Sugarcane Molasses for Ethanolic Fermentation Using *Saccharomyces cerevisiae*

Additive/Supplement	Function	Remarks	References
H_2SO_4	Precipitation of calcium salts; pH adjustment	Removal of calcium salts avoids scaling during distillation	Murtagh (1995)
Synthetic zeolites	Removal of inhibitory substances; changes in flocculation behavior of the yeast	1–20 g/L medium	SivaRaman et al. (1994)
Unsaturated lipids, soy flour, *Aspergillus oryzae* proteolipids	Protection against the inhibitory effects of substrate and product		SivaRaman et al. (1994)
Chitin	Protection against the inhibitory effects of substrate and product	2 g/L medium; reduction of fermentation time	Cachot and Pons (1991); Patil and Patil (1989)
Skim milk powder	Protection against the inhibitory effects of substrate and product	2 g/L medium	Cachot and Pons (1991); Patil and Patil (1989); Patil et al. (1985)
EDTA, potassium ferrocyanide, sodium potassium tartarate	Binding of inhibitory metal ions	EDTA: 50 mg/L $K_4Fe(CN)_6$, NaK tartarate: 1.5 g/L	Pandey and Agarwal (1993)
Alumina beads	Growth promoting effect	2 g/L medium	Cachot and Pons (1991)
Yeast extract	Source of amino acids	2 g/L medium	Cachot and Pons (1991)
Urea, diammonium phosphate	Sources of nitrogen and phosphorous		Murtagh (1995)
Enzymatic complex of amylases, cellulases and amylopectinases	Conversion of nonfermentable substances into assimilable compounds	Commercial preparation *Rhizozyme*: 35 ppm	Acevedo et al. (2003)

Source: Adapted from Sánchez, Ó.J., and C.A. Cardona. 2008. *Bioresource Technology* 99:5270–5295. Elsevier Ltd.

contents using *Saccharomyces bayanus*. In this way, higher ethanol concentrations have been achieved (Castellar et al., 1998). When immobilized cells for ethanol production from molasses are used, different impurities (inorganic salts, nonfermentable sugars, sulfated ash, colored substances) are fixed onto the surface of the biocatalyst decreasing its productivity. One of the strategies to diminish this negative effect is the removal of impurities by microfiltration before the fermentation process that can increase the amount of produced ethanol to 18.1% (Kaseno and Kokugan, 1997).

Several substances employed as nutritive supplements in media based on molasses are also presented in Table 4.1. In general, cane molasses has most of the nutrients needed for yeast growth. However, the addition of small amounts of the nitrogen and phosphorous sources is required. In this case, the addition of nitrogen in the form of ammonium sulfate is not recommended since this salt contributes to the generation of incrustation problems. The addition of gaseous ammonium is also not desirable because it can significantly increase the pH of the medium. Urea is the most used compound as a nitrogen source in alcoholic fermentations.

Diammonium phosphate is employed as a phosphorous source that implies the reduction in the required amount of urea (Murtagh, 1995). The usage of yeast extract as a nitrogen source has been proposed as well but this material is expensive (Cachot and Pons, 1991). Molasses supplementation can be enhanced with the addition of some hydrolytic enzymes converting some biopolymers and non-fermentable substances contained in molasses into compounds assimilable by the yeasts, such as monosaccharides or free amino acids (Acevedo et al., 2003).

4.2   PRETREATMENT OF STARCHY MATERIALS

When cereals are used for producing fuel ethanol, the feedstock enters the process in the form of grains, which need to undergo some preliminary operations like washing and milling. As pointed out in Chapter 3, the milling of cereal grains like corn, wheat, and barley can be carried out by either the wet-milling or dry-milling process. In wet-milling technology only the starch enters the process for fuel ethanol production once it is separated from the rest of the grain components. In the case of corn, these components represent value-added products employed mostly for animal feed and human food. Moreover, part of the starch can be deviated toward the production of sweetening syrups like high-fructose corn syrup (HFCS). This type of milling technology was discussed in Chapter 3 for corn. Currently, 67% of ethanol produced in the United States is obtained in plants using the corn wet-milling technology. The ethanol yields for this process reach 403.1 L EtOH/ton corn (Gulati et al., 1996).

During the dry milling of cereals, the whole grain enters the ethanol production line, which means that the rest of its components are processed along with starch. The nonutilized components are accumulated in the bottom of the first distillation column and are concentrated as a co-product employed in animal feed. In the United States, 33% of ethanol is produced in plants employing the dry milling of corn, although other grains are used to a lesser degree. While the

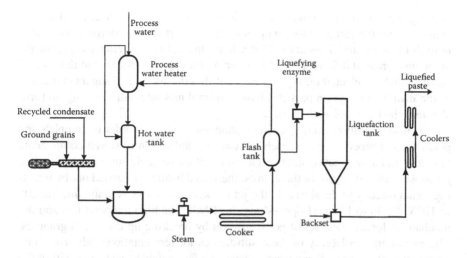

FIGURE 4.1 Scheme of cooking and liquefaction steps of ground corn grains.

dry-milling process generates co-products with lower value than the wet-milling process, this technology offers higher ethanol yields that can be in the range 419.4 to 460.6 L/ton (Gulati et al., 1996). Furthermore, the dry milling has lower capital and labor costs.

The dry milling of corn is made up of washing and milling the grains until they reach 3 to 5 mm. Then, some impurities are removed (Cardona et al., 2005). The milling is generally accomplished with the use of hammer mills. The starch from the ground grain should be gelatinized since the granules of native starch are not susceptible to enzymatic attack. To carry out this process, a starch suspension containing no more than 45% solids is prepared and cooked. To start the cooking process, the suspension is prepared in hot water (88°C). Madson and Monceaux (1995) emphasize that one of the key issues during the design and operation of production processes for ethanol production from starch is the elimination of contaminating bacteria which requires the maintenance of sterile conditions along the production line until the fermentation step. The decisive factor during the design of cooking systems is not that the starch cooks itself, but the elimination of bacteria. The overall process for cooking and liquefying ground corn grains is presented in Figure 4.1, which is based on information provided by Madson and Monceaux (1995).

Considering that the conditions needed to reach the sterility are different from the cooking conditions, the preliminary cooking of the suspension of ground grains should be accomplished with minimum solubilization of the potential fermentable substances in order to avoid undesired reactions. These substances should be released only during the subsequent steps of liquefaction, saccharification, and fermentation. This is explained taking into account that a premature solubilization can lead to the risk of undesired reactions involving fermentable substances, such as sugars contained in the corn fiber. These secondary reactions can provoke the retrogradation of starch (crystallization of soluble starch after cooling the gelatinized starch) or reactions between carbohydrates and amino acids causing the fermentable compounds to convert into nonfermentable compounds (Madson and Monceaux, 1995). In addition, these reactions also contribute to an increased infection risk. Considering these issues, the size of the ground grain should be selected taking into account that the starch and sugars contained in the matrix of the grain particles have minimal mobility, but ensuring, in turn, a suitable hydration of such particles.

The resulting slurry undergoes instantaneous cooking in order to complete the gelatinization process, i.e., to reach the total solubilization of starch components (amylose and amylopectin) and the release of all fermentable substances. For this, jet cookers are employed. In these units, the initial heating is carried out by injecting steam directly to the slurry. In the jet cookers, the slurry is maintained at 105 to 110°C for 10 to 15 sec (López-Munguía, 1993). The high temperatures and the mechanical forces allow a fast gelatinization by breaking up the starch granules. The maximum availability of these substances for fermentation with yeasts can be achieved if the process manages to keep the fermentable substances within the matrix of grain particles just until the moment in which the liquefaction is started.

During the liquefaction, thermoresistant α-amylase is used in order to hydrolyze the starch slurry in a preliminary way that allows abruptly decreasing the viscosity and improving the system operation. This process is carried out in a tank at 80 to 90°C. Moreover, this hydrolysis process allows for the avoidance of starch retrogradation, which is a latent danger since the amylose and amylopectin are dissolved in the system. With the aim of preventing undesired secondary reactions, the liquefaction is accomplished in such a way that a minimum amount of starch is hydrolyzed, thereby avoiding its conversion into other nonfermentable substances before the fermentation is started (Madson and Monceaux, 1995). Among the products of the liquefaction step are the dextrins, which are oligosaccharides formed as a consequence of the partial breakdown of amylose and amylopectin chains. The stream leaving the liquefaction tank undergoes cooling to reach the optimal temperature for the following steps (saccharification and fermentation).

Another variant of the process involves a preliquefaction step that is carried out during the cooking step. For this, 10% of the α-amylase dosage is added to the cooking tank. When the slurry is sent to the jet cooker, while the starch granules are broken down, part of the starch chains begin their hydrolysis process. The gelatinized (cooked) corn slurry is cooled to 80 to 90°C, then the remaining dosage of α-amylase is added and the hydrolysis is kept at least 30 min in the liquefaction tank (Bothast and Schlicher, 2005). When the corn wet-milling technology is used, the starch obtained undergoes cooking and liquefaction in a similar way as the ground corn grains, though process conditions can vary slightly.

From the viewpoint of process synthesis, the different schemes for pretreatment of starchy materials can cause the evaluation of multiple alternative process configurations derived from a set of several combinations of methods and technological procedures. For this, the simulation tools are of paramount importance. Unfortunately, mathematical models describing starch cooking and liquefaction have not been effectively published in the open literature. This fact limits the quality of the related simulations and the possibility of involving a more detailed description of these steps during the application of process synthesis procedures to elucidate the best conditions for fuel ethanol production.

4.3 PRETREATMENT OF LIGNOCELLULOSIC BIOMASS

The lignocellulosic biomass represents the most abundant source of fermentable sugar in nature, not only for fuel ethanol production, but also for producing a wide range of fermentation products, such as additives for the food industry, industrial chemicals, components of balanced animal feed, and pharmaceuticals. But to utilize this renewable resource in the production of most of these products, the lignocellulosic biomass must be pretreated, i.e., the lignocellulosic materials should be suitably processed in such a way that their constituent sugars and polysaccharides are susceptible to the action of hydrolytic enzymes as well as of fermenting microorganisms.

In the previous chapter, the complexity of the structure of lignocellulosic biomass was recognized considering that the lignin and hemicellulose form a sort of seal covering the polysaccharide with the highest potential to release glucose, the cellulose. In addition, it should be emphasized that most of cellulose in biomass has a crystalline structure derived from the longitudinal alignment of its linear chains. In the crystalline cellulose, the polysaccharide–polysaccharide interactions are favored and the polysaccharide–water interactions are reduced so this biopolymer is insoluble in water. A minor fraction of cellulose has an amorphous structure (Figure 4.2). The hemicellulose chains establish hydrogen bonds with the cellulose microfibers forming a matrix reinforced with lignin. The lignin presence makes it so the lignocellulosic complex cannot be directly hydrolyzed with enzymes. In this way, factors such as the crystallinity degree of cellulose, available surface area (porosity of the material), protection of cellulose by the lignin, pod-type cover offered by the hemicellulose to the cellulose, and heterogeneous character of the biomass particles contribute to the recalcitrance of the lignocellulosic materials to the hydrolysis. In addition, the relationship between the biomass structure

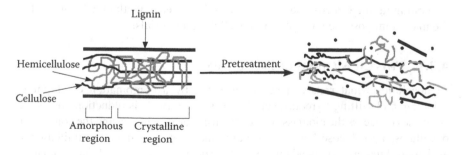

FIGURE 4.2 Schematic diagram of the pretreatment process of lignocellulosic biomass.

and its composition adds a factor implying even more variability exhibited by these materials regarding their digestibility (Mosier et al., 2005b). Therefore, the pretreatment step of the lignocellulosic complex has the following goals:

- Breakdown of the cellulose–hemicellulose matrix
- Reduction of the crystallinity degree of cellulose and increase of the fraction of amorphous cellulose
- Hydrolysis of hemicellulose
- Release and partial degradation of lignin
- Increase of the biomass porosity

In addition, the pretreatment should contribute to the formation of sugars (hexoses and pentoses) through the hemicellulose hydrolysis or to the ability to form glucose during the subsequent enzymatic hydrolysis of cellulose. The pretreatment should also avoid the formation of by-products inhibiting the subsequent bioprocesses. As a complement, the pretreatment avoids the need of reducing the biomass particle size, a very energy-consuming process. The efficiency of this process is evidenced by the fact that, when the biomass is not pretreated, glucose yields during the following cellulose hydrolysis step are less than 20% of theoretical yields, whereas the yields after the pretreatment often exceed 90% of the theoretical yields (Lynd, 1996). In this way, the pretreatment is a crucial step during the overall process for fuel ethanol production from lignocellulosic materials. However, the pretreatment is one of the most expensive steps: the unit costs of pretreatment can reach 30 cents per gallon of produced ethanol (about US\$0.08/L EtOH), according to Mosier et al. (2005b). Nevertheless, the improvement of pretreatment has a great potential to reduce its costs and increase the efficiency of the overall process.

Different methods have been developed for pretreatment of lignocellulosic biomass, which can have physical, chemical, physical–chemical, or biological nature (Sánchez and Cardona, 2008; Sun and Cheng, 2002). The evaluation of each one of these methods is related to whether it meets all the goals mentioned above, in addition to other features involving technoeconomic criteria, such as the cost of the agent or catalyst employed, the possibility of recycling the agents or catalysts involved, the degree of technological maturity of each method, the possibility of generating lignin as a co-product, and the ability of each method to be applied to the maximum possible amount of lignocellulosic materials.

4.3.1 PHYSICAL METHODS OF PRETREATMENT

Lignocellulosic materials can be comminuted by a combination of chipping, grinding, and milling to reduce cellulose crystallinity. This reduction allows the cellulases to access the biomass surface in an easier way increasing the conversion of cellulose into glucose. The energy requirements of mechanical comminution of agricultural materials depend on the final particle size and waste biomass characteristics (Sánchez and Cardona, 2008). In some specific cases, it has been demonstrated that wet or dry crushing used as a sole pretreatment method provokes the conversion of biomass into glucose in the following process steps of 56 to 60% for rice straw. For cane bagasse, a 25% conversion using wet crushing and 49.2% conversion for dry crushing have been observed (Rivers and Emert, 1988). In these cases, particle size plays a crucial role in the efficiency of the method. An important amount of data has been presented on the yields during the hydrolysis of timothy grass and alfalfa for different fractions of milled material with different particle sizes (Alvo and Belkacemi, 1997). Thus, 56.4% yield for timothy grass (53 to 106 μm) and 62.5% for alfalfa (53 to 106 μm) during 24 h of pretreatment using roller mills have been achieved. In contrast, the hydrolysis yields for nonmilled materials were 51.4% for timothy grass and 38% for alfalfa.

The energy requirements of the mechanical comminution of agricultural materials depend on the final particle size and biomass properties (Sun and Cheng, 2002) and are usually very elevated. Cadoche and López (1989) reported that power supply for mechanical comminution in a plant processing different lignocellulosic residues should be maintained below 30 kW/ton wastes by ensuring a final particle size in the range of 3 to 6 mm. On the other hand, it was demonstrated that milling and sieving of such residues as wheat straw could lead to an increased efficiency of the pretreatment using dilute acid due to the removal of noncarbohydrate biomass components before the physical–chemical pretreatment (Papatheofanous et al., 1998). Milling with vibratory balls is an effective method of transferring the energy input into size reduction and altering the crystalline cellulose structure. Although mechanical pretreatment methods increase cellulose reactivity toward enzymatic hydrolysis, they are unattractive due to their high energy and capital costs (Ghosh and Ghose, 2003).

Besides comminution, pyrolysis has been tested as a physical method for pretreatment of lignocellulosic biomass since cellulose rapidly decomposes when treated at high temperature. In particular, it has been reported that this method can be improved in the presence of oxygen. When zinc chloride or calcium carbonate is added as a catalyst, the decomposition of pure cellulose can occur at a lower temperature (Sun and Cheng, 2002). For instance, the pyrolysis of 100 g of waste cotton produces 80 g of a highly viscous pyrolyzate with a high content of levoglucosan (43% by weight), an intramolecular glucoside whose hydrolysis using dilute acid forms significant concentrations of glucose (Yu and Zhang, 2003). Main physical methods employed for pretreatment of lignocellulosic materials are presented in Table 4.2.

4.3.2 Physical–Chemical Methods of Pretreatment

Physical–chemical methods of pretreatment are remarkably more effective than physical ones. The most employed physical–chemical methods are presented in Table 4.3. Steam explosion (autohydrolysis) is the most commonly used method along with dilute-acid process for pretreatment of lignocellulosic materials. The use of saturated steam at high pressure causes autohydrolysis reactions in which part of the hemicellulose and lignin are converted into soluble oligomers with the help of some acids released from the biomass itself during the process. The

TABLE 4.2
Physical Methods for Pretreatment of Lignocellulosic Biomass for Ethanol Production

Methods	Procedure/Agents	Remarks	Examples of Pretreated Materials	References
Mechanical comminution	Chipping, grinding, milling	Milling: vibratory ball mill (final size: 0.2–2 mm), knife or hammer mill (final size: 3–6 mm)	Wood and forestry wastes (hardwood, straw) Corn stover, cane bagasse	Alvo and Belkacemi (1997); Cadoche and López (1989); Papatheofanous et al. (1998); Rivers and Emert (1988); Sun and Cheng (2002)
Pyrolysis	T >300°C, then cooling and condensing	Formation of volatile products and char Residues can undergo mild dilute-acid hydrolysis (1N H_2SO_4, 2.5 h, T = 97°C) to produce 80–85% reducing sugars (>50% glucose) Can be carried out under vacuum (400°C, p = 1 mm Hg, 20 min)	Timothy, alfalfa Wood Waste cotton, corn stover	Khiyami et al. (2005); Prosen et al. (1993); Sun and Cheng (2002); Yu and Zhang (2003)

Source: Adapted from Sánchez, Ó.J., and C.A. Cardona. 2008. *Bioresource Technology* 99:5270–5295. Elsevier Ltd.

TABLE 4.3
Physical–Chemical Methods for Pretreatment of Lignocellulosic Biomass for Ethanol Production

Methods	Procedure/Agents	Remarks	Examples of Pretreated Materials	References
Steam explosion	Saturated steam at 160–290°C, p = 0.69–4.85 MPa for several sec or min, then decompression until atm. pressure	It can handle high solids loads Size reduction with lower energy input compared to comminution 80–100% hemicellulose hydrolysis, destruction of a portion of xylan fraction, 45–65% xylose recovery Inhibitors formation Addition of H_2SO_4, SO_2, or CO_2 improves effect of further enzyme hydrolysis Cellulose depolymerization occurs at certain degree Lignin is not solubilized, but is redistributed	Poplar, aspen, eucalyptus Softwood (Douglas fir) Bagasse, corn stalk, wheat straw, rice straw, barley straw, sweet sorghum bagasse, *Brassica carinata* residue, olive pits Timothy grass, alfalfa, reed canary grass	Ballesteros et al. (2001, 2002, 2004); Belkacemi et al. (1997, 2002); De Bari et al. (2002) Dekker and Wallis (1983); Hamelinck et al. (2005) Heitz et al. (1987); Kaar et al. (1998) Lynd et al. (2002); Moniruzzaman (1996) Nakamura et al (2001); Negro et al. (2003) Shevchenko et al (1999); Söderström et al. (2003) Sun and Cheng (2002)
Liquid hot water (LHW)	Pressurized hot water, p >5 MPa, T = 170–230°C, 1–46 min; solids load <20%	80–100% hemicellulose hydrolysis, 88–98% xylose recovery, >50% olygomers Low or no formation of inhibitors Cellulose depolymerization occurs at certain degree Further cellulose conversion >90% Partial solubilization of lignin (20–50%)	Bagasse, corn stover, olive pulp Alfalfa fiber	Ballesteros et al. (2002); Koegel et al. (1999); Laser et al. (2002); Lynd (1996); Lynd et al. (2002); Mosier et al. (2005a); Negro et al. (2003); Ogier et al. (1999); Sreenath et al. (2001)

Continued

TABLE 4.3 (Continued)
Physical–Chemical Methods for Pretreatment of Lignocellulosic Biomass for Ethanol Production

Methods	Procedure/Agents	Remarks	Examples of Pretreated Materials	References
Ammonia fiber explosion (AFEX)	1–2 kg ammonia/kg dry biomass, 90°C, 30 min, p = 1.12–1.36 MPa	Ammonia recovery is required; 0–60% hemicellulose hydrolysis in dependence on moisture, >90% olygomers; No inhibitors formation; Cellulose depolymerization occurs at certain degree; Further cellulose conversion can be >90%, for biomass with high lignin content <50%; ~10–20% lignin solubilization	Aspen wood chips; Bagasse, wheat straw, barley straw, rice hulls, corn stover; Switchgrass, coastal Bermuda grass, alfalfa; Newsprint; MSW	Dale et al. (1996); Holtzapple et al. (1994); Lynd et al. (2002); Sun and Cheng (2002)
CO_2 explosion	4 kg CO_2/kg fiber, p = 5.62 MPa	No inhibitors formation; Further cellulose conversion can be >75%	Bagasse; Alfalfa; Recycled paper	Sun and Cheng (2002)

Source: Adapted from Sánchez, Ó.J., and C.A. Cardona. 2008. *Bioresource Technology* 99:5270–5295. Elsevier Ltd.

factors affecting steam explosion pretreatment are residence time, temperature, chip size, and moisture content (Sánchez and Cardona 2008). In this process, the combined action of temperature and contact time of steam with the biomass is achieved. To quantify this effect, the severity index has been defined (Shahbazi et al., 2005; Söderström et al., 2003), and the pretreatment severity is described as a function of time t (in min) and temperature T (in degrees Celsius) related to a reference temperature of 100°C:

$$\log R_0 = \log\left(t\exp\left(\frac{T - T_{ref}}{14.75}\right)\right) \qquad (4.1)$$

If the steam explosion process is carried out under acidic conditions (which increases the efficiency of cellulose hydrolysis), an additional term should be introduced in Equation (4.1) to consider the effect of pH on the combined severity (CS):

$$CS = \log R_0 - pH \qquad (4.2)$$

The pH can be calculated from the amount of sulfuric acid added to the material and its water content (Söderström et al., 2003).

An important factor to be considered when pretreatment methods are used (including steam explosion) is the particle size of the lignocellulosic materials. The conventional mechanical methods require 70% more energy than steam explosion in order to achieve the same reduction in the particle size. During steam explosion, some inhibitors of the subsequent biological processes (enzymatic hydrolysis, fermentation) are formed. Inhibitors formation during the pretreatment requires the washing of pretreated biomass. This decreases the global yield of the saccharification due to the removal of sugars generated during the hemicellulose hydrolysis. Typically, 20 to 25% of the initial dry matter is removed by the washing water (Sun and Cheng, 2002). For this reason, the use of very small particles in some cases (e.g., herbaceous wastes) is not desirable if taking into account the economy of the process (Ballesteros et al., 2002).

Steam explosion is recognized as one of the most efficient methods for hardwood (poplar, oak, birch, maple) and agroindustrial residues, but it is less efficient for softwood (pine, cedar; Sánchez and Cardona, 2008). For instance, an increase of 90% in the efficiency of the subsequent enzymatic hydrolysis of poplar chips pretreated by steam explosion was reported compared to 15% efficiency when pretreatment of chips is not carried out (Sun and Cheng, 2002). In the case of cane bagasse pretreatment, Kaar et al. (1998) determined the conditions that maximize sugars concentration by varying the temperature within the range of 188 to 243°C and residence time within the range of 0.5 to 44 min. These authors concluded that these conditions strongly depend on the composition of the lignocellulosic material and it demonstrated the formation of furfural. On the other hand, it has been reported that the susceptibility of the pretreated substrate to the action of cellulases is highly influenced by the steam pressure and vaporization time during the pretreatment, as was demonstrated for rice straw. In particular, a steam pressure of 3.53 MPa during a short vaporization time (2 min) significantly increases the enzymatic hydrolysis without any observable inhibitory effect (Moniruzzaman, 1996).

For the case of softwood that has an increased lignin content and is more difficult to degrade, a two-stage steam pretreatment has been proposed. In the first stage, the operating conditions are defined in such a way that the maximum amount of sugars derived from hemicellulose is obtained. In the second stage, more severe conditions are employed to degrade the solid fraction resulting from the first stage achieving a partial hydrolysis of cellulose. In both stages, the softwood sawdust is impregnated with dilute sulfuric acid. Shahbazi et al. (2005) propose a fractionation procedure for softwood based on steam explosion and alkaline delignification in order to produce ethanol and related coproducts. An analogous fractionation procedure was utilized by Belkacemi et al. (2002) where the captured hemicellulose-rich liquor was enzymatically treated to produce xylose-rich solutions. Regarding the enzymatic hydrolysis of hemicellulose, Saha (2003) points out that there are no suitable commercial hemicellulase preparations that can efficiently hydrolyze feedstocks like corn fiber to monomeric sugars. This author also briefly reviews the microorganisms and enzymes that could be useful for degrading hemicellulose. Other analogous schemes involve two hydrolysis stages (Nguyen et al., 1999) or an initial treatment by steam explosion followed by an acid hydrolysis to completely degrade the xylans with a further acid recovery (Saska and Ozer, 1995).

One of the methods with better indexes is the pretreatment with liquid hot water (LHW) or thermohydrolysis. Laser et al. (2002) mention that under optimal conditions, this method is comparable to the dilute acid pretreatment, but without the addition of acids or production of neutralization wastes. In addition, this method presents elevated recovery rates of pentoses and does not generate inhibitors (Ogier et al., 1999). Nevertheless, solid load for this method is much less than for the steam explosion method, which is usually greater than 50%.

Another physical–chemical method is the ammonia fiber explosion (AFEX) process whose function is similar to steam explosion. The pretreatment with ammonium does not generate inhibitors for subsequent biological processes, so the washing with water is not necessary. In addition, a small particle size is not required. For this method, Dale et al. (1996) report experimental data correlating conditions of the AFEX process, enzyme doses during cellulose hydrolysis, and the corresponding yields for several agricultural and lignocellulosic residues. Similarly to the AFEX method and steam explosion, CO_2 explosion uses the same principle, but the yields are relatively low compared to the other methods (Sánchez and Cardona, 2008; Sun and Cheng, 2002).

4.3.3 Chemical Methods of Pretreatment

Chemical pretreatments utilize different chemical agents such as ozone, acids, alkalis, peroxide, and organic solvents (Table 4.4). The ozonolysis can be used to degrade lignin and hemicellulose in many lignocellulosic materials. For the case of poplar sawdust pretreated with ozone, the enzymatic hydrolysis yield is increased from 0 to 57% while the lignin content is reduced from 29 to 8%. Despite its advantages, this method requires high amounts of ozone for an effective pretreatment making the process quite expensive (Sun and Cheng, 2002).

Inorganic acids, especially sulfuric and hydrochloric acids, are the most used agents for biomass pretreatment using acid catalysts. These acids are toxic, dangerous, and require reactors resistant to corrosion. Moreover, if concentrated acids are employed, their recycling should be implemented by economic considerations. The pretreatment using dilute acids, especially sulfuric acid, has been developed in a successful way to process different lignocellulosic materials whereby high reaction rates can be attained and the subsequent cellulose hydrolysis can be significantly improved. Nevertheless, the dilute-acid pretreatment costs are usually high related to those of steam explosion or AFEX process (Sánchez and Cardona, 2008; Sun and Cheng, 2002) . The pretreatment of corn stover at pilot plant scale has been studied using dilute sulfuric acid (0.5 to 1.4% p/v) in a continuous reactor for processing 1 ton/day of feedstock (Schell et al., 2003). In this case, high solids load (20%) was utilized unlike those reported in the open literature. Xylose yield reached 77% at 190°C. The digestibility of the pretreated material was evaluated by a simultaneous saccharification and fermentation (SSF) process attaining values of 87%.

Like steam explosion, dilute-acid pretreatment can be combined with other pretreatment methods to carry out a two-stage process. In particular, lignin removal can be greater if the biomass is treated with a dilute acid in the first stage followed by the addition of a concentrated acid plus ethanol in the second stage. In this way, the biomass fractionation can be accomplished, i.e., the separation of its three main components:

1. Sugars generated from hemicellulose hydrolysis that remain in the liquid fraction after the first stage,
2. Cellulose with a higher susceptibility to the enzymatic attack that remains in the solid fraction, and
3. Oligomers of lignin that result from the combined action of the concentrated acid and ethanol and that are solubilized in the second stage (delignification) releasing the cellulosic fiber.

These oligomers can be precipitated after biomass fractionation (Papatheofanous et al., 1995). Another variant of the two-stage dilute-acid pretreatment consists of conducting the hemicellulose hydrolysis at 140°C during 15 min in a first stage to reduce the formation of furans and carboxylic acids, and then increasing the temperature to 190°C for 10 min to make the cellulose more accessible to the cellulase attack (Saha et al., 2005a, 2005b; Sánchez and Cardona, 2008). If dilute-acid pretreatment is performed at a lower temperature (121°C), the degradation of sugars into furans (furfural and hydroxymethylfurfural), which may have an inhibitory effect on the fermentation, can be prevented, but sugar yields are reduced.

TABLE 4.4

Chemical Methods for Pretreatment of Lignocellulosic Biomass for Ethanol Production

Methods	Procedure/Agents	Remarks	Examples of Pretreated Materials	References
Ozonolysis	Ozone, room temperature, and pressure	No inhibitors formation Further cellulose conversion can be >57% Lignin degradation	Poplar sawdust Pine Bagasse, wheat straw, cotton straw, green hay, peanut	Sun and Cheng (2002)
Dilute-acid hydrolysis	0.75–5% H_2SO_4, HCl, or HNO_3, p~1 MPa; continuous process for low solids loads (5–10 wt.% dry substrate/mixture): T = 160–200°C; batch process for high solids loads (10–40 wt.% dry substrate/mixture): T = 120–160°C	pH neutralization is required that generates gypsum as a residue 80–100% hemicellulose hydrolysis, 75–90% xylose recovery Cellulose depolymerization occurs at certain degree High temperature favors further cellulose hydrolysis Lignin is not solubilized, but it is redistributed	Poplar wood Bagasse, corn stover, wheat straw, rye straw, rice hulls Switchgrass, Bermuda grass	Esteghlalian et al. (1997); Hamelinck et al. (2005); Lynd et al. (2002); Martinez et al. (2000); Rodriguez-Chong et al. (2004); Saha et al. (2005a, 2005b); Schell et al. (2003); Sun and Cheng (2002); Sun and Cheng (2005); Wooley et al. (1999)
Concentrated-acid hydrolysis	10–30% H_2SO_4, 170–190°C, 1:1.6 solid–liquid ratio 21–60% peracetic acid, silo-type system	Acid recovery is required Residence time greater compared to dilute-acid hydrolysis Peracetic acid provokes lignin oxidation	Poplar sawdust Bagasse	Cuzens and Miller (1997); Teixeira et al. (1999a, 1999b)

Method	Conditions	Notes/Effects	Feedstock	References
Alkaline hydrolysis	Dilute NaOH, 24 h, 60°C; $Ca(OH)_2$, 4 h, 120°C; it can be complemented by adding H_2O_2 (0.5–2.15 vol.%) at lower temperature (35°C)	Reactor costs are lower compared to acid pretreatment >50% hemicellulose hydrolysis, 60–75% xylose recovery. Low inhibitors formation. Cellulose swelling. Further cellulose conversion can be >65%. 24–55% lignin removal for hardwood, lower for softwood	Hardwood. Bagasse, corn stover, straws with low lignin content (10–18%), cane leaves	Hamelinck et al. (2005); Hari Krishna et al. (1998); Kaar and Holtzapple (2000); Lynd et al. (2002); Rivers and Emert (1988); Saha and Cotta (2006); Sun and Cheng (2002); Teixeira et al. (1999b)
Oxidative delignification	Peroxidase and 2% H_2O_2, 20°C, 8 h	Almost total solubilization of hemicellulose. Further cellulose conversion can be 95%. 50% lignin solubilization	Bagasse	Sun and Cheng (2002)
Wet oxidation	1.2 MPa oxygen pressure, 195°C, 15 min; addition of water and small amounts of Na_2CO_3 or H_2SO_4	Solubilization of major part of hemicellulose. Inhibitors formation. Lignin degradation	Corn stover, wheat straw	Bjerre et al. (1996); Varga et al. (2004)
Organosolv process	Organic solvents (methanol, ethanol, acetone, ethylene glycol, triethylene glycol) or their mixture with 1% of H_2SO_4 or HCl; 185–198°C, 30–60 min, pH = 2.0–3.4	Solvent recovery required. Almost total hydrolysis of hemicellulose, high yield of xylose. Almost total lignin solubilization and breakdown of internal lignin and hemicellulose bonds	Poplar wood. Mixed softwood (spruce, pine, Douglas fir)	Lynd et al. (2002); Pan et al. (2005); Rezzoug and Capart (1996); Sun and Cheng (2002)

Source: Adapted from Sánchez, Ó.J., and C.A. Cardona. 2008. Bioresource Technology 99:5270–5295. Elsevier Ltd.

Pretreatment, using concentrated acids, for fuel ethanol production also has been proposed. Arkenol, Inc. (USA) has reported the development of a fuel ethanol production process from cane bagasse through pretreatment with concentrated sulfuric acid, which has been patented (Farone and Cuzens, 1996). This technology requires the retrofitting of sugar mills in order to produce ethanol and improve energetic indexes of this kind of processes (Cuzens and Miller, 1997; Sánchez and Cardona, 2008). This company is evaluating *Zymomonas* bacteria for use in its concentrated-acid process (Mielenz, 2001). An alternative approach has been tested by Teixeira et al. (1999a, 1999b), which employed a silo-type system that introduced the feedstock (bagasse or hybrid poplar) into plastic bags to which a peracetic acid solution was added with a concentration range from 0 to 60% (by weight). To enhance the process efficiency, sodium or ammonium hydroxide was added before the acid treatment, which allowed the use of lower amounts of peracetic acid. Cellulose conversion of pretreated material reached 93.1% in 120 h using 21% acid concentration or in 24 h using 60% acid concentration. This system requires low energy because the process is carried out at room temperature. Other methods involve the conversion of both cellulose and hemicellulose into fermentable sugars, which eliminates the necessity of adding cellulases, but the operation conditions are far from economically viable (Iranmahboob et al., 2002). In addition, there exists an additional problem related to the oxidation of glucose, which is obtained because of the high acid concentration and relatively prolonged times for biomass heating.

Alkaline pretreatment is based on the effects of the addition of dilute bases on the biomass: increase of internal surface by swelling, decrease of polymerization degree and crystallinity, destruction of links between lignin and other polymers, and breakdown of lignin. The effectiveness of this method depends on the lignin content of the biomass (Sun and Cheng, 2002). This type of pretreatment has been applied to corn stover obtaining 60 to 80% delignification efficiency employing 2.5 to 20% ammonium at 170°C for 1 h (Sun and Cheng, 2002), as well as to sugarcane bagasse and rice straw (Rivers and Emert, 1988). The addition of an alkali can be combined with the addition of hydrogen peroxide as reported by Hari Krishna et al. (1998). Lignin degradation can also be carried out using the peroxidase enzyme in the presence of hydrogen peroxide through a process called oxidative delignification (see Table 4.4). Another pretreatment method involving lignin degradation is wet oxidation that is based on the addition of oxygen and water at high temperatures and pressure leading to the opening of crystalline cellulose and the breakdown of lignin into simpler compounds, such as CO_2, water, and carboxylic acids (Bjerre et al., 1996). Lignin oxidation can also be carried out with $KMnO_4$, although with low cellulose conversions (below 50% for rice straw and cane bagasse; Rivers and Emert, 1988). In general, the utilization of bases like sodium hydroxide or solvents like ethanol or methanol (organosolv process) allows the dissolution of lignin, but their costs are so high that these methods are not competitive for large-scale plants (Lynd et al., 1999).

4.3.4 BIOLOGICAL METHODS OF PRETREATMENT

Biological pretreatment has low energy requirements and mild environmental conditions (Table 4.5). However, most of these processes are too slow, which limits their application at an industrial level for ethanol production process. Fungi are enzyme producers when they grow on the surface of wood and other lignocellulosic materials. Brown-rot fungi mostly attack the cellulose, while white- and soft-rot fungi attack both cellulose and lignin. As many white-rot fungi degrade the lignin, they have been employed for production of ligninases and degradation of lignocellulosics. Evans et al. (1994) describe how enzymes released by these fungi attack lignocellulosic materials and emphasized the role of small molecular agents involved in this process. These authors point out that the most important fungi of this class are *Phanerochaete chrysosporium* and *Phlebia radiate*, which synthesize significant amounts of extracellular peroxidases. Lee (1997) reports on the main microorganisms producing lignin-degrading enzymes and suggests the fermentation processes for producing them through both submerged culture and solid-state fermentation. In fact, the fungus *P. chrysosporium* has been proposed in the patent of Zhang (2006) for degrading lignin in a biomass-to-ethanol process scheme involving the separate fermentation of pentoses and hexoses (Sánchez and Cardona, 2008). The cellulases required by the process for bioethanol production can be obtained from lignocellulosic materials in which the fungi grow. Tengerdy and Szakacs (2003) and Kang et al. (2004) highlight the viability of producing cellulases and hemicellulases by solid-state fermentation compared to conventional submerged fermentation (Warzywoda et al., 1992).

4.3.5 ROLE OF PRETREATMENT DURING PROCESS SYNTHESIS

One of the main problems during the pretreatment and hydrolysis of lignocellulosic biomass lies in the big differences found in its content of both lignin and hemicellulose. This content depends not only on the plant species from which the lignocellulosic materials were obtained, but also on crop age, method of harvesting, etc. This means that none of the pretreatment methods could be applied in a generic way for the great amount of potential feedstocks (Claassen et al., 1999). This justifies the need for a detailed analysis of all pretreatment options for different materials, conditions, and regions. With this aim, process synthesis can provide the necessary tools for discarding, in a preliminary way, the less promising pretreatment options and considering new procedures, schemes, and alternatives proposed during the conceptual design involving all the steps of the biomass processing as well.

The significant variety of pretreatment methods of biomass has led to the development of many flowsheet options for ethanol production (Cardona and Sánchez, 2007). Von Sivers and Zacchi (1995) analyze three pretreatment processes for ethanol production from pine: concentrated acid hydrolysis, two-stage hydrolysis by steam explosion using SO_2, and dilute acid and steam explosion using SO_2 followed by the enzymatic hydrolysis. Through sensitivity analysis, these authors show that

TABLE 4.5
Biological Methods for Pretreatment of Lignocellulosic Biomass for Ethanol Production

Methods	Procedure/Agents	Remarks	Examples of Pretreated Materials	References
Fungal pretreatment	Brown-, white- and soft-rot fungi Cellulase and hemicellulase production by solid-state fermentation of biomass	Fungi produces cellulases, hemicellulases, and lignin-degrading enzymes: ligninases, lignin, peroxidases, polyphenoloxidases, laccase and quinone-reducing enzymes Very slow process: *Pleurotus ostreatus* converts 35% of wheat straw into reducing sugars in 5 weeks Brown-rot fungi degrades cellulose White- and soft-rot fungi degrade cellulose and lignin	Corn stover, wheat straw	Sun and Cheng (2002); Tengerdy and Szakacs (2003)
Bioorganosolv pretreatment	*Ceriporiopsis subvermispora* for 2–8 weeks followed by ethanolysis at 140–200°C for 2 h	Fungi decompose the lignin network Ethanol action allows hemicellulose hydrolysis Biological pretreatment can save 15% of the electricity needed for ethanolysis Ethanol can be reused; environmentally friendly process	Beechwood	Itoh et al (2003)

Source: Adapted from Sánchez, Ó.J., and C.A. Cardona. 2008. *Bioresource Technology* 99:5270–5295. Elsevier Ltd.

none of the processes can be discarded as less reliable. Milling has been suggested as a sole pretreatment method before the cellulose hydrolysis since the required equipment is less expensive than the equipment needed for other pretreatment methods such as steam explosion or ammonia fiber explosion (AFEX) process, which can account for 6 to 20% of the capital costs of the process. In contrast, milling equipment accounts for about 1% of these costs. However, it is considered that milling has elevated energy costs. Alvo and Belkacemi (1997) point out that milling of perennial grasses requires much less energy that milling of wood. These authors consider that milling as a unique pretreatment method should not be discarded as an option taking into account the advantages of this configuration: toxic products of degradation are not formed, soluble carbohydrates of the initial biomass are not destroyed, and many rural communities can acquire an easier way to mill in comparison to other expensive pretreatment equipment. This alternative should be evaluated in depth, utilizing simulation and optimization tools in the design step.

Dilute-acid pretreatment is the most studied method in the world along with steam explosion since they have a major probability of being implemented at an industrial scale in the near future. In fact, the utilization of dilute acids is considered one of the most mature technologies compared to the rest of biomass pretreatment methods. The NREL (National Renewable Energy Laboratory) of the U.S. Department of Energy, which is one of the institutions leading the research and industrial development of technologies for fuel ethanol production from lignocellulosic materials in the United States, has chosen the dilute-acid pretreatment as the best model process to have been developed in the past years and offered to the industry (Aden et al., 2002; Wooley et al., 1999). The main advantage of this method compared to steam explosion is the higher recovery of sugars derived from the hydrolyzed hemicellulose. In the case of hardwood, about 80% of sugars can be recovered using dilute sulfuric acid while this recovery does not reach 65% when steam explosion is used (Lynd, 1996). The higher the sugars recovery, the greater the monosaccharide content in the liquid fraction resulting from the pretreatment. This liquid fraction can be employed as a culture medium for pentose-assimilating yeasts, or can be added to the bioreactor where the fermentation of cellulose hydrolyzates is accomplished as an additional sugar source (see the following chapters).

Another prospective method is the pretreatment by LHW. In particular, steam explosion and LHW processes have been compared in the case of poplar biomass obtaining better results for the latter method (Negro et al., 2003). In general, it is considered that the most efficient and promising methods are dilute-acid pretreatment, steam explosion with the addition of acid catalysts, and the LHW method (Ogier et al., 1999). To this regard, the evaluation of the global process for fuel ethanol production from lignocellulosic materials has been performed by Hamelinck et al. (2005) and the above-mentioned promising pretreatment methods should be noted. These authors selected one of the three pretreatment methods and diverse biological conversion technologies according to three different stages of technological maturity and development of the operations involved. For this, they employed spreadsheets and commercial simulators to study the

configuration of the process flowsheet corresponding to each one of three analyzed scenarios (short-, mid-, and long-term variants). For the pretreatment step
and short-term scenario (five years), the dilute-acid pretreatment was selected
considering that this method is the technology offering the highest efficiency and
reliability at the moment, while for the mid-term scenario (10 to 15 years), the
steam explosion was chosen. For the long-term scenario, the LHW method was
analyzed due to its comparative advantages and considering that this technology,
for a period of time greater than 15 years, will be completely developed and that
the current drawbacks will be overcome. Undoubtedly, the evaluation through the
simulation of these alternatives will provide more insight for the selection of the
best configuration of the overall process flowsheet. This will allow, in turn, the
definition of the main research and development directions for the design of more
effective pretreatment methods.

Process simulation requires suitable models for describing the studied processes. Considering process synthesis procedures, the mathematical modeling can
allow a deeper insight into the pretreatment methods and make possible the definition of operating parameters for which the system attains a better performance.
In particular, mathematical modeling is a valuable tool for planning and executing different trials at pilot and industrial scales (Cardona and Sánchez, 2007).
For instance, through a kinetic model of cane bagasse pretreatment using nitric
acid, the best conditions for increasing sugar yields were predicted. In this specific case, the model considered the formation of inhibitory compounds (furfural
and acetic acid). The results obtained were better than when sulfuric acid was
used (Rodríguez-Chong et al., 2004) or when no acid was used (Jacobsen and
Wyman, 2002). This type of kinetic study also has been done for poplar wood,
switchgrass, and corn stover treated with sulfuric acid (Esteghlalian et al., 1997),
as well as for wheat straw using hydrochloric acid (Jiménez and Ferrer, 1991). The
kinetic model of corn stover pretreatment at pilot scale was also used to determine
the process conditions leading to the maximization of xylose yield (Schell et al.,
2003). Malester et al. (1988) studied the kinetics of dilute-acid pretreatment of
municipal solid waste (MSW) since the cellulosic materials are the main component of this potential feedstock for ethanol production. For this case, the major
difficulty lies in the resistance of cellulose to be converted into fermentable sugars in the presence of acids. This is explained by the neutralizing capacity of the
MSW over the acid that imposes an additional difficulty to the measurement of
kinetic parameters of the process. Therefore, these authors proposed to measure
this effect based on pH and not on the concentration of the employed acid. This
capacity is inherent to other lignocellulosic materials like corn stover and sawdust, so the acid concentration should be adjusted slightly in order to maintain the
efficiency of this type of pretreatment (Esteghlalian et al., 1997).

4.4   DETOXIFICATION OF PRETREATED BIOMASS

During the pretreatment of lignocellulosic biomass and along with fermentable
sugars, a significant amount of compounds that can seriously inhibit the subsequent

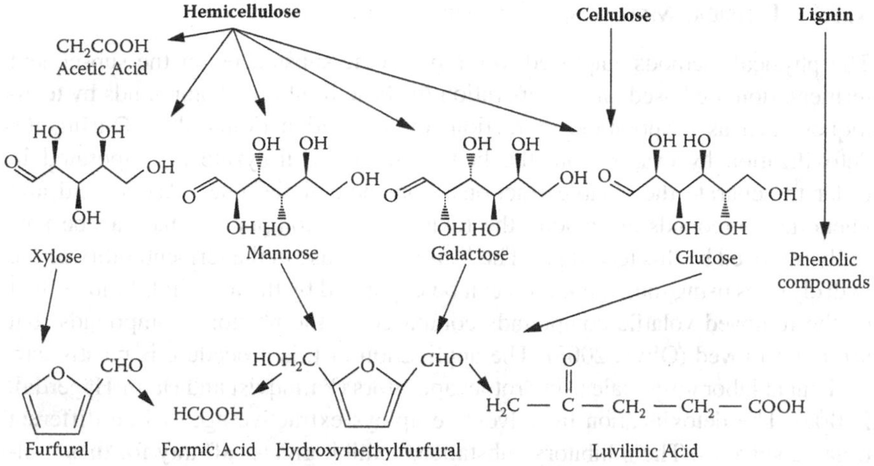

FIGURE 4.3 Main inhibitors formed during the pretreatment of lignocellulosic biomass.

fermentation process are formed (Sánchez and Cardona, 2008). Inhibitory substances are generated as a result of the hydrolysis of extractive components, organic and sugar acids esterified to hemicellulose (acetic, formic, glucuronic, galacturonic), and solubilized phenolic derivatives. In the same way, inhibitors are produced from degradation products of soluble sugars (furfural, hydroxymethylfurfural [HMF]) and lignin (cinnamaldehyde, p-hydroxybenzaldehyde, syringaldehyde), and as a consequence of corrosion (metal ions; Körner et al., 1984; Lynd, 1996; Palmqvist and Hahn-Hägerdal, 2000b), as shown in Figure 4.3.

In general, the liquid stream resulting from pretreatment undergoes detoxification. This stream basically represents the hemicellulose hydrolyzates. The solid fraction containing cellulose and lignin is washed with water in order to remove all the soluble substances including the inhibitors. The resulting washing water is mixed with the hemicellulose hydrolyzates to obtain a liquid medium that can be employed for alcoholic fermentation or for producing other substances by microorganisms. If this liquid medium is employed to produce ethanol, the fermentation of the pentoses and hexoses released from the hemicellulose hydrolysis during the pretreatment can be accomplished in a parallel way to the fermentation of glucose formed during the cellulose hydrolysis. Another option is the unification of the cellulose hydrolyzate and the hemicellulose hydrolyzate for their joint fermentation. In both cases, the previous detoxification of the hemicellulose hydrolyzate is required especially when parallel fermentations are carried out. This is explained by the fact that pentose-assimilating yeasts are very sensitive to the presence of inhibitors.

The methods used for detoxification can be physical, chemical, or biological. As reported by Palmqvist and Hahn-Hägerdal (2000a), these methods cannot be directly compared because they vary in their degree of neutralization of the inhibitors. In addition, the fermenting microorganisms have different tolerances to the inhibitors that increase the complexity of this process.

4.4.1 Physical Methods of Detoxification

The physical methods employed to remove toxic substances for the subsequent fermentation are based on the separation of these inhibiting compounds by techniques such as evaporation, extraction, or adsorption (Table 4.6). During the detoxification by evaporation, the hemicellulose hydrolyzate is evaporated in order to separate the volatile fractions from nonvolatile ones. Acetic acid and phenolic compounds are among the toxic volatile compounds that can be partially removed by this technique. The limited increase in the fermentability of the hydrolyzates using the evaporation can be explained by the lower inhibition effect of the removed volatile compounds compared to the phenolic compounds that are not removed (Oliva, 2003). The application of this procedure is mostly carried out at laboratory scale using rotoevaporators (Palmqvist and Hahn-Hägerdal, 2000a). The detoxification by solvents employs extractive agents like different organic solvents. The inhibitory substances exhibit greater affinity for these solvents than for the aqueous medium of the hydrolyzate allowing their migration to the organic phase. Detoxification with supercritical fluids has been proposed for extracting the inhibitors as well (Persson et al., 2002b). In the case of the detoxification by adsorption, the inhibitors are retained on the surface of a solid material (adsorbent) decreasing their concentration in the hemicellulose hydrolyzate. However, it is possible that useful substances like the glucose also are adsorbed in the adsorbent bed. The usage of molecular sieves as an adsorbing material for partial removal of acetic acid, furfural, and soluble lignin has been proposed (Olsson and Hahn-Hägerdal, 1996).

4.4.2 Chemical Methods of Detoxification

The chemical methods of detoxification are based on the addition of certain chemical compounds that vary the conditions of the aqueous medium provoking changes in the pH, formation of precipitates, or the direct transformation of the toxic compounds. Among these methods, the ionic interaction of the ionic exchange resins can be included in this group of detoxification methods. The most employed chemical methods of detoxification are shown in Table 4.7. By neutralization, the solubility of many inhibitory substances is changed. This allows their removal by a later filtration or adsorption. However, the addition of alkali up to very high pH values (alkaline detoxification) leads to the formation of a significant amount of precipitate composed by calcium salts (if lime is used), which entrains the inhibitory compounds or causes them to settle. In addition, many inhibitors are unstable at pH higher than 9. Alkaline treatment is considered one of the best detoxification methods since a high percentage of substances such as furaldehydes and phenolic compounds can be removed by this method, improving the fermentability of the resulting liquid medium especially when biomass hydrolyzates pretreated with dilute acid are employed (Persson et al., 2002a). The addition of calcium hydroxide (overliming) or ammonium has shown better results than the use of sodium or potassium hydroxide. Some methods to determine the optimal

TABLE 4.6
Physical Methods for Detoxification of Pretreated Biomass

Methods	Procedure/Agents	Examples	Microorganism	Remarks	References
Evaporation	Evaporation, separation of volatile and nonvolatile fractions and dilution of nonvolatile fraction	Willow hz.	*Saccharomyes cerevisiae*	Reduction of acetic acid and phenolic compounds in nonvolatile fraction; roto-evaporation	Palmqvist and Hahn-Hägerdal (2000a)
		Aspen hz.	*Pichia stipitis*	93% yield of ref. fermn.; removal: 54% acetic acid, 100% furfural, 29% vanillin; roto-evaporation	Palmqvist and Hahn-Hägerdal (2000a)
Extraction	Organic solvents, 3:1 org. phase: aqueous phase volumetric ratio	Spruce hz.	*S. cerevisiae*	Solv.: diethyl ether (solv.); yield comparable to ref. fermn.; removal of acetic, formic, and levulinic acids, furfural, HMF	Palmqvist and Hahn-Hägerdal (2000a)
		Aspen hz.	*P. stipitis*	Solv.: ethyl acetate; 93% yield of ref. fermn.; removal: 56% acetic acid, 100% furfural, 100% vanillin, 100% hydroxybenzoic acid	Palmqvist and Hahn-Hägerdal (2000a)
		Pine hz.	*S. cerevisiae*	Solv.: ethyl acetate; removal of low molecular phenolic compounds	Palmqvist and Hahn-Hägerdal (2000a)
		Steam-exploded poplar	*S. cerevisiae*	Solv.: ethyl acetate; EtOH yield (SSF): detoxified hz. 0.51 g/g, undetox. hz. 0 g/g; high degree of phenolic removal	Cantarella et al. (2004)

Continued

TABLE 4.6 (*Continued*)
Physical Methods for Detoxification of Pretreated Biomass

Methods	Procedure/Agents	Examples	Microorganism	Remarks	References
	Supercritical solvent in countercurrent with the hydrolyzate, 20 MPa, 40°C; then, depressurization	Dilute-acid spruce hz.	*S. cerevisiae*	Solv.: supercritical CO_2; 98% yield of ref. fermn.; removal: 93% furfural, 10% HMF	Persson et al. (2002b)
Adsorption	Activated carbon, 0.05–0.20 g/g glucose	Steam-exploded concentr. oak hz.	*S. cerevisiae*	Detoxified hz. with 140–170 g/L initial glucose was utilized; undetox. hz. with 100 g/L initial glucose could not be completely utilized	Lee et al. (1999)
	Amberlite hydrophobic polymeric adsorbent XAD-4, 8% (w/v), 1.5 h, 25°C; regeneration with EtOH; then, neutralized with lime	LHW pretreated corn fiber	Recombinant *Escherichia coli*	Reduction of furfural conc. from 1–5 to <0.01 g/L; 90% yield of theoretical; sugars are not adsorbed	Weil et al. (2002)

Source: Adapted from Sánchez, Ó.J., and C.A. Cardona. 2008. *Bioresource Technology* 99:5270–5295. Elsevier Ltd.

Note: HMF = hydroxymethylfurfural; hz. = hydrolyzate. Reference fermentation (ref. fermn.) refers to fermentation carried out in a glucose-based medium without inhibitors.

TABLE 4.7

Chemical Methods for Detoxification of Pretreated Biomass

Methods	Procedure/Agents	Examples	Microorganism	Remarks	References
Neutralization	Ca(OH)₂ or CaO, pH = 6, then membrane filtration or adsorption	Acid hz. of cotton waste pyrolysate	*Saccharomyces cerevisiae*, *Pichia* sp.	Precipitation or removal of toxic compounds; 10% lower yield for *Pichia* sp.	Yu and Zhang (2003)
		Steam-exploded poplar	*S. cerevisiae*	EtOH yield (SSF); detoxified hz. 0.86 g/g, undetox. hz. 0 g/g	Cantarella et al. (2004)
Alkaline detoxification (overliming)	Ca(OH)₂, pH = 9–10.5, then pH adjustment to 5.5–6.5 with H₂SO₄ or HCl	Dilute-acid hz. of spruce		Yield comparable to ref. fermn.; 20% removal of furfural and HMF	Palmqvist and Hahn-Hägerdal (2000a)
		Steam-exploded bagasse	Recombinant *S. cerevisiae*	Removal of acid acetic, furfural and part of phenolic compounds	Martín et al. (2002)
		Acid hz. of cotton waste pyrolysate	*S. cerevisiae*, *Pichia* sp.	7.5% lower yield for *Pichia* sp.	Yu and Zhang (2003)
		Rice hulls hz.	Recombinant *E. coli*	39% reduction in fermentation time	Saha et al. (2005a)
		Wheat straw hz.	Recombinant *E. coli*	Reduction in fermn. time: SSF −18%, SHF −67%	Saha et al. (2005b)
		Dilute-acid bagasse hz.	Recombinant *E. coli*	Removal: 51% furfural, 51% HMF; 41% phenolic compounds, 0% acetic acid; overliming at 60°C or 25°C, at high temperature, the required amounts of lime and acid are reduced	Martinez et al. (2000, 2001)

Continued

TABLE 4.7 (Continued)
Chemical Methods for Detoxification of Pretreated Biomass

Methods	Procedure/Agents	Examples	Microorganism	Remarks	References
Combined alkaline detoxification	KOH, pH = 10, then pH adjustment to 6.5 with	Bagasse hz.	Pichia stipitis	Reduction of ketones and aldehydes, removal of volatile compounds when hydrolyzate is heated at 90°C	Palmqvist and Hahn-Hägerdal (2000a)
	HCl and addition of 1% sodium sulfite	Dilute-acid hz. of spruce	S. cerevisiae		Palmqvist and Hahn-Hägerdal (2000a)
		Willow hz.	Recombinant E. coli		Palmqvist and Hahn-Hägerdal (2000a)
Ionic exchange	Weak base resins Amberlyst A20, regenerated with ammonia	Dilute-acid poplar	Recombinant Zymomonas mobilis	Removal: 88% acetic acid, 100% H_2SO_4; 100% sugars recovery	Wooley et al. (1999)
		Dilute-acid hz. of spruce	S. cerevisiae	Removal: >80% phenolic compounds, ~100% levulinic, acetic and formic acids, 70% furfural; considerable loss of fermentable sugars	Palmqvist and Hahn-Hägerdal (2000a)
	Poly(4-vinyl pyridine)	Corn stover hz.	Recombinant S. cerevisiae	Sugars eluted earlier than all tested inhibitors; ferment. results were similar to that using pure sugars	Xie et al. (2005)

Source: Adapted from Sánchez, Ó.J., and C.A. Cardona. 2008. Bioresource Technology 99:5270–5295. Elsevier Ltd.
Note: Reference fermentation (ref. fermn.) refers to fermentation carried out in a glucose-based medium without inhibitors; hz = hydrolyzate.

amount of lime to be added in dependence on the acid content of the hydrolyzate have been developed (Martinez et al., 2001). The positive effects of alkaline treatment on the hydrolyzate fermentability cannot be explained only by the removal of inhibitors. It has been postulated that this detoxification method may have possible stimulating effects on ethanol-producing microorganisms (Persson et al., 2002a).

Ionic exchange also has been studied as a detoxification method showing a high efficiency for removing inhibitors. This method can be considered as a special case of adsorption because ionized groups of the ionic exchange resin (the adsorbent) interact electrostatically with the charged molecules of inhibitors. In particular, some anionic exchange resins are used to eliminate phenolic compounds as a consequence of the strong bonds formed between quaternary ammonium groups of the resin (positively charged) and phenols (negatively charged). The rest of substances that do not interact with the resin pass through the adsorbent leading to a detoxified hydrolyzate. Besides the high resin cost, one drawback of this method lies in the fact that the content of fermentable sugars in the hydrolyzate can be reduced (Oliva, 2003). In the model process developed for NREL, the ionic exchange was proposed as a detoxification method for ethanol production process using poplar wood as the feedstock (Wooley et al., 1999). The biomass pretreated with dilute sulfuric acid at 190°C and high pressure is cooled by flashing, which removes 61% furfural and HMF as well as 6.5% of the acetic acid released from hemicellulose. The liquid fraction of the pretreated biomass is sent to an ionic exchange column whose effluent undergoes overliming to enhance the detoxification efficiency. Then, sulfuric acid is added to remove the calcium and suspended solids by forming a precipitate of calcium sulfate (gypsum). This design was based on data obtained at pilot scale level using a column with a diameter of 20 mm and a length of 1 m containing Amberlyst A20, a weak base resin. The regeneration of the resin is accomplished by passing the eluent (ammonium) through the column. Xie et al. (2005), in turn, demonstrated the successful detoxification of corn stover hydrolyzate using a polymeric adsorbent without the need of a subsequent alkalinization. In contrast, the newer model process for ethanol production for corn stover designed for NREL only suggested the overliming as the detoxification method (Aden et al., 2002).

Different methods of detoxification that combine physical and chemical principles have been proposed, such as the neutralization with CaO or Ca(OH)$_2$ followed by the addition of activated carbon and filtration to remove the acetic acid (Olsson and Hahn-Hägerdal, 1996). For lignocellulosic materials pretreated by pyrolysis and hydrolyzed with dilute acid, the utilization of several adsorbents, such as activated carbon, diatomite, bentonite, and zeolites, after the treatment by neutralization has been also studied (Yu and Zhang, 2003).

4.4.3 BIOLOGICAL METHODS OF DETOXIFICATION

Enzymatic detoxification is one of the new methods tested for inhibitors removal. For this, the phenoloxidase laccase is employed. This enzyme oxidizes the phenolic compounds derived from lignin. Martín et al. (2002) have evaluated the fermentation behavior of bagasse hydrolyzates pretreated by steam explosion and treated with commercial cellulases or acids. The hydrolyzates were treated with phenoloxidase laccase and then underwent alkalinization. The specific productivity (measured as g EtOH/g bagasse × h) of the detoxified hydrolyzate was almost four times greater than the specific productivity of the nondetoxified hydrolyzate. The fact that the inhibitory effect was remarkably reduced when the phenolic compounds were specifically removed by the laccase suggests that this enzyme could be used to neutralize the inhibitory effect of low molecular weight soluble phenolic compounds instead of using nonspecific chemical methods (Palmqvist and Hahn-Hägerdal, 2000a). Other alternative biological methods involving microorganisms (microbial detoxification) have been proposed. For instance, in the case of dilute solutions resulting from biomass pretreated by pyrolysis, a biofilm reactor with a mixed culture of aerobic bacterial cells naturally immobilized on a plastic composite support has been used (Khiyami et al., 2005). The most employed biological methods of detoxification are shown in Table 4.8.

TABLE 4.8

Biological Methods for Detoxification of Pretreated Biomass

Methods	Procedure/Agents	Examples	Microorganism	Remarks	References
Enzymatic detoxification	Laccase (phenol oxidase) and lignin peroxidase from *Trametes versicolor*: 30°C, 12 h	Willow hz.	*Saccharomyces cerevisiae*	Two- to three-fold increase of EtOH productivity compared to undetox. hz.; laccase selectively removes phenolic low molecular weight compounds and phenolic acids	Palmqvist and Hahn-Hägerdal (2000b); Jönsson et al. (1998)
		Steam-exploded bagasse	Recombinant *S. cerevisiae*	80% removal of phenolic compounds	Martín et al. (2002)
Microbial detoxification	*Trichoderma reesei*	Steam-exploded willow	*S. cerevisiae*	Three-fold increase of EtOH productivity compared to undetox. hz.; four-fold increase of yield; removal of acetic acid, furfural and benzoic acid derivatives	Palmqvist and Hahn-Hägerdal (2000a)
				Aerobic bacteria oxidize aromatic compounds	
	Immobilized to PCS mixed culture of *Pseudomonas putida* and *Streptomyces setonii* cells (biofilm reactor: PCS tubes attached to CSTR agitator shaft)	Diluted pyrolysate of corn stover		Detoxification of 10 and 25 vol.% of pyrolysate medium, and partially detoxification of 50 vol.% of pyrolysate medium	Khiyami et al., (2005)

Source: Adapted from Sánchez, Ó.J., and C.A. Cardona. 2008. *Bioresource Technology* 99:5270–5295. Elsevier Ltd.

Note: hz = hydrolyzate.

REFERENCES

Acevedo, A., R. Godoy, and G. Bolaños. 2003. Incremento de la producción de alcohol en fermentación de melazas mediante la utilización del complejo enzimático Rhyzozyme (Increase of ethanol production during molasses fermentation through the utilization of the enzymatic complex Rhyzozyme, in Spanish). Paper presented at the XXII Congreso Colombiano de Ingeniería Química, Bucaramanga, Colombia.

Aden, A., M. Ruth, K. Ibsen, J. Jechura, K. Neeves, J. Sheehan, B. Wallace, L. Montague, A. Slayton, and J. Lukas. 2002. *Lignocellulosic biomass to ethanol process design and economics utilizing co-current dilute acid prehydrolysis and enzymatic hydrolysis for corn stover*. Technical Report NREL/TP-510-32438, National Renewable Energy Laboratory, Washington, D.C.

Alvo, P., and K. Belkacemi. 1997. Enzymatic saccharification of milled timothy (*Phleum pretense* L.) and alfalfa (*Medicago sativa* L.). *Bioresource Technology* 61:185–198.

Ballesteros, I., J.M. Oliva, M.J. Negro, P. Manzanares, and M. Ballesteros. 2002. Enzymic hydrolysis of steam exploded herbaceous agricultural waste (*Brassica carinata*) at different particule sizes. *Process Biochemistry* 38:187–192.

Ballesteros, I., J.M. Oliva, F. Sáez, and M. Ballesteros. 2001. Ethanol production from lignocellulosic byproducts of olive oil extraction. *Applied Biochemistry and Biotechnology* 91–93:237–252.

Ballesteros, M., J.M. Oliva, M.J. Negro, P. Manzanares, and I. Ballesteros. 2004. Ethanol from lignocellulosic materials by a simultaneous saccharification and fermentation process (SFS) with *Kluyveromyces marxianus CECT 10875*. *Process Biochemistry* 39:1843–1848.

Belkacemi, K., G. Turcotte, and P. Savoie. 2002. Aqueous/steam-fractionated agricultural residues as substrates for ethanol production. *Industrial & Engineering Chemistry Research* 41:173–179.

Belkacemi, K., G. Turcotte, P. Savoie, and E. Chornet. 1997. Ethanol production from enzymatic hydrolyzates of cellulosic fines and hemicellulose-rich liquors derived from aqueous/steam fractionation of forages. *Industrial & Engineering Chemistry Research* 36:4572–4580.

Bjerre, A.B., A.B. Olesen, T. Fernqvist, A. Plöger, and A.S. Schmidt. 1996. Pretreatment of wheat straw using combined wet oxidation and alkaline hydrolysis resulting in convertible cellulose and hemicellulose. *Biotechnology and Bioengineering* 49:568–577.

Bothast, R.J., and M.A. Schlicher. 2005. Biotechnological processes for conversion of corn into ethanol. *Applied Microbiology and Biotechnology* 67:19–25.

Cachot, T., and M.N. Pons. 1991. Improvement of alcoholic fermentation on cane and beet molasses by supplementation. *Journal of Fermentation and Bioengineering* 71 (1):24–27.

Cadoche, L., and G.D. López. 1989. Assessment of size reduction as a preliminary step in the production of ethanol from lignocellulosic wastes. *Biological Wastes* 30:153–157.

Cantarella, M., L. Cantarella, A. Gallifuoco, A. Spera, and F. Alfani. 2004. Comparison of different detoxification methods for steam-exploded poplar wood as a substrate for the bioproduction of ethanol in SHF and SSF. *Process Biochemistry* 39:1533–1542.

Cardona, C.A., and Ó.J. Sánchez. 2007. Fuel ethanol production: Process design trends and integration opportunities. *Bioresource Technology* 98:2415–2457.

Cardona, C.A., Ó.J. Sánchez, M.I. Montoya, and J.A. Quintero. 2005. Simulación de los procesos de obtención de etanol a partir de caña de azúcar y maíz (Simulation of processes for ethanol production from sugarcane and corn, in Spanish). *Scientia et Technica* 28:187–192.

Castellar, M.R., M.R. Aires-Barros, J.M.S. Cabral, and J.L. Iborra. 1998. Effect of zeolite addition on ethanol production from glucose by *Saccharomyces bayanus*. *Journal of Chemical Technology and Biotechnology* 73:377–384.

Claassen, P.A.M., J.B. van Lier, A.M. López Contreras, E.W.J. van Niel, L. Sijtsma, A.J.M. Stams, S.S. de Vries, and R.A. Weusthuis. 1999. Utilisation of biomass for the supply of energy carriers. *Applied Microbiology and Biotechnology* 52:741–755.

Cuzens, J.C., and J.R. Miller. 1997. Acid hydrolysis of bagasse for ethanol production. *Renewable Energy* 10 (2–3):285–290.

Dale, B.E., C.K. Leong, T.K. Pham, V.M. Esquivel, I. Rios, and V.M. Latimer. 1996. Hydrolysis of lignocellulosics at low enzyme levels: Application of the AFEX process. *Bioresource Technology* 56:111–116.

De Bari, I., E. Viola, D. Barisano, M. Cardinale, F. Nanna, F. Zimbardi, G. Cardinale, and G. Braccio. 2002. Ethanol production at flask and pilot scale from concentrated slurries of steam-exploded aspen. *Industrial & Engineering Chemistry Research* 41:1745–1753.

Dekker, R.F.H., and A.F.A. Wallis. 1983. Enzymic saccharification of sugarcane bagasse pretreated by autohydrolysis-steam explosion. *Biotechnology and Bioengineering* 25:3027–3048.

Ergun, M., S.F. Mutlu, and Ö. Gürel. 1997. Improved ethanol production by *Saccharomyces cerevisiae* with EDTA, ferrocyanide and zeolite X addition to sugar beet molasses. *Journal of Chemical Technology and Biotechnology* 68:147–150.

Esteghlalian, A., A.G. Hashimoto, J.J. Fenske, and M.H. Penner. 1997. Modeling and optimization of the dilute-sulfuric-acid pretreatment of corn stover, poplar and switchgrass. *Bioresource Technology* 59:129–136.

Evans, C.S., M.V. Dutton, F. Guillén, and R.G. Veness. 1994. Enzymes and small molecular mass agents involved with lignocellulose degradation. *FEMS Microbiology Reviews* 13:235–240.

Farone, W.A., and J.E. Cuzens. 1996. Method of strong acid hydrolysis. WIPO Patent No. WO9640970.

Ghosh, P., and T.K. Ghose. 2003. Bioethanol in India: Recent past and emerging future. *Advanced Biochemical Engineering and Biotechnology* 85:1–27.

Gulati, M., K. Kohlman, M.R. Ladish, R. Hespell, and R.J. Bothast. 1996. Assessment of ethanol production options for corn products. *Bioresource Technology* 5:253–264.

Hamelinck, C.N., G. van Hooijdonk, and A.P.C. Faaij. 2005. Ethanol from lignocellulosic biomass: techno-economic performance in short-, middle- and long-term. *Biomass and Bioenergy* 28:384–410.

Hari Krishna, S., K. Prasanthi, G. Chowdary, and C. Ayyanna. 1998. Simultaneous saccharification and fermentation of pretreated sugar cane leaves to ethanol. *Process Biochemistry* 33 (8):825–830.

Heitz, M., F. Carrasco, M. Rubio, A. Brown, E. Chornet, and R.P. Overend. 1987. Physicochemical characterization of lignocellulosic substrates via autohydrolysis: An application to tropical woods. *Biomass* 13:255–273.

Holtzapple, M.T., E.P. Ripley, and M. Nikolaou. 1994. Saccharification, fermentation, and protein recovery from low-temperature AFEX-treated coastal bermudagrass. *Biotechnology and Bioengineering* 44:1122–1131.

Iranmahboob, J., F. Nadim, and S. Monemi. 2002. Optimizing acid-hydrolysis: A critical step for production of ethanol from mixed wood chips. *Biomass and Bioenergy* 22:401–404.

Itoh, H., M. Wada, Y. Honda, M. Kuwahara, and T. Watanabe. 2003. Bioorganosolve pretreatments for simultaneous saccharification and fermentation of beech wood by ethanolysis and white rot fungi. *Journal of Biotechnology* 103:273–280.

Jacobsen, S.E., and C.E. Wyman. 2002. Xylose monomer and oligomer yields for uncatalyzed hydrolysis of sugarcane bagasse hemicellulose at varying solids concentration. *Industrial & Engineering Chemistry Research* 41:1454–1461.

Jiménez, L., and J.L. Ferrer. 1991. Saccharification of wheat straw for the production of alcohols. *Process Biochemistry* 26:153–156.

Jönsson, L.J., E. Palmqvist, N.-O. Nilvebrant, and B. Hahn-Hägerdal. 1998. Detoxification of wood hydrolysates with laccase and peroxidase from the white-rot fungus *Trametes versicolor*. *Applied Microbiology and Biotechnology* 49:691–697.

Kaar, W.E., C.V. Gutiérrez, and C.M. Kinoshita. 1998. Steam explosion of sugarcane bagasse as a pretreatment for conversion to ethanol. *Biomass and Bioenergy* 14 (3):277–287.

Kaar, W.E. , and M.T. Holtzapple. 2000. Using lime pretreatment to facilitate the enzymic hydrolysis of corn stover. *Biomass and Bioenergy* 18:189–199.

Kang, S.W., Y.S. Park, J.S. Lee, S.I. Hong, and S.W. Kim. 2004. Production of cellulases and hemicellulases by *Aspergillus niger* KK2 from lignocellulosic biomass. *Bioresource Technology* 91:153–156.

Kaseno, M. and T. Kokugan. 1997. The effect of molasses pretreatment by ceramic microfiltration membrane on ethanol fermentation. *Journal of Fermentation and Bioengineering* 83 (6):511–582.

Khiyami, M.A., A.L. Pometto, III, and R.C. Brown. 2005. Detoxification of corn stover and corn starch pyrolysis liquors by *Pseudomonas putida* and *Streptomyces setonii* suspended cells and plastic compost support biofilms. *Journal of Agricultural and Food Chemistry* 53:2978–2987.

Koegel, R.G., H.K. Sreenath, and R.J. Straub. 1999. Alfalfa fiber as a feedstock for ethanol and organic acids. *Applied Biochemistry and Biotechnology* 77–79:105–115.

Körner, H.-U., D. Gottschalk, J. Wiegel, and J. Puls. 1984. The degradation pattern of oligomers and polymers from lignocelluloses. *Analytica Chimica Acta* 163:55–66.

Laser, M., D. Schulman, S.G. Allen, J. Lichwa, M.J. Antal, Jr., and L.R. Lynd. 2002. A comparison of liquid hot water and steam pretreatments of sugar cane bagasse for bioconversion to ethanol. *Bioresource Technology* 81:33–44.

Lee, J. 1997. Biological conversion of lignocellulosic biomass to ethanol. *Journal of Biotechnology* 56:1–24.

Lee, W.G., J.S. Lee, C.S. Shin, S.C. Park, H.N. Chang, and Y.K. Chang. 1999. Ethanol production using concentrated oak wood hydrolysates and methods to detoxify. *Applied Biochemistry and Biotechnology* 77–79:547–559.

López-Munguía, A. 1993. Edulcorantes (Sweeteners, in Spanish). In *Biotecnología Alimentaria*, ed. M. García, R. Quintero, and A. López-Munguía. Centro, México: Limusa.

Lynd, L.R. 1996. Overview and evaluation of fuel ethanol from cellulosic biomass: Technology, economics, the environment, and policy. *Annual Review of Energy and the Environment* 21:403–465

Lynd, L.R., P.J. Weimer, W.H. van Zyl, and I.S. Pretorious. 2002. Microbial cellulose utilization: Fundamentals and biotechnology. *Microbiology and Molecular Biology Reviews* 66 (3):506–577.

Lynd, L.R., C.E. Wyman, and T.U. Gerngross. 1999. Biocommodity engineering. *Biotechnology Progress* 15:777–793.

Madson, P.W., and D.A. Monceaux. 1995. Fuel ethanol production. In *The Alcohol Textbook*, ed. T. P. Lyons, D. R. Kelsall, and J. E. Murtagh. Nottingham, U.K.: University Press.

Malester, A.I., M. Green, S. Kimchie, and G. Shelef. 1988. The effect of the neutralizing capacity of cellulosic materials on the kinetics of cellulose dilute acid hydrolysis. *Biological Wastes* 26:115–124.

Martín, C., M. Galbe, C.F. Wahlbom, B. Hahn-Hägerdal, and L.J. Jönsson. 2002. Ethanol production from enzymatic hydrolysates of sugarcane bagasse using recombinant xylose-utilising *Saccharomyces cerevisiae*. *Enzyme and Microbial Technology* 31:274–282.

Martinez, A., M.E. Rodriguez, M.L. Wells, S.W. York, J.F. Preston, and L.O. Ingram. 2001. Detoxification of dilute acid hydrolysates of lignocellulose with lime. *Biotechnology Progress* 17:287 -293.

Martinez, A., M.E. Rodriguez, S.W. York, J.F. Preston, and L.O. Ingram. 2000. Effects of $Ca(OH)_2$ treatments ("overliming") on the composition and toxicity of bagasse hemicellulose hydrolysates. *Biotechnology and Bioengineering* 69 (5):526 -536.

Mielenz, J.R. 2001. Ethanol production from biomass: Technology and commercialization status. *Current Opinion in Microbiology* 4:324–329.

Moniruzzaman, M. 1996. Saccharification and alcohol fermentation of steam-exploded rice straw. *Bioresource Technology* 55:111–117.

Mosier, N., R. Hendrickson, N. Ho, M. Sedlak, and M.R. Ladisch. 2005a. Optimization of pH controlled liquid hot water pre-treatment of corn stover. *Bioresource Technology* 96:1986–1993.

Mosier, N., C. Wyman, B. Dale, R. Elander, Y.Y. Lee, M. Holtzapple, and M. Ladisch. 2005b. Features of promising technologies for pretreatment of lignocellulosic biomass. *Bioresource Technology* 96:673–686.

Murtagh, J.E. 1995. Molasses as feedstock for alcohol production. In *The Alcohol Textbook*, ed. T.P. Lyons, D.R. Kelsall, and J.E. Murtagh. Nottingham, U.K.: University Press.

Nakamura, Y., T. Sawada, and E. Inoue. 2001. Enhanced ethanol production from enzymatically treated steam-exploded rice straw using extractive fermentation. *Journal of Chemical Technology and Biotechnology* 76:879–884.

Negro, M.J., P. Manzanares, I. Ballesteros, J.M. Oliva, A. Cabañas, and M. Ballesteros. 2003. Hydrothermal pretreatment conditions to enhance ethanol production from poplar biomass. *Applied Biochemistry and Biotechnology* 105–108:87–100.

Nguyen, Q.A., M.P. Tucker, F.A. Keller, D.A. Beaty, K.M. Connors, and F.P. Eddy. 1999. Dilute acid hydrolysis of softwoods. *Applied Biochemistry and Biotechnology* 77–79:133–142.

Ogier, J.-C., D. Ballerini, J.-P. Leygue, L. Rigal, and J. Pourquié. 1999. Production d'éthanol à partir de biomasse lignocellulosique (Ethanol production from lignocellulosic biomass, in French). *Oil & Gas Science and Technology—Revue de l'IFP* 54 (1):67–94.

Oliva, J.M. 2003. Efecto de los productos de degradación originados en la explosión por vapor de biomasa de chopo sobre *Kluyveromyces marxianus*, (Effect of the degradation products generated during the steam explosion of poplar biomass on *Kluyveromyces marxianus*, in Spanish), Ph.D. Thesis, Departamento de Microbiología III, Universidad Complutense de Madrid, Spain.

Olsson, L., and B. Hahn-Hägerdal. 1996. Fermentation of lignocellulosic hydrolysates for ethanol production. *Enzyme and Microbial Technology* 18:312–331.

Palmqvist, E., and B. Hahn-Hägerdal. 2000a. Fermentation of lignocellulosic hydrolysates. I: Inhibition and detoxification. *Bioresource Technology* 74:17–24.

Palmqvist, E., and B. Hahn-Hägerdal. 2000b. Fermentation of lignocellulosic hydrolysates. II: Inhibitors and mechanisms of inhibition. *Bioresource Technology* 74:25–33.

Pan, X., C. Arato, N. Gilkes, D. Gregg, W. Mabee, K. Pye, Z. Xiao, X. Zhang, and J. Saddler. 2005. Biorefining of softwoods using ethanol organosolv pulping: Preliminary evaluation of process streams for manufacture of fuel-grade ethanol and co-products. *Biotechnology and Bioengineering* 90 (4):473–481.

Pandey, K., and P.K. Agarwal. 1993. Effect of EDTA, potassium ferrocyanide, and sodium potassium tartarate on the production of ethanol from molasses by *Saccharomyces cerevisiae*. *Enzyme and Microbial Technology* 15:887–898.

Papatheofanous, M.G., E. Billa, D.P. Koullas, B. Monties, and E.G. Koukios. 1995. Two-stage acid-catalyzed fractionation of lignocellulosic biomass in aqueous ethanol systems at low temperatures. *Bioresource Technology* 54:305–310.

Papatheofanous M.G., E. Billa, D.P. Koullas, B. Monties, and E.G. Koukios. 1998. Optimizing multisteps mechanical-chemical fractionation of wheat straw components. *Industrial Crops and Products* 7:249–256.

Patil, S.G., D.V. Gokhale, and B.G. Patil. 1985. Enhancement in ethanol production from cane molasses by skim milk supplementation. *Enzyme and Microbial Technology* 8:481–484.

Patil, S.G., and B.G. Patil. 1989. Chitin supplement speeds up the ethanol production in cane molasses fermentation. *Enzyme and Microbial Technology* 11:38–43.

Persson, P., J. Andersson, L. Gorton, S. Larsson, N.-O. Nilvebrant, and L.J. Jönsson. 2002a. Effect of different forms of alkali treatment on specific fermentation inhibitors and on the fermentability of lignocellulose hydrolysates for production of fuel ethanol. *Journal of Agricultural and Food Chemistry* 50:5318–5325.

Persson, P., S. Larsson, L.J. Jönsson, N.-O. Nilvebrant, B. Sivik, F. Munteanu, L. Thörneby, and Lo. Gorton. 2002b. Supercritical fluid extraction of a lignocellulosic hydrolysate of spruce for detoxification and to facilitate analysis of inhibitors. *Biotechnology and Bioengineering* 79 (6):694–700.

Prosen, E.M., D. Radlein, J. Piskorz, D.S. Scott, and R.L. Legge. 1993. Microbial utilization of levoglucosan in wood pyrolysate as a carbon and energy source. *Biotechnology and Bioengineering* 42 (4):538–541.

Rezzoug, S.-A., and R. Capart. 1996. Solvolysis and hydrotreatment of wood to provide fuel. *Biomass and Bioenergy* 11 (4):343–352.

Rivers, D.B., and G.H. Emert. 1988. Factors affecting the enzymatic hydrolysis of bagasse and rice straw. *Biological Wastes* 26:85–95.

Rodríguez-Chong, A., J.A. Ramírez, G. Garrote, and M. Vázquez. 2004. Hydrolysis of sugar cane bagasse using nitric acid: A kinetic assessment. *Journal of Food Engineering* 61 (2):143–152.

Saha, B.C. 2003. Hemicellulose conversion. *Journal of Industrial Microbiology & Biotechnology* 30:279–291.

Saha, B.C., and M.A. Cotta. 2006. Ethanol production from alkaline peroxide pretreated enzymatically saccharified wheat straw. *American Chemical Society and American Institute of Chemical Engineers*: A-E, Washington, D.C.

Saha, B.C., L.B. Iten, M.A. Cotta, and Y.V. Wu. 2005a. Dilute acid pretreatment, enzymatic saccharification, and fermentation of rice hulls to ethanol. *Biotechnology Progress* 21:816–822.

Saha, B.C., L.B. Iten, M.A. Cotta, and Y.V. Wu. 2005b. Dilute acid pretreatment, enzymatic saccharification and fermentation of wheat straw to ethanol. *Process Biochemistry* 21:816–822.

Sánchez, Ó.J., and C.A. Cardona. 2008. Trends in biotechnological production of fuel ethanol from different feedstocks. *Bioresource Technology* 99:5270–5295.

Saska, M., and E. Ozer. 1995. Aqueous extraction of sugarcane bagasse hemicellulose and production of xylose syrup. *Biotechnology and Bioengineering* 45:517–523.

Schell, D.J., J. Farmer, M. Newman, and J.D. McMillan. 2003. Dilute-sulfuric acid pretreatment of corn stover in pilot-scale reactor. Investigation of yields, kinetics, and enzymatic digestibilities of solids. *Applied Biochemistry and Biotechnology* 105 (1–3):69–85.

Shahbazi, A., Y. Li, and M.R. Mims. 2005. Application of sequential aqueous steam treatments to the fractionation of softwood. *Applied Biochemistry and Biotechnology* 121–124:973–987.

Shevchenko, S.M., R.P. Beatson, and J.N. Saddler. 1999. The nature of lignin from steam explosion/enzymatic hydrolysis of softwood. *Applied Biochemistry and Biotechnology* 77–79:867–876.

SivaRaman, H., A. Chandwadkar, S.A. Baliga, and A.A. Prabhune. 1994. Effect of synthetic zeolite on ethanolic fermentation of sugarcane molasses. *Enzyme and Microbial Technology* 16:719–722.

Söderström, J., L. Pilcher, M. Galbe, and G. Zacchi. 2003. Two-step steam pretreatment of softwood by dilute H_2SO_4 impregnation for ethanol production. *Biomass and Bioenergy* 24:475–486.

Sreenath, H.K., R.G. Koegel, A.B. Moldes, T.W. Jeffries, and R.J. Straub. 2001. Ethanol production from alfalfa fiber fractions by saccharification and fermentation. *Process Biochemistry* 36:1199–1204.

Sun, Y., and J.J. Cheng. 2002. Hydrolysis of lignocellulosic materials for ethanol production: A review. *Bioresource Technology* 83:1–11.

Sun, Y., and J.J. Cheng. 2005. Dilute acid pretreatment of rye straw and Bermuda grass for ethanol production. *Bioresource Technology* 96 (14):1599–1606.

Teixeira, L.C., J.C. Linden, and H.A. Schroeder. 1999a. Alkaline and peracetic acid pretreatments of biomass for ethanol production. *Applied Biochemistry and Biotechnology* 77–79:19–34.

Teixeira, L.C., J.C. Linden, and H.A. Schroeder. 1999b. Optimizing peracetic acid pretreatment conditions for improved simultaneous saccharification and co-fermentation (SSCF) of sugar cane bagasse to ethanol fuel. *Renewable Energy* 16:1070–1073.

Tengerdy, R.P., and G. Szakacs. 2003. Bioconversion of lignocellulose in solid substrate fermentation. *Biochemical Engineering Journal* 13:169–179.

Varga, E., H.B. Klinkle, K. Réczey, and A.B. Thomsen. 2004. High solid simultaneous saccharification and fermentation of wet oxidized corn stover to ethanol. *Biotechnology and Bioengineering* 88 (5):567–574.

von Sivers, M., and G. Zacchi. 1995. A techno-economical comparison of three processes for the production of ethanol from pine. *Bioresource Technology* 51:43–52.

Warzywoda, M., E. Larbre, and J. Pourquié. 1992. Production and characterization of cellulolytic enzymes from *Trichoderma reesei* grown on various carbon sources. *Bioresource Technology* 39:125–130.

Weil, J.R., B. Dien, R. Bothast, R. Hendrickson, N.S. Mosier, and M.R. Ladisch. 2002. Removal of fermentation inhibitors formed during pretreatment of biomass by polymeric adsorbents. *Industrial & Engineering Chemistry Research* 41:6132–6138.

Wooley, R., M. Ruth, J. Sheehan, K. Ibsen, H. Majdeski, and A. Galvez. 1999. *Lignocellulosic biomass to ethanol process design and economics utilizing co-current dilute acid prehydrolysis and enzymatic hydrolysis. Current and futuristic scenarios.* Technical Report NREL/TP-580-26157, National Renewable Energy Laboratory, Washington, D.C.

Xie, Y., D. Phelps, C.-H. Lee, M. Sedlak, N. Ho, and N.-H.L. Wang. 2005. Comparison of two adsorbents for sugar recovery from biomass hydrolyzate. *Industrial & Engineering Chemistry Research* 44:6816–6823.

Yu, Z., and H. Zhang. 2003. Pretreatments of cellulose pyrolysate for ethanol production by *Saccharomyces cerevisiae, Pichia* sp. YZ-1 and *Zymomonas mobilis. Biomass and Bioenergy* 24:257–262.

Zhang, B.S. 2006. Process for preparing fuel ethanol by using straw fiber materials. China Patent CN1880416.

5 Hydrolysis of Carbohydrate Polymers

The hydrolysis of glucans is a significant source of fermentable sugars for fuel ethanol production. The most important glucans in ethanol industry are starch and cellulose. In this chapter, the features of starch and cellulose hydrolyses are discussed. In particular, the difficulties related to the enzymatic hydrolysis of cellulose are analyzed taking into consideration the enzyme complexes used and the presence of solid particles in the reaction mixture. Some process engineering aspects emphasizing the kinetic modeling of these processes are also discussed.

5.1 STARCH SACCHARIFICATION

In Chapter 4, the use of α-amylase during starch liquefaction was mentioned. It is worth emphasizing here some general issues related to the enzymatic hydrolysis of starch. The utilization of enzymes to break down the starch has some advantages over the hydrolysis using acids. In the latter case, strong conditions are required to achieve the degradation of starch (150°C, pH of 1.5 to 1.8). The amylolytic enzymes work under milder conditions (temperatures lower than 110°C, neutral pH) with the corresponding energy savings. In addition, the enzymatic process does not generate compounds resulting from degradation or oxidation of sugars due to the very high specificity of the enzymes (Lopez-Munguía, 1993). For these reasons, the sweeteners industry does not use the acid hydrolysis of starch. This process has been taken over by the fuel ethanol industry in order to obtain the fermentable sugars needed for yeast cultivation.

Main amylases employed for starch hydrolysis at the industrial level are from bacterial and fungal origin though some plant enzymes are eventually used (Table 5.1). The enzyme α-amylase obtained from thermophilic bacteria or produced by a recombinant microorganism using the gene obtained from a thermophilic organism is one of the most used amylases. This enzyme randomly hydrolyzes the α(1,4) glycosidic bonds within the chains of both amylose and amylopectin. For this enzyme to attack the starch, the previous gelatinization of starch should have been carried out. Thus, the broken starch granules release the amylose and amylopectin and the enzymatic action can be started. The α-amylase can support high temperatures of up to 110°C while keeping its activity, thus it is ideal for starch liquefaction process. The molecular weights and optimum temperatures of enzymatic activity of the α-amylases most commonly employed for starch processing are shown in Table 5.2. Apar and Özbek (2004) provide information about the effects of operating conditions on the enzymatic hydrolysis of corn starch using commercial α-amylase. In general, the optimum pH of

TABLE 5.1
Main Enzymes Used for Starch Hydrolysis

Enzymes	Source	Bond Hydrolyzed	Products	Mechanism
α-amylase	Bacillus licheniformis B. subtilis	α(1,4)	Dextrines, maltodextrines, maltose	Endoenzyme
β-amylase	Asergillus oryzae Bacillus cereus Barley	α(1,4) from nonreducing ends	Dextrines, maltose	Exoenzyme
Glucoamylase (amyloglucosidase)	A. niger Rhizopus sp.	α(1,4) from nonreducing ends and α(1,6)	Glucose	Exoenzyme
Pullulanase	B. acidopullulyticus Klebsiella pneumoniae	α(1,6)	Maltodextrines	Endoenzyme

TABLE 5.2
Origin and Properties of Different α-Amylases and Glucoamylases

Microbial Source	Molecular Weight	Optimum Temperature/°C	Optimum/pH	References
α-amylases				Nigam and Singh
Bacillus subtilis	54,780	80	5.6	(1995); Pandey et
Bacillus amyloliquefaciens	49,000	70		al. (2000a)
Bacillus licheniformis	62,000	90	6.0–6.5	
Glucoamylases				Nigam and Singh
Aspergillus awamori	83,700–88,000	60	4.5	(1995); Pandey et
Aspergillus niger I	99,000	60	4.5–5.0	al. (2000a)
Aspergillus niger II	112,000	60	4.4	
Aspergillus oryzae I	76,000	60	4.5	
Aspergillus oryzae II	38,000	50	4.5	
Aspergillus oryzae III	38,000	40	4.5	
Aspergillus saitoi	90,000		4.5	
Cephalosporium eichhormonie	26,850	45–62		
Lipomyces kononenkoae	811,500	50		
Mucor rouxianus I	59,000	55	4.7	
Mucor rouxianus II	49,000	55	4.7	
Penicillium oxalicum I	84,000	55–60	4.6	
Penicillium oxalicum II	86,000	60	4.6	
Rhizophus delemar	100,000	40	4.5	

the α-amylase is about 6. For this reason, the pH of the cooked starch should be adjusted before the enzyme addition during the liquefaction process.

Glucoamylase (amyloglucosidase) is the other most employed enzyme in starch-to-ethanol process. This enzyme is generally obtained from *Aspergillus niger* or a species of *Rhizopus* genus (Labeille et al., 1997; Nigam and Singh, 1995; Shigechi et al., 2004). The glucoamylase is an exo-enzyme that hydrolyzes the α(1,4) bonds from the nonreducing ends of amylose or amylopectin chains forming glucose. Unlike α-amylase, most glucoamylases have the ability to hydrolyze the α(1,6) bonds in branching points of the amylopectin, though the hydrolysis rate of this bond is 15 times lower than for the α(1,4) bonds (López-Munguía, 1993; Pandey et al., 2000b). This feature allows this enzyme to convert the dextrins formed during the liquefaction step into glucose. The optimum temperatures of enzymatic activity for glucoamylases are lower than those of α-amylases employed during starch liquefaction (see Table 5.2). Consequently, the cooling of liquefied starch is needed to ensure an appropriate conversion toward glucose. In some commercial formulations of amylases, the enzyme pullulanase, which specifically hydrolyzes the α(1,6) bonds, is also employed. Finally, the β-amylase is mostly employed in the brewing industry and for production of maltose syrups.

The process of hydrolysis (or saccharification) of the stream exiting the liquefaction tank is aimed at obtaining a glucose-rich solution for its later fermentation. This stream is adjusted at a pH of 4.5 and cooled down to 65°C in order to ensure the optimum conditions of the hydrolysis process (Bothast and Schlicher, 2005). The saccharification product is called corn mash if this cereal is the feedstock employed or saccharified starch if starch is the feedstock employed, as in the case of wet milling.

The steps related to the starch degradation are responsible for 10 to 20% of the energy consumption of the ethanol process in the case of fuel ethanol produced from starchy materials. One potential option to minimize this high amount of energy is the substitution of enzymatic hydrolysis technologies in liquid media at high temperatures with technologies involving the starch hydrolysis using amylases working at low temperatures in solid phase. This approach would make possible the "cold hydrolysis" of the native starch (Cardona and Sánchez, 2007). For this, the discovery and characterization of new enzymes displaying these properties are required (Robertson et al., 2006). Some microorganisms, such as the bacterium *Clostridium thermohydrosulfuricum,* have the ability to digest nongelatinized starch and convert it into ethanol at 66°C. However, the productivities attained are too low (Mori and Inaba, 1990).

5.2   HYDROLYSIS OF CELLULOSE

The yeast *Saccharomyces cerevisiae* and the bacterium *Zymomonas mobilis* are not able to directly utilize the cellulose for ethanol production. In general, these microorganisms have the highest probability to be used in an industrial process for conversion of lignocellulosic biomass into fuel ethanol. For this reason, a step for cellulose hydrolysis (saccharification) to obtain a fermentable solution of glucose is required in an analogous way to the starch saccharification. As in the case of starch, the cellulose can be hydrolyzed with the help of acids either dilute or concentrated. If dilute acids are used, temperatures of 200 to 240°C at 1.5% acid concentrations are needed in order to hydrolyze the crystalline cellulose, but the degradation of glucose into hydroxymethylfurfural (HMF) and other nondesired products is unavoidable under these conditions. In a similar way, xylose is degraded into furfural and other compounds (Wyman, 1994). H_2SO_4 and HCl have been historically used for these purposes (Jones and Semrau, 1984; Song and Lee, 1984). As mentioned in Chapter 4, the pretreatment and cellulose hydrolysis of lignocellulosic biomass can be carried out in two stages using dilute acids. Thus, a first stage under mild conditions (190°C, 0.7% acid, 3 min) is carried out to recover pentoses, while the remaining solids undergo harsher conditions (215°C, 0.4% acid, 3 min) to recover hexoses from cellulose hydrolysis in the second stage. In this way, 50% glucose yield is obtained (Hamelinck et al., 2005). One variant of the acid hydrolysis is the employment of extremely low acid and high temperature conditions during batch processes (autohydrolysis approach) that have been applied to sawdust (Ojumu and Ogunkunle, 2005; Sánchez and Cardona, 2008). The degradation of cellulose with near-critical water has been proposed, although the obtained glucose yield is low (40%) and the fermentability of resulting hydrolyzate is limited due to the formation of unknown inhibitors (Sakaki et al., 1996).

The hydrolysis of cellulose with concentrated acids allows achieving glucose yields near 90%, but in this case, the recovery of the acid is a key factor in the process economy (Hamelinck et al., 2005). Several configurations for separation of formed glucose and recovery of employed acid have been proposed. Among the early procedures suggested, the lime addition, ionic exclusion columns based on commercial resins (Neuman et al., 1987), or electrodialysis (Baltz et al., 1982) can be highlighted. Concentrated acid processes using 30 to 70% H_2SO_4 have a higher glucose yield (90%) and are relatively fast (10 to 12 h), although the amount of acid used is a critical economic factor. By continuous ionic exchange, it is possible to recover over 97% of the acid (Hamelinck et al., 2005). In general, the acid hydrolysis of cellulose implies high energy costs and the construction of reactors resistant to the corrosion, which increase the capital costs.

5.2.1 ENZYME SYSTEMS FOR CELLULOSE HYDROLYSIS

The microbial cellulolytic enzymes (cellulases) can overcome the disadvantages of acid hydrolysis of cellulose. Being specific biological catalysts, secondary products of degradation are not formed and working under milder conditions (temperatures up to 60°C, pH of 2.5 to 5.5), exigencies to the enzymatic bioreactors are not very strict. Nevertheless, the reaction times are more prolonged than in the case of acid hydrolysis.

A significant number of microorganisms have the ability to biosynthesize cellulases. In general, the anaerobic bacteria degrade the cellulose through high-molecular-weight complex systems with cellulolytic activity named cellulosomes, as in the case of *Clostridium thermocellum*. The cellulosomes are extracellular structures consisting of spherical polypeptidic complexes that include an enzymatic package with cellulase activity, which group together the enzymes through the action of one polypeptide with no hydrolytic activity involved in the adhesion of cellulosomes to the substrate. The cellulosomes increase the magnitude of the enzymatic action on the plant particles rich in cellulose due to the close contact of the bacterium with its enzymes and such particles (Fondevila, 1998). *C. thermocellum* cellulosomes are distributed not only in the culture medium, but also on the surface of bacterial cells, though several bacterial strains do not release appreciable amounts of these cellulolytic complexes to the medium. In this case, the cellulosomes are concentrated in the cell wall surface or inside its structure. These complexes have a high effectiveness in the degradation of crystalline cellulose (Lynd et al., 2002). In particular, anaerobic thermophilic bacteria exhibit high growth rates on cellulose and have enzymes with a high stability. Moreover, their culture requires less agitation energy. These bacteria could be potential candidates for the direct conversion of cellulose to ethanol (Lee, 1997).

In contrast, aerobic cellulose-degrading microorganisms, including bacteria and fungi, break down this polysaccharide through the production of significant amounts of individual extracellular cellulases, although some enzymatic complexes can be occasionally found on the cell surface. Due to their individual nature, these cellulases exhibit a synergic action on the cellulose (Lynd et al., 2002). At an industrial level, a great variety of cellulases are produced and employed in the formulation of detergents, in the textile and food industries, and during the production of paper pulp and paper (Bhat, 2000). Most of the commercial cellulases are obtained from *Trichoderma reesei*, though a small portion is obtained from *A. niger*. Several reports can be found on the features of cellulase aerobic production by *T. reesei*. See, for example, the work of Marten et al. (1996). *T. reesei* releases a mixture of cellulases, among which at least two cellobiohydrolases, five endoglucanases, β-glucosidases, and hemicellulases can be found. The two cellobiohydrolases and one of the endoglucanases represent approximately 92% of total production of cellulases (Zhang and Lynd, 2004). Cellobiohydrolases break down β(1,4) linkages from nonreducing or reducing ends of the cellulose chain releasing cellobiose or even glucose, whereas endoglucanases hydrolyze these same linkages randomly inside the chain. The action of cellobiohydrolases causes a gradual decrease in the polymerization degree, while endoglucanases cause the rupture of cellulose into smaller chains reducing rapidly the polymerization degree. Endoglucanases especially act on amorphous cellulose, whereas cellobiohydrolases are capable of acting on crystalline cellulose as well (Lynd et al., 2002), as illustrated in Figure 5.1.

Although *T. reesei* produces some β-glucosidases, which is the result of hydrolyzing formed cellobiose into two molecules of glucose, their activities are not very high. Unfortunately, cellobiohydrolases are inhibited by cellobiose. Therefore, β-glucosidase from other sources needs to be added in order to complement the action of the cellulases from this fungus (Sánchez and Cardona, 2008). Dekker and Wallis (1983) have determined a ratio of FPU (filter paper

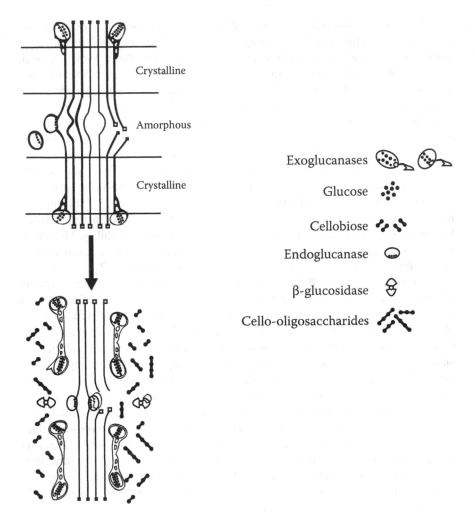

Exoglucanases

Glucose

Cellobiose

Endoglucanase

β-glucosidase

Cello-oligosaccharides

FIGURE 5.1 Schematic representation of cellulose hydrolysis using individual cellulolytic enzymes.

units) to β-glucosidase units of enzymatic activity as the minimum required to achieve more than 80% conversion of cellulose into glucose. The combined action of these enzymes is synergic leading to the conversion of cellulose into glucose (Béguin and Aubert, 1994; Walker et al. 1993). Factorial optimization techniques have been applied to the design of mixtures of cellulases from different sources along with β-glucosidase in order to maximize the yield of glucose produced (Kim et al., 1998). It has been suggested to develop a multicellulase plasmid in which the different cellulase genes could be expressed to produce cellulases with an optimum ratio from a single cultivation.

Cellulases should be adsorbed onto the surface of substrate particles before hydrolysis of insoluble cellulose takes place. The three-dimensional structure of these particles in combination with their size and shape determines if β-glycosidic bonds are or are not accessible to enzymatic attack (Zhang and Lynd, 2004). This makes the hydrolysis process slower in relation to the enzymatic degradation of other biopolymers. For comparison, the hydrolysis rate of starch by amylases is 100 times faster than the hydrolysis rate of cellulose by cellulases under industrial processing conditions. Apparently, this difference in hydrolysis rates can be explained to a greater extent by the higher accessibility of the enzymes to the substrate, which is more limited in the case of the cellulose, than by the fact that the β(1,4) bond of cellulose is more difficult to hydrolyze than the α(1,4) bond of starch. In the case of the pretreated lignocellulosic complex, the cellulases can bind in a reversible way not only to the cellulose particles, but also to the lignin that reduces their effectiveness. In addition, the cellulases can bind in an irreversible way to the substrate provoking a progressive loss of enzymatic activity. It has been postulated that the addition of surfactants to the reaction mixture can improve the effectiveness of enzymatic cellulose hydrolysis due to the reduction of enzyme loss through irreversible binding to the substrate. The surfactant increases the rate of hydrolysis as well as prolongs the enzyme life. This allows the reduction of enzyme dosage to 50%. It has been reported that the usage of Tween-80 improved sugar production for any given particle size of cellulose when milled newsprint was used as a feedstock. Similarly, the use of sophorolipid increases by 67% the hydrolysis of steam-exploded wood (Duff et al., 1995; Helle et al., 1993).

One nonconventional approach for saccharification that demonstrates the diversity of trends for research and development of cellulose hydrolysis is the enzymatic hydrolysis in biphasic media, by which higher glucose concentrations can be attained. The goal of replacing part of the water with organic substances is explained by the need of ensuring the necessary rheological properties to accomplish saccharification using higher substrate loads taking into account that glucose does not migrate to the organic phase. Cantarella et al. (2001) employed this approach for saccharification of steam-exploded wheat straw employing a medium with 25% (by volume) aqueous phase and 75% organic phase (acetates). Higher glucose concentrations (measured in the aqueous phase) were obtained compared to the case when only the aqueous phase was used. These concentrations reached about 150 g/L. However, when the aqueous phase was fermented using *S. cerevisiae*, an increase in the lag-phase and a small reduction in ethanol yield were observed (nearly 86 to 92% of the theoretical maximum). Another approach to cellulose hydrolysis consists in the design of hydrolyzing agents, which are different from mineral acids (that degrade the formed glucose) or enzymes. Mosier et al. (2002) have systematically studied several organic acids in order to assess their cellulose hydrolysis and glucose degradation characteristics in comparison to sulfuric acid. Their results indicate that maleic acid presented the best characteristics. In particular, this acid does not degrade the formed glucose. These studies are aimed at developing a nonprotein catalyst that mimics the action of the cellulases. In this way, maleic acid is a suitable catalytic domain for the synthesis of such enzymatic mimicry.

5.2.2 CONVERSION OF CELLULOSE TO GLUCOSE

From the viewpoint of the conversion process, the most important factors to be taken into account for hydrolysis of cellulose contained in lignocellulosic materials are the reaction time, temperature, pH, enzyme dosage, and substrate load (Sánchez and Cardona, 2008). By testing lignocellulosic material from sugarcane leaves, Hari Krishna et al. (1998) have found the best values of all these parameters varying in each experimental series the value of one of the factors while fixing the other ones. Sixty-five to seventy percent cellulose conversion was achieved at 50°C and pH of 4.5. Although enzyme doses of 100 FPU*/g cellulose caused almost a 100% hydrolysis, this amount is not economically justifiable. Hence, a 40 FPU/g cellulose dosage was proposed for which a 13% reduction in conversion was observed. Regarding the substrate concentration, solids loads of 10% were defined as the most adequate considering rising mixing difficulties and accumulation of inhibitors in the reaction medium. Though these conditions were determined for a specific material pretreated by overliming and extrapolations to other lignocellulosic feedstocks are risky, found values are within the reported ones in the literature for a wide range of materials. Hydrolysis tests for steam-pretreated spruce also indicate the need of high enzyme loadings of both cellulases and β-glucosidase in order to achieve cellulose conversions greater than 70% due to the lower degradability of the softwood (Tengborg et al., 2001). Similar studies were carried out for saccharification of dilute-acid pretreated Douglas fir showing that enzyme dosage has a significant effect on glucose yield (Schell et al., 1999). Saha and Cotta (2006) obtained 96.7% yield of monomeric sugars using an enzymatic cocktail of cellulase, β-glucosidase, and xylanase for saccharification of wheat straw pretreated by alkaline peroxide method. An ethanol concentration of 18.9 g/L and a yield of 0.46 g/g available sugars were achieved in the subsequent fermentation using a recombinant *Escherichia coli* strain capable of assimilating both hexoses and pentoses. Jeffries and Schartman (1999) propose the use of sequential enzyme addition in a countercurrent mode for the saccharification of fiber fines from a paper recycling plant. Sequential addition of enzyme to the pulp in small aliquots produced a higher overall sugar yield per activity unit of enzyme than the addition of the same total amount of enzyme in a single dose. During the succeeding fermentation employing different yeasts, 78% ethanol yield of the theoretical maximum was obtained.

---

* Filter paper units (FPU) allow the indirect quantification of enzymatic activity of cellulases. One FPU is equivalent to the cellulase amount needed to form 1 μmol glucose in one minute measured as reducing sugars (or to form 0.10 mg glucose in one minute) during the cellulose hydrolysis reaction employing as the substrate Whatman No. 1 paper filter under determined conditions (pH = 4.8, 0.05 M sodium citrate buffer, 50°C). See Ghose (1987).

5.2.3   CELLULOSE HYDROLYSIS ASSESSMENT FOR PROCESS SYNTHESIS

5.2.3.1   Efficiency of Cellulases

Cellulase utilization plays a crucial role considering the global costs of the bio-mass-to-ethanol process. The cellulases available for ethanol industry account for 36 to 45% of the costs of bioethanol produced from lignocellulosic materials. According to different evaluations (Cardona and Sánchez, 2007; Reith et al., 2002), a 30% reduction in capital costs and 10-fold decrease in the cost of current cellulases are required for this process to be competitive in relation to ethanol produced from starchy materials. These analyses evidence the need of improving the cellulase performance in the following aspects:

- Increase of thermal stability
- Improvement of the binding to cellulose
- Increase of specific activity
- Reduction of the nonspecific binding to lignin

The increase of the thermal stability of cellulases is important since the temperature increase implies an increase of the cellulose hydrolysis rate (Mielenz, 2001). The augment of specific activity, in turn, can lead to significantly reduced costs. It is estimated that a 10-fold increase in specific activity could lead to nearly 16 cents (U.S.) savings per liter of ethanol produced. Among the strategies to attain the increase of specific activity are the increase in the efficiency of active sites through protein engineering or random mutagenesis, augment of thermal tolerance, improvement in the degradation of the crystalline structure of cellulose, enhancement of the synergism among the cellulases from different sources, and reduction of nonspecific bindings (Cardona and Sánchez, 2007; Sheehan and Himmel, 1999).

In general, the costs of cellulases are considered high. According to preliminary evaluations of the National Renewable Energy Laboratory (NREL) cited by Tengerdy and Szakacs (2003), the cost of cellulase production *in situ* by submerged culture is U.S.$0.38/100,000 FPU. Hence, cellulase costs make up 20% of ethanol production costs, assuming them at U.S.$1.5/gallon. On the other hand, commercial cellulase cost (U.S.$16/100,000 FPU) is prohibitive for this process. In contrast, these authors indicate that the cost for producing cellulases by solid-state fermentation of corn stover would be U.S.$0.15/100,000 FPU that would correspond to U.S.$0.118/gal EtOH, i.e., nearly 8% of total costs. The analysis of process integration via simulation of not only the technological scheme, but also the costs structure, can provide key elements allowing a deep evaluation of all of these alternatives in order to choose the more convenient option for *in situ* cellulase production. On the other hand, the mathematical description of cellulase production can allow the definition of useful relationships to assess the performance and quality of the production of such enzymes using different process analysis approaches. These approaches undoubtedly will contribute to the definition of strategies to lower the costs of cellulases. This is the case of cellulase production by *Trichoderma reesei* using kinetic and neural networks approaches (Tholudur et al., 1999), which allowed the optimization of operating conditions considering two performance indexes based on the estimated protein value and volumetric productivity.

5.2.3.2 Modeling of Cellulose Hydrolysis

A significant number of papers and works related to cellulose hydrolysis have been published. For instance, the reader can consult the reviews of Lynd et al. (2002) and Zhang and Lynd (2004) on this theme. The description of cellulose degradation, in particular, the kinetics of cellulose hydrolysis, is difficult due to the influence of factors such as (Bernardez et al., 1993; Kadam et al., 2004; Meunier-Goddik and Penner, 1999; Philippidis and Hatzis, 1997; Philippidis et al., 1993; Zhang and Lynd, 2004):

- Adsorption of the enzymes to the substrate particles.
- Presence of interactions between the lignin and enzymes that do not lead to sugars formation.
- Decrease of cellulose hydrolysis as substrate conversion progresses.
- Enzyme dosage.
- Synergistic action of endoglucanases and cellobiohydrolases.
- Need of supplementing the cellulases with β-glucosidase to diminish cellulase inhibition by cellobiose formed.
- β-glucosidase inhibition by the glucose formed.
- Solids load to the reactor, among others.

The importance of cellulose hydrolysis description is recognized considering that the mathematical modeling of this process allows for developing appropriate simulation tools to be used during process synthesis procedures. Such institutions as the NREL have funded projects regarding the modeling of cellulose breakdown, sugar formation, and ethanolic fermentation in order to assess several alternative technological configurations of the biomass-to-ethanol process. This was the case of the work of South et al. (1995) dealing with the hydrolysis of cellulose contained in pretreated wood that, in turn, used the experimental data obtained in a previous work by these same authors using commercial fungal cellulase (South et al., 1993). In this study, a kinetic model, considering the cellulose conversion and the formation and disappearance of cellobiose and glucose, was developed. In addition, a Langmuir-type model taking into account the adsorption of cellulases on the solid particles of cellulose and lignin and expressions describing the dependence of cellulose conversion on the residence time of nonsoluble solid particles of biomass were considered.

The adsorption of cellulases to the particles present in the solid fraction of pretreated lignocellulosic biomass should be modeled in a suitable way to obtain results useful during the design of cellulosic ethanol production. This is not the case for starch saccharification considering that the amylases attack their substrate in a soluble form after starch cooking. In contrast, during the first steps of ligno-cellulosic hydrolysis, the cellulases are added to a suspension of cellulose and lignin particles. The Langmuir model is widely used for description of adsorption processes involving cellulases taking into account the good adjustment to experimental data in most cases (Cardona and Sánchez, 2007). In addition, it represents a simple mechanistic model that can be used to compare kinetic properties of various cellulase–cellulose systems. Kadam et al. (2004) employed a Langmuir-type isotherm to describe the enzymatic hydrolysis of cellulose for the case of dilute-acid pretreated corn stover. In this model, the inhibition effect on cellulases of other sugars present in the biomass hydrolyzate as xylose was considered as well as the effect of temperature (through the Arrhenius equation) and the dosage of β-glucosidase. In an early work, Bernardez et al. (1993) studied the adsorption process of complexed cellulase systems (cellulosomes) released by the anaerobic thermophilic bacterium *Clostridium thermocellum* onto crystalline cellulose, pretreated wood, and lignin employing the Langmuir description. Nevertheless, some experimental data indicate that the negative effect of lignin content in the hydrolyzate is not principally due to the enzyme partitioning between cellulose and lignin, suggesting that lignin hinders saccharification by physically limiting the enzyme accessibility of the cellulose (Meunier-Goddik and Penner, 1999). Hence, more structure-oriented modeling is required to gain insight on biomass hydrolyzate's hydrolysis and its optimal operating conditions. Other models have been proposed since the union of the cellulases to the cellulose does not meet all the assumptions inherent to the Langmuir model. To this end, two-site adsorption models, Freundlich isotherms, and combined Langmuir–Freundlich isotherms have been proposed (Zhang and Lynd, 2004). Lynd et al. (2002) present in their wide review about the microbial cellulose utilization, a compilation of values of adsorption parameters for cellulases isolated from different microorganism and for diverse substrates. In that work, the kinetic constants for cellulose utilization by different microorganisms are reported as well. On the other hand, it has been shown that the intensity of the agitation in batch reactors has little effect over cellulose hydrolysis when cellulose particles are suspended. Based on the analysis of the kinetic constants and on experimental data, it was concluded that the external mass transfer is not a limiting factor of the global process of hydrolysis. However, when the internal area is much greater than the external one, as in the case of most cellulosic substrates, it is probable that cellulases can remain entrapped in the pores provoking lower hydrolysis rates (Zhang and Lynd, 2004). These considerations are essential when mathematical representations of cellulose saccharification are developed. On the other hand, some kinetic studies for cellulose hydrolysis highlight the significant effect the enzyme dosage has on glucose yield. In particular, Schell et al. (1999) accomplished the saccharification of dilute-acid pretreated Douglas fir and obtained data from which was derived a useful empirical model to calculate the glucose yield as well as to formulate a kinetic cellulose hydrolysis model.

   Zhang and Lynd (2004) reviewed in detail the works concerning the modeling of cellulose hydrolysis and point out that most of proposed models for the design of industrial systems fall in the category of semimechanistic models, i.e., models taking into account the substrate concentration or one of the enzymatic activities as a state variable. These models meet the requirement of including the minimum of necessary information for the description of the process (Cardona and Sánchez, 2007). These authors emphasize that most kinetic models do not consider the changes in the hydrolysis rate during the course of the reaction, and that those models that do this, are based mainly on empirically adjusted parameters and not on a mechanistic approach. For instance, the model of a simultaneous saccharification and fermentation (SSF, to be analyzed in Chapter 9) process developed for the case of nonpretreated wastepaper using commercial cellulases and *S. cerevisiae* for both batch and two-stage continuous regimes (Philippidis and Hatzis, 1997) made use of an exponential decay term to describe the time-dependent decline in the rate of cellulose hydrolysis. With the help of an exhaustive sensitivity analysis, the model showed that further improvements in the fermentation stage do not have great influence on ethanol yield. In contrast, the digestibility of substrate (as a result of pretreatment), cellulase dosage, specific activity, and composition have a great effect on ethanol yield. This confirms that major research efforts should be oriented to the development of more effective pretreatment methods and the production of cellulases with higher specific activity (Cardona and Sánchez, 2007).

Half of the mechanistic models cited by Zhang and Lynd (2004) are based on the Michaelis–Menten model, which is valid when the limiting substrate is in excess relative to the enzyme. In addition, competitive inhibition is the mechanism most found in the literature, although a combination of both noncompetitive and competitive mechanisms for different inhibition effects can be found, as are analyzed in the work of Philippidis et al. (1993). Due to the importance of modeling, these authors highlight the need for developing functional models that include the adsorption process, several state variables for substrate besides the concentration (e.g., polymerization degree or amount of amorphous cellulose), and multiple enzymatic activities.

REFERENCES

Apar, D.K., and B. Özbek. 2004. σ-Amylase inactivation during cornstarch hydrolysis process. *Process Biochemistry* 39 (12):1877–1892.

Baltz, R.A., A.F. Burcham, O.C. Sitton, and N.L. Book. 1982. The recycle of sulfuric acid and xylose in the prehydrolysis of corn stover. *Energy* 7 (3):259–265.

Béguin, P., and J.P. Aubert. 1994. The biological degradation of cellulose. *FEMS Microbiology Reviews* 13:25–58.

Bernardez, T.D., K. Lyford, D.A. Hogsett, and L.R. Lynd. 1993. Adsorption of *Clostridium thermocellum* cellulases onto pretreated mixed hardwood, avicel, and lignin. *Biotechnology and Bioengineering* 42:899–907.

Bhat, M.K. 2000. Cellulases and related enzymes in biotechnology. *Biotechnology Advances* 18:355–383.

Bothast, R.J., and M.A. Schlicher. 2005. Biotechnological processes for conversion of corn into ethanol. *Applied Microbiology and Biotechnology* 67:19–25.

Cantarella, M., F. Alfani, L. Cantarella, A. Gallifuoco, and A. Saporosi. 2001. Biosaccharification of cellulosic biomass in inmiscible solvent–water mixtures. *Journal of Molecular Catalysis B: Enzymatic* 11:867–875.

Cardona, C.A., and Ó.J. Sánchez. 2007. Fuel ethanol production: Process design trends and integration opportunities. *Bioresource Technology* 98:2415–2457.

Dekker, R.F.H., and A.F.A. Wallis. 1983. Enzymic saccharification of sugarcane bagasse pretreated by autohydrolysis-steam explosion. *Biotechnology and Bioengineering* 25:3027–3048.

Duff, S.J.B., J.W. Moritz, and T.E. Casavant. 1995. Effect of surfactant and particle size reduction on hydrolysis of deinking sludge and nonrecyclable newsprint. *Biotechnology and Bioengineering* 45:239–244.

Fondevila, M. 1998. Procesos implicados en la digestión microbiana de los forrajes de baja calidad. (Processes involved in the microbial digestion of low-quality forages, in Spanish.) *Revista de la Facultad de Agronomía Universidad del Zulia* 15:87–106.

Ghose, T.K. 1987. Measurement of cellulase activities. *Pure & Applied Chemistry* 59 (2): 257–268.

Hamelinck, C.N., G. van Hooijdonk, and A.P.C. Faaij. 2005. Ethanol from lignocellulosic biomass: Techno-economic performance in short-, middle- and long-term. *Biomass and Bioenergy* 28:384–410.

Hari Krishna, S., K. Prasanthi, G. Chowdary, and C. Ayyanna. 1998. Simultaneous saccharification and fermentation of pretreated sugar cane leaves to ethanol. *Process Biochemistry* 33 (8):825–830.

Helle, S.S., S.J.B. Duff, and D.G. Cooper. 1993. Effect of surfactants on cellulose hydrolysis. *Biotechnology and Bioengineering* 42:611–617.

Jeffries, T.W., and R. Schartman. 1999. Bioconversion of secondary fiber fines to ethanol using counter-current enzymatic saccharification and co-fermentation. *Applied Biochemistry and Biotechnology* 77–79:435–444.

Jones, J.L., and K.T. Semrau. 1984. Wood hydrolysis for ethanol production—Previous experience and the economics of selected processes. *Biomass* 5:109–135.

Kadam, K.L., E.C. Rydholm, and J.D. McMillan. 2004. Development and validation of a kinetic model for enzymatic saccharification of lignocellulosic biomass. *Biotechnology Progress* 20:698–705.

Kim, E., D.C. Irwin, L.P. Walker, and D.B. Wilson. 1998. Factorial optimization of a six-cellulase mixture. *Biotechnology and Bioengineering* 58 (5):494–501.

Labeille, P., J.L. Baret, Y. Beaux, and F. Duchiron. 1997. Comparative study of wheat flour saccharification and ethanol production with two glucoamylase preparations. *Industrial Crops and Products* 6:291–295.

Lee, J. 1997. Biological conversion of lignocellulosic biomass to ethanol. *Journal of Biotechnology* 56:1–24.

López-Munguía, A. 1993. Edulcorantes (Sweeteners, in Spanish). In *Biotecnología Alimentaria*, ed. M. García, R. Quintero, and A. López-Munguía. Centro, México: Limusa.

Lynd, L.R., P.J. Weimer, W.H. van Zyl, and I.S. Pretorious. 2002. Microbial cellulose utilization: Fundamentals and biotechnology. *Microbiology and Molecular Biology Reviews* 66 (3):506–577.

Marten, M.R., S. Velkovska, S.A. Khan, and D.F. Ollis. 1996. Rheological, mass transfer, and mixing characterization of cellulase-producing *Trichoderma reesei* suspensions. *Biotechnology Progress* 12:602–611.

Meunier-Goddik, L., and M.H. Penner. 1999. Enzyme-catalyzed saccharification of model celluloses in the presence of lignacious residues. *Journal of Agricultural and Food Chemistry* 47:346–351.

Mielenz, J.R. 2001. Ethanol production from biomass: Technology and commercialization status. *Current Opinion in Microbiology* 4:324–329.

Mori, Y., and T. Inaba. 1990. Ethanol production from starch in a pervaporation membrane bioreactor using *Clostridium thermohydrosulfuricum*. *Biotechnology and Bioengineering* 36:849–853.

Mosier, N.S., C.M. Ladisch, and M.R. Ladisch. 2002. Characterization of acid catalytic domains for cellulose hydrolysis and glucose degradation. *Biotechnology and Bioengineering* 79 (6):610–618.

Neuman, R.P., S.R. Rudge, and M.R. Ladisch. 1987. Sulfuric acid-sugar separation by ion exclusion. *Reactive Polymers* 5:55–61.

Nigam, P., and D. Singh. 1995. Enzyme and microbial systems involved in starch processing. *Enzyme and Microbial Technology* 17:770–778.

Ojumu, T.V., and O.A. Ogunkunle. 2005. Production of glucose from lignocellulosic under extremely low acid and high temperature in batch process, auto-hydrolysis approach. *Journal of Applied Sciences* 5 (1):15–17.

Pandey, A., P. Nigam, C.R. Soccol, V.T. Soccol, D. Singh, and R. Mohan. 2000a. Advances in microbial amylases. *Biotechnology and Applied Biochemistry* 31:135–152.

Pandey, A., C. Soccol, C. Nigam, and V. Soccol. 2000b. Biotechnological potential of agro-industrial residues. I: Sugarcane bagasse. *Bioresource Technology* 74:69–80.

Philippidis, G.P., and C. Hatzis. 1997. Biochemical engineering analysis of critical process factors in the biomass-to-ethanol technology. *Biotechnology Progress* 13:222–231.

Philippidis, G.P., T.K. Smith, and C.E. Wyman. 1993. Study of the enzymatic hydrolysis of cellulose for production of fuel ethanol by the simultaneous saccharification and fermentation process. *Biotechnology and Bioengineering* 41:846–853.

Reith, J.H., H. den Uil, H. van Veen, W.T.A.M. de Laat, J.J. Niessen, E. de Jong, H.W. Elbersen, R. Weusthuis, J.P. van Dijken, and L. Raamsdonk. 2002. Co-production of bioethanol, electricity and heat from biomass residues. Paper presented at the 12th European Conference and Technology Exhibition on Biomass for Energy, Industry and Climate Protection, Amsterdam, The Netherlands.

Robertson, G.H., D.W.S. Wong, C.C. Lee, K. Wagschal, M.R. Smith, and W.J. Orts. 2006. Native or raw starch digestion: A key step in energy efficient biorefining of grain. *Journal of Agricultural and Food Chemistry* 54:353–365.

Saha, B.C., and M.A. Cotta. 2006. Ethanol production from alkaline peroxide pretreated enzymatically saccharified wheat straw. *American Chemical Society and American Institute of Chemical Engineers*: A-E, Washington, D.C.

Sakaki, T., M. Shibata, T. Miki, H. Hirosue, and N. Hayashi. 1996. Decomposition of cellulose in near-critical water and fermentability of the products. *Energy & Fuels* 10:684–688.

Sánchez, Ó.J., and C.A. Cardona. 2008. Trends in biotechnological production of fuel ethanol from different feedstocks. *Bioresource Technology* 99:5270–5295.

Schell, D.J., M.F. Ruth, and M.P. Tucker. 1999. Modeling the enzymatic hydrolysis of dilute-acid pretreated Douglas fir. *Applied Biochemistry and Biotechnology* 77–79:67–81.

Sheehan, J., and M. Himmel. 1999. Enzymes, energy, and the environment: A strategic perspective on the U.S. Department of Energy's research and development activities for bioethanol. *Biotechnology Progress* 15:817–827.

Shigechi, H., Y. Fujita, J. Koh, M. Ueda, H. Fukuda, and A. Kondo. 2004. Energy-saving direct ethanol production from low-temperature-cooked cornstarch using a cell-surface engineered yeast strain co-displaying glucoamylase and α-amylase. *Biochemical Engineering Journal* 18:149–153.

Song, S.K., and Y.Y. Lee. 1984. Acid hydrolysis of wood cellulose under low water condition. *Biomass* 6:93–100.

South, C.R., D.A.L. Hogsett, and L.R. Lynd. 1993. Continuous fermentation of cellulosic biomass to ethanol. *Applied Biochemistry and Biotechnology* 39/40:587–600.

South, C.R., D.A.L. Hogsett, and L.R. Lynd. 1995. Modeling simultaneous saccharification and fermentation of lignocellulose to ethanol in batch and continuous reactors. *Enzyme and Microbial Technology* 17:797–803.

Tengborg, C., M. Galbe, and G. Zacchi. 2001. Influence of enzyme loading and physical parameters on the enzymatic hydrolysis of steam-pretreated softwood. *Biotechnology Progress* 17:110–117.

Tengerdy, R.P., and G. Szakacs. 2003. Bioconversion of lignocellulose in solid substrate fermentation. *Biochemical Engineering Journal* 13:169–179.

Tholudur, A., W.F Ramírez, and J.D. McMillan. 1999. Mathematical modelling and optimization of cellulase protein production using *Trichoderma reesei* RL-P37. *Biotechnology and Bioengineering* 66 (1):1–16.

Walker, L.P., C.D. Belair, D.B. Wilson, and D.C. Irwin. 1993. Engineering cellulase mixtures by varying the mole fraction of *Thermomonospora fusca* E5 and E3, *Trichoderma reesei* CBHI, and *Caldocellum saccharolyticum* β-glucosidase. *Biotechnology and Bioengineering* 42:1019–1028.

Wyman, C.E. 1994. Ethanol from lignocellulosic biomass: Technology, economics, and opportunities. *Bioresource Technology* 50:3–16.

Zhang, Y.-H.P., and L.R. Lynd. 2004. Toward an aggregated understanding of enzymatic hydrolysis of cellulose: Noncomplexed cellulose systems. *Biotechnology and Bioengineering* 8 (7):797–824.

6 Microorganisms for Ethanol Production

Microorganisms are the key component for conversion of different raw materials into ethyl alcohol during the fermentation step. The success of the overall fuel ethanol production process depends on the selected microbial strains. In this chapter, some metabolic features of microorganisms used for ethanol biosynthesis are discussed. The development of efficient, low-cost, and environmentally friendly processes for fuel ethanol production requires the selection of suitable microorganisms that contribute to achieving such goals. In addition, the development of microbial strains allowing innovative designs and making use of the wide availability of feedstock resources, especially lignocellulosic materials, is required. For this reason, the main strategies for genetic modification of microorganisms with better performance during ethanol production from different feedstocks are discussed in this chapter.

6.1 METABOLIC FEATURES OF ETHANOL PRODUCING MICROORGANISMS

Ethyl alcohol is obtained from coal or natural gas using a synthetic process. South Africa is the major producer of this type of ethanol. Ethanol is produced by oxidation of olefins as well. However, 95% of ethanol worldwide is obtained by fermentation from carbohydrate-containing feedstocks (Berg, 2001). As a result of the metabolism of microorganisms employed in fermentative processes, their growth and maintenance under conditions favoring the attainment of high ethanol concentrations is accomplished. The metabolism can be considered as the overall sum of chemical reactions generally catalyzed by enzymes taking place in biological systems (cells) aimed at maintaining their life. Microscopic organisms possess certain peculiarities that make them ideal for production of a wide variety of industrial products, especially in the case of bacteria, filamentous fungi (molds), and yeasts.

Among the main metabolic features of microorganisms, the great diversity of these substances that can be utilized as carbon and energy sources should be emphasized. This feature has a paramount importance at the moment of choosing different technological options for fuel ethanol production. The new ethanol production processes are oriented to the utilization of highly available lower-cost feedstocks, such as lignocellulosic materials. In this regard, conventional microorganisms used during ethanolic fermentation, the yeasts, display a lower versatility regarding the utilization of a wide range of substances. Thus, the yeast

Saccharomyces cerevisiae can assimilate only some hexoses and a much-reduced amount of disaccharides.

In general, bacteria and filamentous fungi provide a wider variety of substances (compared to yeasts) that can be potentially utilized to obtain metabolites with economic importance. For instance, the bacterium *Clostridium thermocellum* assimilates not only hexoses, but also the cellulose polysaccharide as a substrate during the biosynthesis of ethanol.

The microorganisms also feature a much more intense metabolism than that of higher multicellular organisms. Due to their small size, microbial cells present a very high surface/volume ratio, which makes possible the active input of many substances to the cytoplasm. In addition and due to the presence of a resistant cell wall, the microorganisms can build up many substances in high concentrations. This allows a high rate of metabolic reactions, such as a very fast fermentation and high cell division rate. This intense metabolism permits the development of continuous fermentation processes at an industrial level by which a bioreactor is fed with fresh nutritive medium, while permanently withdrawing the effluent stream containing the target product, in this case ethanol. This is possible because the cell growth rate offsets the rate at which the cells are removed from the bioreactor with the effluent. In fact, the alcoholic fermentation is one of the few microbiological processes that have been successfully implemented in continuous regime at the industrial level.

During the development of microbial cells, it is possible to clearly differentiate the biosynthesis of substances needed to ensure cell growth and reproduction. These substances are called *primary metabolites* and are relatively similar for all kinds of microorganisms, such as the adenosine triphosphate (ATP), carboxylic acids involved in the Krebs cycle, polysaccharides constituting the bacterial cell wall, and the ethanol itself. Ethanol is a compound that, like the lactic acid, is formed in the cytoplasm as a result of the metabolic processes intended to produce energy under anoxic conditions. These anoxic processes taking place in yeasts and some bacteria allow the production of energy required for cell growth, suggests the biosynthesis of ethyl alcohol in a parallel way. The similarity of the metabolic pathways for ethanol biosynthesis allows broadening the possibilities to choose different microorganisms during the selection of the process bioagent for industrial alcoholic fermentations. At this point, the wide range of yeast species able to uptake both hexoses and pentoses for their conversion into ethanol should be noted. This is particularly important to design fermentation processes using the liquid fraction of the pretreated lignocellulosic biomass that contains a high content of xylose and hexoses.

Another feature of some microorganisms related to ethanol production is their ability to "predigest" the available food resources. Thus, these microorganisms release into the culture medium not only end products of the metabolism, but also intermediate metabolites such as certain hydrolytic enzymes. These enzymes hydrolyze complex polymeric structures outside the cell membrane in order to assimilate the nutrients resulting from their degradation. In the case of *C. thermocellum*, the action of cellulosomes in the outer side of the cell wall allows the formation of glucose from cellulose. The glucose formed is assimilated by this bacterium for its growth and the production of ethanol.

In general, ethanol-producing microorganisms exhibit a fermentative type of metabolism, which is characterized by the utilization of organic compounds (e.g., sugars) with an intermediate degree of oxidation without an end electron acceptor. In turn, the products of this type of metabolism are other compounds (such as ethanol or lactic acid) with an intermediate degree of oxidation. In this process, the oxygen does not participate, thus the involved microorganisms are strict anaerobes, aero tolerant anaerobes, or facultative anaerobes. The alcoholic fermentation occurs when microorganisms such as the yeasts utilize sugars, obtain energy for their growth, and generate ethanol through a fermentative metabolism. It is worth noting that the term *fermentation* in biochemistry refers to this kind of metabolism, while *fermentation* in industrial microbiology refers to the conversion of substrates into specific products by the action of microorganisms. This latter process is accomplished in biological reactors called *fermenters*. The fermenters for fermentative (in the biochemical sense) processes are simpler than those used for aerobic processes (e.g., for processes using certain microorganisms with a respiratory type of metabolism) because no aeration and, sometimes, agitation devices are required, which reduces the corresponding capital costs. The respiratory type of metabolism, in turn, can be aerobic (the end electron acceptor is the oxygen) or anaerobic (with end electron acceptors other than oxygen such as nitrates or sulfates). Besides fermentative and respiratory types of metabolism, the microorganisms can exhibit either methanogenic or phototrophic types of metabolism (Table 6.1).

In the specific case of ethanolic fermentation, most of producing microorganisms obtain their energy under low or null oxygen concentrations through the transformation of one glucose molecule into two molecules of ATP forming, in addition, two molecules of ethyl alcohol. Due to the absence of oxygen, the pyruvate formed during the glycolysis cannot be linked to the Krebs cycle, electron transfer chain, or oxidative phosphorylation (characteristic metabolic processes of the aerobic respiratory type of metabolism). This implies that the pyruvate is

TABLE 6.1
Main Types of Microbial Metabolism according to Their Catabolic Features

Metabolism	Energy Source	Carbon Source	Feed Kind
Fermentative	Organic compounds	Organic compounds	Organotrophic
Respiratory	Organic compounds	Organic compounds,	Organotrophic
	Inorganic compounds	CO_2	Autotrophic
Methanogenic	Organic compounds	Organic compounds,	Organotrophic
	Inorganic compounds, H_2	CO_2	Autotrophic
Phototrophic	Solar radiation	Organic compounds,	Photoorganotrophic
		CO_2	Photoorganotrophic

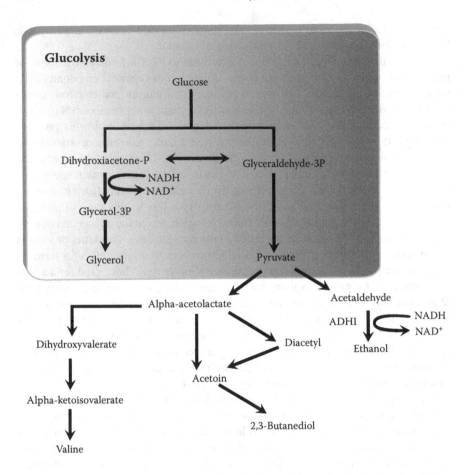

FIGURE 6.1 Main metabolic pathways involved in the biosynthesis of ethanol and other related metabolites during alcoholic fermentation using *Saccharomyces cerevisiae* (ADHI = alcohol dehydrogenase).

decarboxylated to form acetaldehyde, which, in turn, is finally reduced to ethanol, an energy-rich product. The main metabolic pathways involved in the transformation of glucose into ethanol using *S. cerevisiae* are depicted in Figure 6.1 where, in addition, the overall pathways for formation of other metabolites related to ethanolic fermentation are also shown.

6.2 NONGENETICALLY MODIFIED MICROORGANISMS FOR ETHANOL PRODUCTION

There exists a significant number of species of yeasts and bacteria having the ability to synthesize ethanol. Main ethanol-producing microorganisms that are currently used in the industry or have potential utilization in the future are shown in Table 6.2. Among the main criteria in choosing a microorganism producing

TABLE 6.2
Main Ethanol-Producing Microorganisms

Microorganism	Fermentable Substrates	Fermentation Conditions	Ethanol Tolerance	Remarks	References
Saccharomyces cerevisiae	Glucose, fructose, maltose, sucrose	Anaerobic, 30–37°C	150 g/L	Up to 95% of theor. yield	Claassen et al. (1999)
Schizosaccharomyces pombe	Glucose, fructose, maltose, sucrose	Anaerobic, 30–35°C		High osmotic tolerance	Bullock (2002); Goyes and Bolaños (2005)
Kluyveromyces marxianus	Glucose	Anaerobic, 40–45°C		80–90% of theor. yield	Kádár et al (2004)
Candida shehatae	Glucose, xylose	Microaerophilic, 20–31°C	30–45 g/L	94% of theor. yield on xylose; produces moderate amounts of xylitol	Olsson and Hahn-Hägerdal (1996); Jeffries and Jin (2000)
Pichia stipitis	Glucose, xylose	Microaerophilic, 26–35°C	35–47 g/L	92% of theor. yield on xylose	Olsson and Hahn-Hägerdal (1996); Jeffries and Jin (2000)
Pachysolen tannophilus	Glucose, xylose, glycerol	Microaerophilic	37.5–75 g/L	47–55% of theor. yield on xylose; produces large amounts of xylitol	Olsson and Hahn-Hägerdal (1996); Jeffries and Jin (2000)
Zymomonas mobilis	Glucose, fructose, sucrose	Anaerobic, 30°C	100 g/L	Up to 97% of theor. yield	Claassen et al. (1999)
Clostridium thermocellum	Glucose, cellulose	Anaerobic, 55–65°C	10–30 g/L	Produces acetic acid	Claassen et al. (1999); South et al. (1993)
Clostridium thermosaccharolyticum	Glucose, xylose	Anaerobic, 60°C	<30 g/L	41% of theor. yield on xylose	Olsson and Hahn-Hägerdal (1996)

ethanol, the ability to assimilate a wide range of substrates should be emphasized. Unfortunately, alcoholic fermentation presents an end-product inhibition, so one of the most desired features of the potential microorganisms is a high ethanol tolerance. Similarly, the cultivation conditions allow defining some desirable features of the microorganisms to be used. For instance, elevated temperatures allow the reduction of cooling costs as well as the acceleration of the metabolic processes (although this can imply an increase in the inhibitory effect of ethanol). Increased temperatures can be advantageous when the microorganisms are cultivated along with hydrolytic enzymes in simultaneous hydrolysis and fermentation processes of either starch or cellulose since these enzymes have, in general, optimum temperatures of enzymatic activity higher than fermentation temperature. Finally, one of the most important selection criteria is the achieved ethanol yield on substrate because the economy of the overall process directly depends on this parameter.

6.2.1 Yeasts

The most employed microorganisms for fuel ethanol production are the yeasts of the *Saccharomyces cerevisiae* species that convert hexoses, such as glucose and fructose, into pyruvate through glycolysis, which is finally reduced to ethanol generating two moles of ATP for each molecule of consumed hexose under anaerobic conditions (Claassen et al., 1999) as shown in Figure 6.1. This microorganism also has the ability to convert hexoses into CO_2 by aerobic respiration. One of the two processes may be favored depending on the oxygen concentration in the culture medium and the carbon source. In the latter case, mainly biomass is formed and it is the base for large-scale production of baker's yeast. In addition to their ability to be grown under anaerobic conditions, yeasts have the advantage of tolerating relatively high concentrations of ethanol (up to 150 g/L). Below a concentration of 30 g/L the inhibition is negligible (Kargupta et al., 1998). The yeast *Schizosaccharomyces pombe* has the additional advantage of tolerating high osmotic pressures (high amounts of salts) and high solids content (Bullock, 2002; Goyes and Bolaños, 2005). In fact, a fermentation process using a wild strain of this yeast has been patented (Carrascosa, 2006).

Other yeasts having the capability of growing under thermophilic conditions have been evaluated from an industrial viewpoint. Increased fermentation temperature accelerates metabolic processes and lowers the refrigeration requirements. For this reason, one of the yeasts that is most studied for ethanol production is *Kluyveromyces marxianus*, which can be cultivated at temperatures higher than 40°C (Ballesteros et al., 2001). This condition makes this yeast very promising in the case of cellulose conversion schemes for ethanol production by the simultaneous accomplishment of hydrolysis and fermentation (Ballesteros et al., 2004) because the cellulases have greater activity at temperatures much higher (50 to 60°C) than those of conventional fermentations.

One of the main problems during ethanol production from lignocellulosic materials is that *S. cerevisiae* can ferment only certain mono- and disaccharides, such as glucose, fructose, maltose, and sucrose. Nevertheless, this microorganism cannot assimilate either cellulose or hemicellulose directly. Furthermore, this yeast does not assimilate the pentoses obtained during the pretreatment of lignocellulosic biomass when hemicellulose is hydrolyzed at a higher degree. This hemicellulose hydrolyzate contains pentoses (mostly xylose, though also arabinose) as well as other hexoses (glucose, mannose, and galactose). For this reason, the utilization of pentose-utilizing microorganisms has been proposed similarly to some species of yeasts. Yeasts, such as *Pichia stipitis*, *Candida shehatae*, and *Pachysolen tannophilus* can assimilate both pentoses and hexoses (Olsson and Hahn-Hägerdal, 1996) as shown in Table 6.2. One key aspect in the metabolism of xylose is its conversion into xylulose that is integrated to the metabolic pathways for pyruvate synthesis (final product of glycolysis) from which ethanol is derived. Pyruvate is also the starting point for the cycle of tricarboxylic acids (Krebs cycle; Figure 6.2). The cultivation of these yeasts requires a thorough control to ensure low levels of oxygen in the medium needed for the oxidative respiratory metabolism.

For pentose-assimilating yeasts, the hexoses are, however, the most readily and rapidly assimilable substrate during ethanol production. This implies a diauxic growth, which means that the hexoses are consumed firstly before the pentoses if the fermentation is extended enough. After a relatively short lag-phase in which the enzymes necessary for pentose metabolism are synthesized, the pentoses are consumed until the end of fermentation. This means that the microorganisms do not utilize the two types of sugar at the same time, which causes a decrease in the

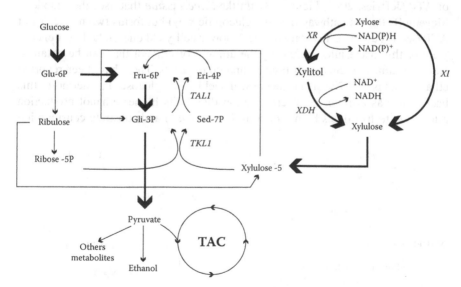

FIGURE 6.2 Main metabolic pathways involved during ethanolic fermentation using microorganisms assimilating both hexoses and pentoses. (Glu-6P = glucose-6-phosphate, Fru-6P = fructose-6-phosphate, Eri-4P = erithrose-4-phosphate, Gly-3P = glyceraldehyde-3-phosphate, Sed-7P = sedoheptulose-6-fosfato, TAC = Krebs cycle, *XR* =: xylose-reductase, *XDH* = xylitol-dehydrogenase, *XI* = xylosa-isomerase, *TAL1* = transaldolase, *TKL1* = transketolase)

biomass utilization rate. As a rule, the microorganisms prefer the glucose over the galactose, followed by the xylose and arabinose (Gong et al., 1999), which is explained by the catabolic repression the glucose exerts on the consumption rate of xylose and other pentoses as in the case of *C. shehatae*. In addition, ethanol productivity achieved using xylose-assimilating yeasts is lower than that of microorganisms fermenting only hexoses. Thus, their ethanol production rate from glucose is at least five times lower than that observed in *S. cerevisiae*. Moreover, their culture requires oxygen and ethanol tolerance that is two to four times lower (Claassen et al., 1999). Most xylose-utilizing yeasts are mesophiles, i.e., they are cultivated at temperatures near 30°C; likewise *S. cerevisiae*, though there exist reports about the methylotrophic yeast *Hansenula polymorpha* cultivated at 37°C in a xylose-containing medium (Ryabova et al., 2003).

6.2.2 BACTERIA

Some bacteria have the capacity to produce ethanol in significant amounts making them potential microorganisms for industrial purposes (see Table 6.2). Among bacteria, the most promising microorganism is *Zymomonas mobilis*. This facultative anaerobe presents higher ethanol yields than yeasts, which is related to the metabolic pathways involved. *Z. mobilis* makes use of the Entner–Doudoroff pathway converting 1 mol of hexose into 2 mol of ethanol, but releasing only 1 mol of ATP (Jeffries, 2005; Figure 6.3), unlike *S. cerevisiae* that uses the Embden–Meyerhoff–Parnas pathway (i.e., the glucolytic way) but forms two molecules of ATP (see Figure 6.1). This fact implies a lower cell yield due to the lower energy yield of this bacterium, increasing the amount of ethanol that can be obtained from the same amount of substrate compared to yeasts. There have been reported ethanol yields of 97% of the theoretical yield from glucose. Furthermore, this bacterium has a more rapid fermentation due to its higher ethanol production rate (three to five times higher than in *S. cerevisiae*) and substrate consumption

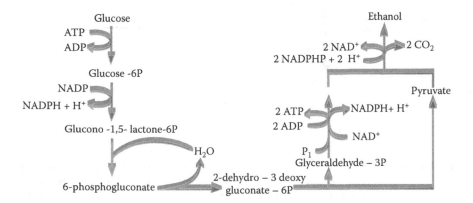

FIGURE 6.3 Main metabolic pathway involved during ethanol biosynthesis by the bacterium *Zymomonas mobilis*.

rate. Additionally, this bacterium has a high ethanol tolerance (100 g/L) and a higher optimal production temperature. Z. *mobilis* requires no controlled addition of oxygen and is much more susceptible to genetic manipulations to enhance its yield or transfer new traits than is the case of yeasts.

Among the drawbacks of Z. *mobilis*, the too narrow range of fermentable substrates (glucose, fructose, and sucrose) should be noted (Claassen et al., 1999; Hawgood et al., 1985). Other disadvantage of the use of this bacterium for fermentation of sugarcane syrup and other sucrose-based media is the formation of the polysaccharide levan (made up of fructose units), which increases the viscosity of fermentation broth, and of sorbitol, a product of fructose reduction that reduces the efficiency of the conversion of sucrose into ethanol (Doelle and Doelle, 1989; Grote and Rogers, 1985; Lee and Huang, 2000). In addition, preculture conditions have a significant influence on bacterium performance, especially on sucrose hydrolysis rate. Hence, the addition of invertase to the culture medium has been proposed (Doelle and Doelle, 1989). Lee and Huang (2000) studied batch ethanolic fermentation using Z. *mobilis* through nonstructured models based on metabolic analysis. These models allowed the use of ethanol and sorbitol formation during cultivation on a medium containing glucose and fructose or on a sucrose medium supplemented with immobilized invertase.

Other bacteria that have been investigated in order to implement processes of direct conversion of lignocellulosic biomass into ethanol are thermophilic and saccharolytic clostridia. *Clostridium thermohydrosulfuricum*, *C. thermosaccharolyticum*, and *C. thermocellum* can synthesize up to 2 mol EtOH/mol hexose. Likewise, these bacteria may transform pentoses and amino acids into ethanol. Having saccharolytic properties, these microorganisms have the ability to grow on a wide variety of nontreated wastes. *C. thermocellum* can even directly convert lignocellulosic materials into ethanol (McMillan, 1997). In this way, ethanol can be directly obtained from pretreated lignocellulosic biomass without the need of adding costly cellulases. Moreover, the cultivation of these microorganisms at high temperatures offers the possibility of an easier ethanol removal by distillation or pervaporation and reduced cooling expenditures (Claassen et al., 1999). *Clostridium thermocellum* has been the most studied thermophilic clostridium because it has the capacity to produce cellulases, hydrolyzing the cellulose and fermenting the glucose forming ethyl alcohol. On the other hand, the possibility of employing *C. thermosaccharolyticum* to produce ethanol from pentoses resulting from hemicellulose degradation during the pretreatment of biomass has been shown (Wyman, 1994).

The main drawback of these bacteria consists in their very low ethanol tolerance compared to yeasts. Consequently the maximum reached ethanol concentrations are lower than 30 g/L. In addition, they exhibit a reduced ethanol yield due to the formation of fermentation by-products like acetic acid and lactate that make the final ethanol concentrations very low and cultivation times prolonged (3 to 12 days) (Baskaran et al., 1995; McMillan, 1997; Szczodrak and Fiedurek, 1996; Wyman, 1994). Process integration can play a crucial role for evaluating the most optimal configurations in order to implement them at an industrial level.

6.3 GENETICALLY MODIFIED MICROORGANISMS FOR ETHANOL PRODUCTION

Due to the need for improving fuel ethanol production processes especially when lignocellulosic biomass is used, the development of microorganism strains with better performance in terms of ethanol yield and productivity is required, particularly concerning the direct conversion of polymeric feedstocks. For instance, a way to improve the technoeconomic indexes of ethanol production processes from starch consists in developing yeasts able to hydrolyze this polysaccharide and then ferment the glucose formed without the addition of amylases. Moreover, the effective utilization of alternative feedstocks to obtain ethanol as the lignocellulosic residue requires microorganisms with traits difficult to find in a single species (cellulase production, pentose assimilation, high ethanol yields, high ethanol tolerance, among others). The native strains of microorganisms cannot meet all these exigencies; therefore, their genetic modification is needed.

Of the thousands of genes contained in the microbial DNA, 90 to 95% are repressed, i.e., at a given moment, the microbial cells require only the expression of a reduced amount of genes for them to accomplish all their metabolic functions directed to cell biomass growth. This implies that most genes are involved in complex regulation processes so that the information contained in them supplies the necessary instructions for the cells to produce myriad metabolites. From an industrial viewpoint, the higher the number of genes expressed, the wider the spectrum of potential value-added products synthesized as well as the greater the number of substrates that can be assimilated by the microorganism. This is accomplished through the synthesis of a higher number of enzymes responsible for that wider assimilation. Similarly, if the natural repression of the genes encoding the production of a given metabolite is eliminated, significant increases in the yield of that metabolite can be reached, though the cell growth may be reduced. In this way, super-producing strains of microorganisms can be developed for production of different products with commercial importance. Therefore, the selection of an industrial strain of microorganisms requires a selection program that can include the genetic modification of native strains of those microorganisms. The genetic modification of industrial microorganisms can mainly be done in two ways: random modification of DNA and directed modification.

6.3.1 MUTAGENESIS

By applying random genetic modification techniques, also called mutagenesis, mutations in the microbial DNA are induced through employing physical or chemical agents (mutagens). The mutations are produced randomly in the DNA chain and lead to the production of mutant microorganisms that can transfer these changes in the genome to the subsequent generations. Among the most used physical agents, ultraviolet radiation (usually with a wave length of 260 nm) is highlighted. This radiation is absorbed by the double bonds of pyrimidines composing the DNA chain. As a result, DNA photoproducts are obtained that provoke drastic changes in the nucleic acids. This process is not directed and leads to the death of most irradiated cells. However, a small number of cells survive and even acquire some desirable traits as a higher product yield. The best mutants are selected by screening procedures and are irradiated again in order to obtain mutants with better yields. Besides ultraviolet radiation, x-rays and ionizing radiations are also employed though these agents can cause the death of all the irradiated cell population. Among the most employed chemical mutagens are the nitrogenated base analogues (5-bromouracile, 2-aminopurine), deaminating or hydroxylating agents (nitrous acid, hydroxyl amine), alkylating agents (mustard gas, ethyl-ethanesulfonate, ethyl-methanesulfonate, nitrosoguanidine), intercalant agents (acridine orange, ethidium bromide, proflavine), and pairing blocking agents (benzopyrene, aflatoxine B_1; Crueger and Crueger, 1993). The selection programs of industrial microorganisms by mutagenesis are very prolonged, tedious, and expensive, but can cause significant increases in the yield of the overall process.

In the case of fuel ethanol production, the development of mutant cells has been oriented, besides increasing ethanol yield, to the enhancement of tolerance to salts and impurities contained in the medium (e.g., for the case of yeasts cultivated on molasses) or to the acquiring of flocculating properties. This latter trait allows the ready separation of yeast cells during schemes of continuous fermentation or by repeated-batch regimes (see Chapter 7) because the cells agglomerate and settle (or float) allowing their rapid removal from the cultivation broth.

6.3.2 RECOMBINANT DNA TECHNOLOGY

The directed modification of DNA through recombinant DNA technology, widely known as genetic engineering, represents a powerful tool not only for improving strains of industrial microorganisms (e.g., by the increase of product yields), but also for transferring specific traits or properties from some species to others, which are very separated in the evolutionary chain. This technology implies the extraction of genes from organisms exhibiting a determined trait in order to introduce (recombine) them in the DNA of a host microorganism (or organism). When the modified microorganism is reproduced, the succeeding generations will inherit the newly acquired trait. In the case of bacteria, the most common vectors used for DNA transfer to the host organism are the plasmids, relatively short autonomous (nonintegrated to the bacterial chromosome) DNA sequences that are autoreplicable, i.e., can form copies of themselves especially during the cell reproduction. It is worth emphasizing that the recombination of the gene with the plasmid is carried out *in vitro* (Figure 6.4). The recombinant vector obtained is introduced into the bacterial cells through diverse techniques that include a temporal increase in the porosity of both cell wall and cell membrane for the vector to be introduced into the cytoplasm attaining, in this way, the transformation of the host cells. The microorganisms having genes of other organisms are called recombinant microorganisms or, in a general way, genetically modified organisms or simply engineered organisms.

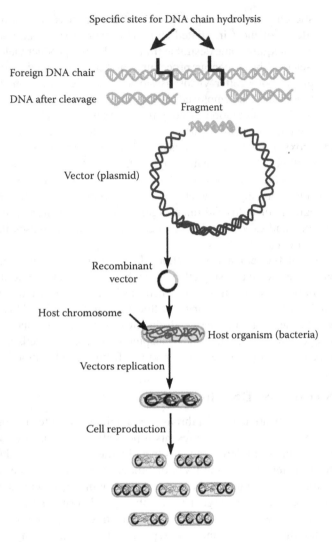

Specific sites for DNA chain hydrolysis

Foreign DNA chair

DNA after cleavage

Fragment

Vector (plasmid)

Recombinant
vector

Host chromosome

Host organism (bacteria)

Vectors replication

Cell reproduction

FIGURE 6.4 Principle of recombinant DNA technology to develop genetically modified bacteria. A plasmid is the vector for transferring the foreign gene to the bacterium.

The introduction of recombinant microorganisms in the ethanol industry can lead to the development of industrial processes radically different from the conventional technologies based on the fermentation of molasses or starch hydrolyzates using native or improved yeast strains. Due to the specific and directed character of the modifications that can be done by genetic engineering, it is possible to "create" microbial strains capable of assimilating alternative feedstocks (first, lignocellulosic materials) or ones with the ability to perform simultaneous transformations in integrated process (e.g., the direct conversion of starch or lignocellulosic biomass). In this regard, when the microorganisms are genetically modified to assimilate more substrates or to form new products, the integration of a series of very complex chemical transformations allowing the execution of several chemical processes in the same unit can be carried out. This kind of integration is verified at cell or even molecular level. These integrated processes represent a new approach in ethanol production and are the base for integral utilization of feedstocks in the so-called biorefineries (Cardona and Sánchez, 2007).

6.3.2.1 Recombinant Microorganisms for Starch Processing

Through recombinant DNA technology, amylolytic yeast strains have been "constructed." This allows for the design of ethanol production processes, excluding the liquefaction and saccharification steps using exogenous enzymes, and the utilization of only one bioagent during the transformation, the yeast (consolidated bioprocessing, CBP). The savings obtained during the commercial implementation of such a process could offset by far the lower growth rates and the longer fermentation times. In this way, a single microorganism can directly convert the starch into ethanol (Cardona and Sánchez, 2007). Some examples of these efforts are shown in Table 6.3. Lynd et al. (2002) mention, among the saccharolytic genes that have been introduced into microorganisms as *S. cerevisiae* and *Klebsiella oxytoca*, those encoding α-amylase, glucoamylase, amylopullanalase, pectate lyase, and polygalacturonase obtained from bacterial and fungal sources.

Many of the investigated recombinant strains have demonstrated the production of ethanol from starch, but in some cases, the results are not definitive. Surprisingly, reduced starch hydrolysis and fermentation rates have been observed for yeast strains expressing a set of genes previously considered as appropriate, as is the case of the work of Knox et al. (2004). This example shows the difficulties that arise during the research using recombinant microorganisms. Although the methods of genetic transformation are relatively developed, the results can be unexpected. This is a key factor when these microorganisms are evaluated from an industrial point of view. For this reason, deep studies on the effects of genetic modifications on engineered strains are required.

The processes with microorganisms modified by genetic engineering involve the optimization not only of microbial physiology parameters, but also of cell culture parameters (retention and stability of plasmids, nutritional factors, cell growth, and protein synthesis). Therefore, the modeling of these processes and the application of the principles of biochemical engineering can be helpful considering the uncertainties and complexities inherent to these biological systems (Cardona and Sánchez, 2007). An example of this type of modeling is the work of Kobayashi and Nakamura (2004), who corroborated experimentally at laboratory scale the higher productivity of the continuous fermentation process from starch using recombinant yeast cells immobilized in calcium alginate beads in comparison with the free cell system.

Another approach employed for modeling this process is the so-called flux balance analysis. Çakır et al. (2004) have employed and experimentally validated this methodology in the case of yeasts. They have determined that if the split ratio in the branch point of the glucose-6-phosphate corresponding to the glucolytic

TABLE 6.3

Some Examples of Recombinant Microorganisms with Potential Use for Fuel Ethanol Production

Host	Expressed Enzymes	Gene Donor	Feedstock/Medium	Remarks	References
Saccharomyces cerevisiae YPG/AB; *S. cerevisiae* YPG/G	α-amylase Glucoamylase	*Bacillus subtilis* *Aspergillus awamori*	Starch-containing medium	Batch and fed-batch cultures; productivity 0.27–0.35 g/(L.h); EtOH conc. 47.5 g/L (fed-batch); yield 0.46 g/g starch	Çakır et al. (2004); Altıntaş et al. (2002); Ülgen et al. (2002)
S. cerevisiae YF207/pGA11/pAA12	α-amylase Glucoamylase	*B. stearothermophilus* *Rhizopus oryzae*	Low-temperature-cooked cornstarch	Batch culture; 36 h cultivation; EtOH conc. 18–30 g/L	Shigechi et al. (2004)
S. cerevisiae Σ1278b	α-amylase Glucoamylase	*Lipomyces kononenkoae* *Saccharomycopsis fibuligera*	Starch-containing medium	Batch culture; 120 h cultivation; EtOH conc. 21 g/L; yield 0.4 g/g starch; *L. kononenkoae* genes included promoter and terminator of *S. cerevisiae* PGK1	Knox et al. (2004)
S. cerevisiae SR96	Glucoamylase	*S. diastaticus*	Starch-containing medium	Immobilized cells in calcium alginate; EtOH conc. 7.2 g/L; 200 h cultivation	Kobayashi and Nakamura (2004)
Klebsiella oxytoca P2(pC46)	α-amylase Pullulanase	*B. stearothermophilus* *Thermoanaerobium brockii*	Starch-containing medium	EtOH conc. 15 g/L; yield 0.36–0.40 g/g starch	dos Santos et al. (1999)

Saccharomyces sp. 1400(pLNH32); S. cerevisiae CEN. PK113-7D	Xylose reductase Xylitol dehydrogenase Xylulokinase	Pichia stipitis P. stipitis S. cerevisiae	Glucose and xylose containing medium; acid pretreated starch industry effluents	Saccharomyces sp. strain is the product of fusion between S. diastaticus and S. uvarum and is capable of fermenting xylose to ethanol and utilizing it for aerobic growth; batch cultures; 48 h (sugars medium), 8 d (starch effluents) cultivation; EtOH conc. 60 g/L; yield 84%	Ho et al. (1998); Zaldivar et al. (2005)
Zymomonas mobilis ZM4(pZB5)	Xylose isomerase Xylulokinase Transketolase Transaldolase	Escherichia coli E. coli E. coli E. coli	Glucose and xylose containing medium	Introduced genes codify xylose assimilation and metabolism; batch culture; 48 h cultivation; EtOH conc. 62 g/L	Leksawasdi et al. (2001)
E. coli ATCC 9637, ATCC 11303, ATCC 15244	Pyruvate decarboxylase Alcohol dehydrogenase	Z. mobilis Z. mobilis	Glucose or xylose containing medium	Hexose and pentose-assimilating E. coli acquires the ability for producing ethanol by introducing ethanol pathway genes; productivity 1.4 g/(L.h) from glucose, 0.64 g/(L.h) from xylose; EtOH conc. 41 g/L from xylose, 57 g/L from glucose; patented strain	Ingram et al. (1991)

Continued

TABLE 6.3 (Continued)

Some Examples of Recombinant Microorganisms with Potential Use for Fuel Ethanol Production

Host	Expressed Enzymes	Gene Donor	Feedstock/Medium	Remarks	References
E. coli FBR3	Pyruvate decarboxylase Alcohol dehydrogenase	Z. mobilis Z. mobilis	Mixed sugars (glucose, xylose, arabinose)	Batch process; 72 h cultivation; EtOH conc. 44 g/L; yield 0.46–0.47 g/g; productivity 1.14 g/(L.h)	Dien et al. (1998)
Klebsiella oxytoca M5A1	Pyruvate decarboxylase Alcohol dehydrogenase	Z. mobilis Z. mobilis	Crystalline cellulose	Native strain can uptake cellobiose and cellotriose; introduced genes codify ethanol pathway; batch SSF with reduced dosage of cellulases; EtOH conc. 47 g/L; yield 0.47 g/g cellulose	Ingram and Doran (1995)
Klebsiella oxytoca M5A1	Pyruvate decarboxylase Alcohol dehydrogenase 2 endoglucanases	Z. mobilis Z. mobilis Erwinia chrysanthemi	Amorphous cellulose	Strain produced over 20,000 U/L endoglucanase and was able to ferment cellulose to ethanol; yield 58–76%; no addition of exogenous cellulases; potential strain for DMC of cellulose	Zhou and Ingram (2001)

S. cerevisiae L2612δGC	Endo/exoglucanase β-glucosidase	Bacillus sp. strain DO4 B. circulans	Crystalline cellulose	Batch SSF; 12 h cultivation; saccharolytic yeast capable of enzyme production; EtOH conc. 20.4 g/L; a considerable amount of commercial enzymes was reduced; potential strain for DMC of cellulose	Cho and Yoo (1999)
Clostridium cellalolyticum	Pyruvate decarboxylase Alcohol dehydrogenase	Z. mobilis	Cellulose	Native cellulolytic strategy	Guedon et al. (2002)
S. cerevisiae	Cellobiohydrolase	Thermoascus aurantiacus	Crystalline cellulose	Recombinant cellulytic strategy	Hong et al. (2003)

metabolic pathway is changed by genetic manipulations, ethanol yield from starch can be considerably affected. This shows that the improvement in ethanol production goes with a rational design of metabolic pathways. In the future, this information could be crucial when different bioprocesses are designed, although it is necessary to analyze the costs and the complexity for acquiring this information in comparison with other "more traditional" procedures of design and control.

Similarly, important parameters, such as the stability of the plasmids used for introducing the desired traits to yeast cells, depend on the definition of the best environmental conditions during the cultivation of recombinant microorganisms. For example, Mete Altıntaş et al. (2002) showed that culture media containing specific salts and yeast extract drastically enhance plasmid stability during fed-batch cultures of yeasts using starch as feedstock.

6.3.2.2 Recombinant Microorganisms for Processing of Lignocellulosic Biomass

In the case of ethanol production from lignocellulosic biomass, the main obstacle to be overcome is the fact that fermenting microorganisms are not able to assimilate all the sugars released during pretreatment and hydrolysis of biomass in an effective way. Genetic engineering has been contributing to the development of microorganisms exhibiting this feature. To face this challenge, there exist two approaches (Aristodou and Penttilä, 2000; Chotani et al., 2000; Zaldivar et al., 2001). The first one consists of modifying microorganisms in such a way that they can assimilate a wide spectrum of substrates, e.g., by introducing the metabolic pathways required for the utilization of xylose or arabinose to good ethanologenic microorganisms such as yeasts and *Z. mobilis*. The second approach is based on the modification of microorganisms that assimilate a great variety of substrates; in this case, genes encoding the conversion of pyruvate into ethanol are introduced to microorganisms, like *Escherichia coli,* capable of assimilating hexoses and pentoses. Some examples illustrating these two approaches are shown in Table 6.3.

Through clonation of genes encoding xylose reductase and xylitol dehydrogenase in *S. cerevisiae*, the conversion of xylose into xylulose via xylitol can be achieved. The xylulose is a pentose, which can be assimilated by the yeasts. Unfortunately, the productivity is low and xylitol is formed as a by-product, which deviates part of the substrate that could be utilized for ethanol synthesis (Claassen et al., 1999). Ingram and Doran (1995) report the development of recombinant strains of gram-negative bacteria (*E. coli, K. oxytoca,* or *Erwinia* sp.) in whose chromosomes have been inserted into the genes of *Z. mobilis* encoding the metabolic pathway for ethanol production. In this way, these strains can efficiently convert all the sugars released during the hydrolysis of cellulose and hemicellulose. The saccharification of cellulose is more complicated, although one of the obtained recombinant strains of *K. oxytoca* has the natural ability for degrading cellobiose that employs a lower dosage of added cellulases for the effective conversion of purified cellulose into ethanol by a simultaneous hydrolysis and fermentation process. In their documented review on the fermentation

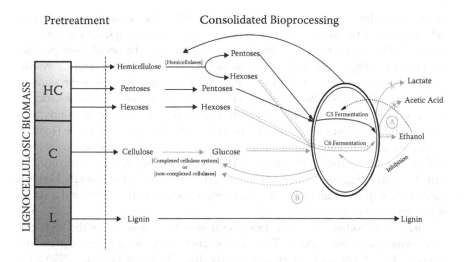

FIGURE 6.5 Conceptual scheme of consolidated bioprocessing for bioethanol production from lignocellulosic biomass. (A) Native cellulolytic strategy. (B) Recombinant cellulolytic strategy. HC = hemicellulose, C = cellulose, L = lignin. Processing pathway for a thermoanaerobic microorganism (e.g., *C. thermocellum*) is indicated by continuous gray lines. Processing pathway for an ethanologenic microorganism (e.g., *Z. mobilis*) is indicated by dotted lines. (From Cardona, C.A., and Ó.J. Sánchez. 2007. *Bioresource Technology* 98:2415–2457. Elsevier Ltd. With permission.)

of lignocellulosic hydrolyzates, Olsson and Hahn-Hägerdal (1996) discuss the performance of different pentose-utilizing microorganisms, among them several recombinant strains. Fermentation indexes for lignocellulosic hydrolyzates pretreated and detoxified by different methods are also provided.

Lynd et al. (2002) point out that the feasibility of a CBP (see Chapter 9, Section 9.2.4) for lignocellulosic ethanol process will be established when a microorganism or microbial consortium can be developed according to one of two strategies. The first of them, called native cellulolytic strategy (A in Figure 6.5), is oriented to engineer microorganisms having a high native cellulolytic activity in order to improve the ethanol production through the increase of their yield or tolerance, i.e., by the improvement of the fermentative properties of a good producer of cellulases. In this case, the modifications in the process microorganism should be aimed at reducing or eliminating the production of by-products, such as acetic acid or lactate, and at increasing the ethanol tolerance and titres (Cardona and Sánchez, 2007). Recent studies have demonstrated the possibility of obtaining ethanol-tolerant strains of *C. thermocellum* growing at ethanol concentrations exceeding 60 g/L, a titer not sufficient to put thermophiles at a disadvantage relative to more conventional ethanol producers (Lynd et al., 2005).

The recombinant cellulolytic strategy (B in Figure 6.5) contemplates the genetic modification of microorganisms that present high ethanol yields and tolerances in such a way that they are capable of utilizing cellulose within a CBP configuration, i.e., making a microorganism with good fermentative properties to produce cellulases (Cardona and Sánchez, 2007). The second strategy is considered more difficult due to the complexity of cellulases system (Béguin and Aubert, 1994). Currently, the production of cellulases by bacterial hosts producing ethanol at high yield such as engineered *E. coli, K. oxytoca,* and *Z. mobilis* and by the yeast *S. cerevisiae* has been studied. For instance, the expression of cellulases in *K. oxytoca* has allowed an increased hydrolysis yield for microcrystalline cellulose and an anaerobic growth on amorphous cellulose. Similarly, several cellobiohydrolases have been functionally expressed in *S. cerevisiae* (Lynd et al., 2005). Undoubtedly, ongoing research on genetic and metabolic engineering will make possible the development of effective and stable strains of microorganisms for converting cellulosic biomass into ethanol. This fact will surely lead to a qualitative improvement in the industrial production of fuel ethanol in the future (Cardona and Sánchez, 2007).

One of the bottlenecks of ethanol production from biomass is the high cost of enzymes, as noted above. The current cost of producing lignocellulolytic enzymes by submerged fermentation mainly using *T. reesei,* is about US$0.40 to 0.60/gal ethanol, but an increase in the specific activity of the enzymes or in the efficiency of their production through genetic engineering can be expected. This would allow an eventual cost reduction of the enzymes to US$0.07/gal ethanol, as suggested by (suggested by Tengerdy and Szakacs, 2003). However, these authors consider that the genetic improvement of fungi producing these enzymes by solid-state fermentation could have a greater potential than the genetic improvement of fungi synthesizing the same enzymes by submerged fermentation, considering the fact that fungi growing on the surface of biomass have enzymatic complexes with optimal characteristics and proportions for the hydrolysis of lignocellulosic materials. On the other hand, the U.S. Department of Energy has contracted with the world enzyme-producing leaders, Novozymes (Denmark) and Genencor (USA), with the aim of developing research for improving cellulase systems and reducing their costs. The research must be oriented not only to the enhancement of yield, stability, and specific activity of cellulases, but also to the development of an enzyme mixture that will remain active under hard conditions related to such steps as acid pretreatment (Mielenz, 2001).

Undoubtedly, if the worldwide use of fuel ethanol uses the development of the technology of lignocellulosic biomass utilization, genetic engineering will be called to supply the "tailored" microorganisms needed to meet the exigencies of this new technology. From this, the importance of the metabolic pathway engineering is inferred. Metabolic pathway engineering is aimed at establishing metabolic pathways and production hosts, which are capable of delivering optimal flow of carbon from substrate to final product at high yields and volumetric productivities. In particular, pathway engineering can achieve the integration of the process at the molecular level through the optimization of the primary metabolic pathways for the synthesis of the targeted product (Chotani et al., 2000).

REFERENCES

Altıntaş, M.M., K.Ö. Ülgen, B. Kırdar, Z.I. Önsan, and S.G. Oliver. 2002. Improvement of ethanol production from starch by recombinant yeast through manipulation of environmental factors. *Enzyme and Microbial Technology* 31:640–647.

Aristodou, A., and M. Penttilä. 2000. Metabolic engineering applications to renewable resource utilization. *Current Opinion in Biotechnology* 11:187–198.

Ballesteros, I., J.M. Oliva, F. Sáez, and M. Ballesteros. 2001. Ethanol production from lignocellulosic byproducts of olive oil extraction. *Applied Biochemistry and Biotechnology* 91-93:237–252.

Ballesteros, M., J.M. Oliva, M.J. Negro, P. Manzanares, and I. Ballesteros. 2004. Ethanol from lignocellulosic materials by a simultaneous saccharification and fermentation process (SSF) with *Kluyveromyces marxianus CECT 10875*. *Process Biochemistry* 39:1843–1848.

Baskaran, S., H.-J. Ahn, and L.R. Lynd. 1995. Investigation of the ethanol tolerance of *Clostridium thermosaccharolyticum* in continuous culture. *Biotechnology Progress* 11:276–281.

Béguin, P., and J.P. Aubert. 1994. The biological degradation of cellulose. *FEMS Microbiology Reviews* 13:25–58.

Berg, C. 2001. *World Fuel Ethanol. Analysis and Outlook*. F.O. Licht. http://www.agra-europe.co.uk/FOLstudies /FOL-Spec04.html (accessed March 2004).

Bullock, G.E. 2002. *Ethanol from sugarcane*, Sugar Research Institute, Brisbane, Australia.

Çakır, T., K.Y. Arga, M.M. Altıntaş, and K.Ö. Ülgen. 2004. Flux analysis of recombinant *Saccharomyces cerevisiae* YPB-G utilizing starch for optimal ethanol production. *Process Biochemistry* 39 (12):2097–2108.

Cardona, C.A., and Ó.J. Sánchez. 2007. Fuel ethanol production: Process design trends and integration opportunities. *Bioresource Technology* 98:2415–2457.

Carrascosa, A.V. 2006. Obtención de etanol en condiciones de alta presión osmótica mediante *Schizosaccharomyces pombe* (CECT 12775). Spain Patent ES2257206.

Cho, K.M., and Y.J. Yoo. 1999. Novel SSF process for ethanol production from microcrystalline cellulose using the δ-integrated recombinant yeast, *Saccharomyces cerevisiae* L2612δGC. *Journal of Microbial Biotechnology* 9 (3):340–345.

Chotani, G., T. Dodge, A. Hsu, M. Kumar, R. LaDuca, D. Trimbur, W. Weyler, and K. Sanford. 2000. The commercial production of chemicals using pathway engineering. *Biochimica et Biophysica Acta* 1543:434–455.

Claassen, P.A.M., J.B. van Lier, A.M. López Contreras, E.W.J. van Niel, L. Sijtsma, A.J.M. Stams, S.S. de Vries, and R.A. Weusthuis. 1999. Utilisation of biomass for the supply of energy carriers. *Applied Microbiology and Biotechnology* 52:741–755.

Crueger, W., and A. Crueger. 1993. *Biotecnología. Manual de microbiología industrial* (*Biotechnology. Handbook of industrial microbiology*, in Spanish). Zaragoza, Spain: Acribia.

Dien, B.S., R.B. Hespell, H.A. Wyckoff, and R.J. Bothast. 1998. Fermentation of hexose and pentose sugars using a novel ethanologenic *Escherichia coli* strain. *Enzyme and Microbial Technology* 23:366–371.

Doelle, M.B., and W. Doelle. 1989. Ethanol production from sugar cane syrup using *Zymomonas mobilis*. *Journal of Biotechnology* 11:25–36.

dos Santos, V.L., E.F. Araújo, E.G. de Barros, and W.V. Guimarães. 1999. Fermentation of starch by *Klebsiella oxytoca* P2, containing plasmids with α-amylase and pullulanase genes. *Biotechnology and Bioengineering* 65:673–676.

Gong, C.S., N.J. Cao, J. Du, and G.T. Tsao. 1999. Ethanol production from renewable resources. *Advances in Biochemical Engineering/Biotechnology* 65:207–241.

Goyes, A., and G. Bolaños. 2005. Un estudio preliminar sobre el tratamiento de vinazas en agua supercrítica (A preliminary study on stillage treatment in supercritical water, in Spanish). Paper presented at the XXIII Congreso Colombiano de Ingeniería Química, Manizales, Colombia.

Grote, W., and P.L. Rogers. 1985. Ethanol production from sucrose-based raw materials using immobilized cells of *Zymomonas mobilis*. *Biomass* 8:169–184.

Guedon, E., M. Desvaux, and H. Petitdemange. 2002. Improvement of cellulolytic properties of *Clostridium cellulolyticum* by metabolic engineering. *Applied and Environmental Microbiology* 68 (1):53–58.

Hawgood, N., S. Evans, and P.F. Greenfield. 1985. Enhanced ethanol production in multiple batch fermentations with an auto-flocculating yeast strain. *Biomass* 7:261–278.

Ho, N.W.Y., Z. Chen, and A.P. Brainard. 1998. Genetically engineered *Saccharomyces* yeast capable of effective co-fermentation of glucose and xylose. *Applied and Environmental Microbiology* 64 (5):1852–1859.

Hong, J., H. Tamaki, K. Yamamoto, and H. Kumagai. 2003. Cloning of a gene encoding thermostable cellobiohydrolase from *Thermoascus aurantiacus* and its expression in yeast. *Applied Microbiology and Biotechnology* 63 (42–50).

Ingram, L.O., T. Conway, and F. Alterthum. 1991. Ethanol production by *Escherichia coli* strains co-expressing *Zymomonas* PDC and ADH genes. U.S. Patent US5000000.

Ingram, L.O. , and J.B. Doran. 1995. Conversion of cellulosic materials to ethanol. *FEMS Microbiology Reviews* 16:235–241.

Jeffries, T.W. 2005. Ethanol fermentation on the move. *Nature Biotechnology* 23 (1):40–41.

Jeffries, T.W., and Y.-S. Jin. 2000. Ethanol and thermotolerance in the bioconversion of xylose by yeasts. *Advances in Applied Microbiology* 47:221–268.

Kádár, Zs., Zs. Szengyel, and K. Réczey. 2004. Simultaneous saccharification and fermentation (SSF) of industrial wastes for the production of ethanol. *Industrial Crops and Products* 20:103–110.

Kargupta, K., S. Datta, and S.K. Sanyal. 1998. Analysis of the performance of a continuous membrane bioreactor with cell recycling during ethanol fermentation. *Biochemical Engineering Journal* 1:31–37.

Knox, A.M., J.C. du Preez, and S.G. Kilian. 2004. Starch fermentation characteristics of *Saccharomyces cerevisiae* strains transformed with amylase genes from *Lipomyces kononenkoae* and *Saccharomycopsis fibuligera*. *Enzyme and Microbial Technology* 34:453–460.

Kobayashi, F., and Y. Nakamura. 2004. Mathematical model of direct ethanol production from starch in immobilized recombinant yeast culture. *Biochemical Engineering Journal* 21:93–101.

Lee, W.-C., and C.-T. Huang. 2000. Modeling of ethanol fermentation using *Zymomonas mobilis* ATCC 10988 grown on the media containing glucose and fructose. *Biochemical Engineering Journal* 4:217–227.

Leksawasdi, N., E.L. Joachimsthal, and P.L. Rogers. 2001. Mathematical modeling of ethanol production from glucose/xylose mixtures by recombinant *Zymomonas mobilis*. *Biotechnology Letters* 23:1087–1093.

Lynd, L.R. , W.H. van Zyl, J.E. McBride, and M. Laser. 2005. Consolidated bioprocessing of cellulosic biomass: An update. *Current Opinion in Biotechnology* 16:577–583.

Lynd, L.R., P.J. Weimer, W.H. van Zyl, and I.S. Pretorious. 2002. Microbial cellulose utilization: Fundamentals and biotechnology. *Microbiology and Molecular Biology Reviews* 66 (3):506–577.

McMillan, J.D. 1997. Bioethanol production: Status and prospects. *Renewable Energy* 10 (2/3):295–302.

Mete Altıntaş, M. , K.Ö. Ülgen, B. Kırdar, Z.I. Önsan, and S.G. Oliver. 2002. Improvement of ethanol production from starch by recombinant yeast through manipulation of environmental factors. *Enzyme and Microbial Technology* 31:640–647.

Mielenz, J.R. 2001. Ethanol production from biomass: Technology and commercialization status. *Current Opinion in Microbiology* 4:324–329.

Olsson, L., and B. Hahn-Hägerdal. 1996. Fermentation of lignocellulosic hydrolysates for ethanol production. *Enzyme and Microbial Technology* 18:312–331.

Ryabova, O.B., O.M. Chmil, and A.A. Sibirny. 2003. Xylose and cellobiose fermentation to ethanol by the thermotolerant methylotrophic yeast *Hansenula polymorpha*. *FEMS Yeast Research* 4:157–164.

Shigechi, H., Y. Fujita, J. Koh, M. Ueda, H. Fukuda, and A. Kondo. 2004. Energy-saving direct ethanol production from low-temperature-cooked cornstarch using a cell-surface engineered yeast strain co-displaying glucoamylase and α-amylase. *Biochemical Engineering Journal* 18:149–153.

South, C.R., D.A. Hogsett, and L.R. Lynd. 1993. Continuous fermentation of cellulosic biomass to ethanol. *Applied Biochemistry and Biotechnology* 39/40:587–600.

Szczodrak, J., and J. Fiedurek. 1996. Technology for conversion of lignocellulosic biomass to ethanol. *Biomass and Bioenergy* 10 (5/6):367–375.

Tengerdy, R.P., and G. Szakacs. 2003. Bioconversion of lignocellulose in solid substrate fermentation. *Biochemical Engineering Journal* 13:169–179.

Ülgen, K.Ö., B. Saygılı, Z.İ. Önsan, and B. Kırdar. 2002. Bioconversion of starch into ethanol by a recombinant *Saccharomyces cerevisiae* strain YPG-AB. *Process Biochemistry* 37:1157–1168.

Wyman, C.E. 1994. Ethanol from lignocellulosic biomass: Technology, economics, and opportunities. *Bioresource Technology* 50:3–16.

Zaldivar, J, C. Roca, C. Le Foll, B. Hahn-Hägerdal, and L. Olsson. 2005. Ethanolic fermentation of acid pre-treated starch industry effluents by recombinant *Saccharomyces cerevisiae* strains. *Bioresource Technology* 96:1670–1676.

Zaldivar, J., J. Nielsen, and L. Olsson. 2001. Fuel ethanol production from lignocellulose: A challenge for metabolic engineering and process integration. *Applied Microbiology and Biotechnology* 56 (1–2):17–34.

Zhou, S., and L.O. Ingram. 2001. Simultaneous saccharification and fermentation of amorphous cellulose to ethanol by recombinant *Klebsiella oxytoca* SZ21 without supplemental cellulase. *Biotechnology Letters* 23 (18):1455–1462.

7 Ethanolic Fermentation Technologies

In this chapter, main fermentation technologies for ethanol production are discussed. The analysis of these technologies is oriented to the description of the different ways by which different sugar solutions resulting from conditioned and pretreated feedstocks are converted into ethyl alcohol. The different regimes for ethanologenic fermentation are reviewed. The role of mathematical modeling of fermentation during the design of ethanol production processes is highlighted and related illustrative examples are presented. Finally, some simulation case studies are disclosed in the framework of process synthesis procedures for fuel ethanol production.

7.1 DESCRIPTION OF MAIN FERMENTATION TECHNOLOGIES FOR ETHANOL PRODUCTION

The fermentation step is central in the overall fuel ethanol production process since it represents the actual transformation of the conditioned and pretreated raw materials into the main product, ethyl alcohol, using bioagents such as yeast or other ethanol-producing microorganism. Ethanolic fermentation is one of the most studied biological processes. Nevertheless, the need of increasing the efficiency of ethanol production including the usage of alternative feedstocks has led to the development of new fermentation methods with better technoeconomic and environmental indicators.

Traditionally, the most used microorganism for ethanolic fermentation is the yeast *Saccharomyces cerevisiae*. This is valid for practically every one of the main types of feedstocks employed for ethanol production: sucrose-based media, starchy materials, and even lignocellulosic materials. However, in the last case, there is a wider variety of process microorganisms employed (e.g., *Zymomonas* bacteria, xylose-assimilating yeasts, or thermophilic clostridia).

7.1.1 FEATURES OF ETHANOLIC FERMENTATION USING *SACCHAROMYCES CEREVISIAE*

As mentioned in the Chapter 6, *S. cerevisiae* converts hexoses into pyruvate through glycolysis, which is decarboxylated to obtain acetaldehyde that is finally reduced to ethanol generating two moles of adenosine triphosphate (ATP) for each mol of consumed hexose under anaerobic conditions. In addition, this microorganism also has the ability to convert hexoses into CO_2 by aerobic respiration favoring the production of yeast cells. Therefore, aeration is an important factor for both cell growth and ethanol production. Although these yeasts have the ability to grow under anaerobic conditions, small amounts of oxygen are needed for synthesis of such substances as fatty acids and sterols. In the case of continuous cultures, cell concentration, cell yield from glucose, and yeast viability are enhanced by increasing air supply while decreasing ethanol concentration under both microaerobic and aerobic conditions. Inhibition of cell growth by ethanol decreases at microaerobic conditions related to fully anaerobic cultivation. Specific ethanol productivity is stimulated with the increase of oxygen percentage in the feed (see, for example, the work of Alfenore et al., 2004). In the case of fed-batch cultures, Alfenore et al. (2004) show that higher ethanol concentrations (147 g/L) can be obtained in cultures without oxygen limitation (0.2 vvm) in only 45 h in comparison to microaerobic conditions. In addition, a 23% increase in the viable cell mass was achieved. Similar studies for fed-batch cultures were performed for assessing the synergistic effect of temperature and ethanol content on the behavior of S. cerevisiae cultures (Aldiguier et al., 2004). Best results were found at 30°C and 33°C for around 120 g/L of ethanol produced in 30 h. Slight benefits for growth at 30°C and for ethanol production at 33°C were observed. These data suggest the possibility of designing two-stage, high-cell-density bioreactors. One proposed method to reduce the inhibition by ethanol of yeast cultures is the preadaptation to a medium with an initial content of ethanol. In this way, yeasts could be acclimatized to acquire ethanol tolerance by serially transferring them into a medium with ethanol. Vriesekoop and Pamment (2005) point out that this approach has not provided important success because which preadaptation to ethanol causes a decrease in the rate of yeast death, it does not prevent it. These authors showed that the addition of acetaldehyde to preadapted cultures of S. cerevisiae eliminates the long lag-phase of yeasts caused by the sudden exposure to initial ethanol concentrations as high as 50 g/L.

7.1.2 FERMENTATION OF SUCROSE-BASED MEDIA

The technology for fermentation of sucrose-based media, mostly sugarcane juice, cane molasses, or beet molasses, can be considered as a mature technology especially if the process is accomplished in batch regime. However, many research efforts are being made worldwide to improve the efficiency of this process, particularly to increase the conversion of these feedstocks into ethanol as well as to gauge its productivity. Ethanolic fermentation can be carried out by discontinuous, semicontinuous, and continuous processes. In general, temperature and medium pH are quite similar in the different types of ethanolic fermentation: about 30°C and pH in the range of 4.0 to 4.5. Nevertheless, each one of these cultivation regimes presents very different performance indicators, especially considering the ethanol volumetric productivity in terms of g EtOH/(L × h). Another performance indicator is ethanol yield defined as the grams of ethanol produced from one gram of substrate consumed (usually the carbon source, i.e., the sugars). Finally, substrate conversion is defined as the ratio between the amount of substrate consumed and the initial amount of substrate either loaded at the start of fermentation for batch regime or contained in the feed stream for continuous regime. This indicator can be expressed in percentage. It is worth emphasizing that substrate conversion implies that the microorganisms consume the substrate not only for ethanol biosynthesis, but also for formation of new cellular biomass and other substances generated as fermentation by-products. *S. cerevisiae* is the most employed microorganism for industrial ethanol production from sucrose-based media.

7.1.2.1 Batch Fermentation

During batch fermentation, a series of operating procedures are periodically repeated to ensure the growth and development of process microorganisms. These procedures can include the washing and disinfection of the fermenter, filling up the fermenter with the culture medium and sterilization of such medium, inoculation of microbial cells, fermentation, and unloading of the bioreactor content at the end of cultivation process. The simplest batch ethanolic fermentation comprises an initial step for yeast propagation in an aerated bioreactor (seed fermenter) where the microbial cells are multiplied in order to attain the appropriate concentration for them to be inoculated into a larger fermenter. In this fermenter, the conditions needed for anaerobic cultivation of yeasts are ensured, favoring in this way, the production of ethanol. The main drawback of this process consists in the operating and feedstock costs needed for each fermentation batch to ensure the yeast propagation until it reaches a concentration high enough to allow for the appropriate cell growth and ethanol production rates. In addition, the yeasts are not reutilized, which indicates one should not employ all the potential of cell biomass formed during the process.

The typical process for fuel ethanol production based on batch fermentation is the so-called Melle-Boinot process. In Brazil, it uses molasses or cane syrup and comprises the weight and sterilization of feedstock, followed by the adjustment of pH with H_2SO_4 and of the degrees Brix to values of 14 to 22. Yeasts ferment obtained wort. The produced wine is decanted, centrifuged, and sent to the separation stage of ethanol, whereas the yeasts are recycled to the fermentation stage in order to reach high cell concentration during cultivation (Kosaric and Velikonja, 1995; Sánchez and Cardona, 2008), as shown in Figure 7.1. A simple and effective method proposed to maintain high cell concentrations is the propagation of yeasts every 13 batches in the fermentation medium, which leads

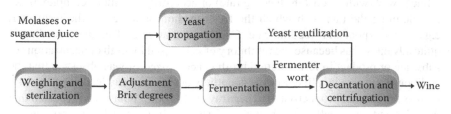

FIGURE 7.1   Schematic sequence of operations during batch industrial process for ethanol production from sugarcane.

to a productivity higher than that obtained for the conventional process with yeast recovery (Navarro et al., 1986). Shojaosadati (1996) highlights that yeast reuse results in a decrease in new growth with more sugar available for conversion to ethanol and a corresponding increase in ethanol yield of 2 to 7%. For ethanol production from beet molasses, this author obtained 8% lower consumption of the feedstock employing cell recycling. Traditional productivities reached by batch fermentation are in the range of 1 to 3 g/(L × h). However, if a significantly high concentration of yeasts (44 g/L) and a high supplementation of yeast extract (28 g/L) at low ethanol content (approximately 60 g/L) are employed in a glucose-based medium, ethanol productivity as high as 21 g/(L × h) can be achieved (Chen, 1981). This value is comparable to those obtained during continuous cultivation, as shown in Table 7.1.

The recirculation of stillage from a previous batch to the current fermentation, as schematically illustrated in Figure 7.2, has also been proposed. The stillage represents one of the distillation product streams during the subsequent ethanol recovery step that contains a significant amount of water and a much-reduced amount of ethanol. The addition of stillage to the culture broth can lead to lower water consumption and the reduction of stillage volume to be treated.

7.1.2.2 Semicontinuous Fermentation

Fed-batch fermentation is one of the most employed cultivation regimes when process microorganisms present catabolic repression, i.e., when high substrate concentrations inhibit specific metabolic processes like those related to cell growth rate. For this reason, the microorganisms grow faster at low substrate concentrations. In fact, the cultivation of *S. cerevisiae* to produce baker's yeasts is accomplished by this process. Nevertheless, the application of fed-batch cultivation to ethanolic fermentation has also offered important results by maintaining low substrate levels as ethanol is accumulated in the medium. This type of cultivation regime along with the cell recycling is the most utilized technology in Brazil for bioethanol production due to the possibility of achieving higher volumetric productivities (Sánchez and Cardona, 2008). To implement such a process, conventional batch fermentation is performed though using a less-concentrated medium. Once the sugars have been consumed, the bioreactor is fed with portions of fresh medium or by adding a small amount of medium permanently until the end of fermentation. This continuous feeding of the medium can be done in a linear way (with a constant feeding rate) or according to a more complex function defining the rate with which the fresh medium is added to the fermenter, e.g., by an exponential feeding rate. Control of flow rate of medium feeding is quite advantageous because the inhibitory effect caused by high concentrations of substrate or product in fermentation broth is neutralized. It was observed that the addition of sucrose in a linear or exponentially decreasing way leads to 10 to 14% increase in ethanol productivity (Echegaray et al., 2000). It has been reported that immobilized yeast cells have better performance in fed-batch cultures regarding ethanol production (Roukas 1996). Aeration is an important factor during this fermentation regime as well. For fed-batch cultures, Alfenore et al. (2004)

TABLE 7.1

Some Fermentation Processes for Ethanol Production from Sugarcane Molasses Using *Saccharomyces cerevisiae*

Regime	Configuration	Ethanol Conc. in Broth/g/L	Productivity/ g/(L.h)	Yield, % of Theor. Max.	References
Batch	Reuse of yeast from previous batches; yeast separation by centrifugation	80–100	1–3	85–90	Claassen et al.,1999
Fed-batch	Stirred tank with variable feeding rate (exponent. depend. with time)	53.7–98.1	9–31	73.2–89	Echegaray et al., 2000
Repeated batch	Stirred tank; flocculating yeast; up to 47 stable batches	92–106	2.5–3.5	80.8–83	Kida et al., 1991; Morimura et al., 1997
	Stirred tank; auto-flocculating yeast separated by settling at the completion of each batch; up to 22 batches; cane juice	87.7	9.1	88.2	Hawgood et al., 1985
Continuous	CSTR; cell recycling using a settler; flocculating yeast; aeration 0.05 vvm	70–80	7–8		Hojo et al., 1999
	Biostill; residence time 3–6 h; cell recycling by centrifugation; recycled stream from distillation column to fermenter	30–70	5–20	94.5	Ehnstroem, 1984; Kosaric and Velikonja 1995
	Cascade of two reactors; flocculating yeast; no cell recycling	81–84	20.3	84–97	Kida et al., 1990
	Tower reactor; flocculating yeast; cell recycling by settling	69.5–70.5	25–28		Kuriyama et al., 1993
	Fluidized bed; highly flocculant yeast; aeration 0.1 vvm	67.8–71.8	5–15	90.2	Wieczorek and Michalski, 1994
Continuous removal of EtOH	Removal by vacuum; cell recycling	50	21–123		Costa et al., 2001; Cysewski and Wilke, 1977; Maiorella et al., 1984
	Removal by membrane	40–50	20–98.4		Maiorella et al., 1984; Shabtai et al., 1991

Source: Modified from Sánchez, Ó.J., and C.A. Cardona. 2008. *Bioresource Technology* 99:5270–5295. Elsevier Ltd.

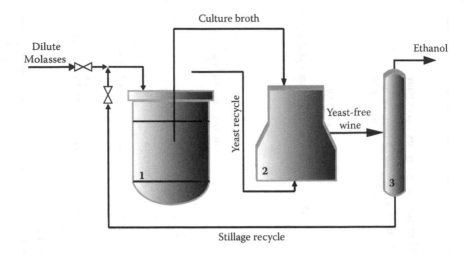

FIGURE 7.2 Reutilization of yeast cells and stillage during batch fermentation: (1) fermenter, (2) separation of cells by centrifugation, (3) distillation column.

have shown that higher ethanol concentrations (147 g/L) could be obtained during cultivation without oxygen limitation (0.2 volumes of air per volume of broth in 1 min or vvm) during only 45 h in comparison to microaerobic conditions.

In the case of multiple or repeated batch fermentation, the use of flocculating strains of yeasts is of great importance. In this type of culture, after starting a conventional batch, the yeasts are decanted in the same vessel where they were cultivated and then the clarified culture broth located in the upper zone of the fermenter is removed. Then, an equal amount of fresh culture medium is added for the following batch. In this way, high cell concentrations are reached and inhibition effect by ethanol is reduced without the need of adding flocculation aids or using separation or recirculation devices. These repeated batches can be accomplished until the moment when the activity and viability of culture are lost as a consequence of a high exposition to fermentation environment. When this occurs, the system should be reinoculated (Sánchez and Cardona, 2008). Some factors, such as agitation, allow flock size to be optimum for reaching higher ethanol concentrations. Even small levels of dissolved oxygen in the medium can facilitate the neutralization of inhibition effect by ethanol, as suggested by Hawgood et al. (1985). Maia and Nelson (1993) point out that the addition of unsaturated fatty acids can reduce or eliminate the need for microaeration because the oxygen requirement is related to the synthesis of these acids. These authors evaluated the supplementation of sucrose-based medium with fatty acids sources (soy and corn flours) in repeated-batch cultures obtaining best results with corn flour and justifying the traditional and empiric use of this component during the fermentation step by the small Brazilian distillers. Some examples of fed-batch and repeated batch fermentations for bioethanol production from sugarcane molasses can be observed in Table 7.1.

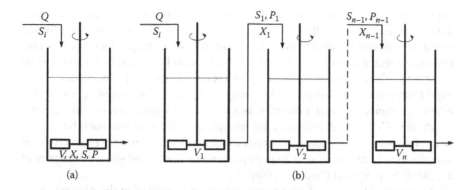

FIGURE 7.3 Schematic diagram of continuous fermentation: (a) single-stage continuous cultivation employing one CSTR, (b) multistage continuous cultivation employing several CSTR. Q = feed stream flow rate; S, X, P = concentrations of substrate, cell biomass, and product, respectively; V = volume of culture broth in fermentation vessel. Subindexes: i, feed stream (influent); $1,2,...n-1,n$, stages of the cascade scheme.

7.1.2.3 Continuous Fermentation

Continuous fermentation consists of the cultivation of cells in a bioreactor to which the fresh medium is permanently added and from which an effluent stream of culture broth is permanently removed, as shown in Figure 7.3. The microorganisms are reproduced within the bioreactor at a grow rate that offsets the cells withdrawal with the effluent achieving the corresponding steady state. To ensure the system homogeneity and reduce concentration gradients in culture broth, continuous stirred-tank reactors (CSTR) are employed. In this way, a constant production of fermented wort can be obtained without the need of stopping the bioreactor operation in order to perform the periodic procedures typical of batch processes, such as filling-up and unloading. This allows a remarkable increase in volumetric productivity compared to discontinuous or semicontinuous processes (see Table 7.1).

The design and development of continuous fermentation systems have allowed the implementation of more effective processes from the viewpoint of the production costs. Continuous processes have a series of advantages in comparison to conventional batch processes mainly due to reduced construction costs of the bioreactors, lower requirements of maintenance and operation, better control of the process, and higher productivities (Sánchez and Cardona, 2008). For those very reasons, 30% of ethanol production facilities in Brazil utilize continuous fermentation processes (Monte Alegre et al., 2003). Most of these advantages are due to the high cell concentration found in this type of bioreactor. Such high densities can be reached by immobilization techniques, recovery and recycling of biomass, or control of cell growth. The major drawback is that cultivation of yeasts during a long time under anaerobic conditions diminishes their ability to produce ethanol. In addition, at high dilution rates (a magnitude proportional to the feed or effluent flow rate) ensuring elevated productivities, the substrate is not completely consumed and yields are reduced. In general, in commercial processes for ethanol production, although the productivity is important, it is more relevant for the substrate conversion considering that the main part of the production costs correspond to feedstocks (Gil et al., 1991). On the other hand, aeration also plays an important role during continuous ethanolic cultivations. Cell concentration, cell yield from glucose, and yeast viability may be enhanced by increasing air supply, whereas ethanol concentration decreases under both microaerobic and aerobic conditions. Cell growth inhibition by ethanol is reduced at microaerobic conditions compared to fully anaerobic cultivation, and specific ethanol productivity is stimulated with the increase of oxygen percentage in the feed stream (Alfenore et al., 2004; Sánchez and Cardona, 2008).

An important feature of continuous processes is related to the diminution of the product inhibition effect. Through cascade of continuous reactors, ethanol obtained in the first reactors is easily transported to the next ones reducing in this way its inhibitory effect (see Figure 7.3b). On the other hand, other configurations employing one fermenter can contribute to the reduction of product inhibition. In particular, the Swedish company Alfa Laval implemented a continuous process for producing 150,000 L EtOH/d in Brazil by Biostill technology (Kosaric and Velikonja, 1995). This process is based on yeast cultivation carried out in a fermentation vessel from which a liquid stream is continuously withdrawn to be sent to a centrifuge. From the centrifuge, one concentrated yeast stream is continuously removed and recycled back to the fermenter. The other yeast-free stream is directed to a distillation tower. From this tower, concentrated solution of ethanol and stillage are removed. A portion of the stillage is also recycled back to the fermenter in order to maintain the mass balance necessary for conserving steady-state conditions according to the original configuration patented by Alfa Laval (Ehnstroem, 1984), which is shown in Figure 7.4. In this process, there is significant savings in process water, which reduces stillage volumes and low residence times (3 to 6 h) in the fermenter can be achieved. A modification to this process using no recirculation stream from a distillation column and reaching yields of 96% of theoretical has been patented as well (Da Silva and Vaz, 1989).

Other alternatives of continuous fermentation have been proposed, but many of them still have not reached the commercial level. Some of them require the use of highly flocculating yeast strains similar to the tower and fluidized-bed reactors. These types of reactors allow much higher cell concentrations (70 to 100 g/L) and ethanol productivities, and have a long-term stability due to the self-replenishing of fresh yeasts. Moreover, these fermenters do not require stirring devices or centrifugation (Gong et al., 1999). *S. uvarum* is one of the most promising yeasts to be employed in these configurations thanks to its flocculating properties. All of these efforts have been directed to the increase of productivity and yield, as can be seen in Table 7.1. Another approach for increasing process productivity is the continuous ethanol removal from culture broth during fermentation by means of a vacuum or membranes. These configurations enhance efficiency of the process remarkably well, but imply an increase in capital costs. The use of vacuum flash coupled with continuous fermenters could eliminate the

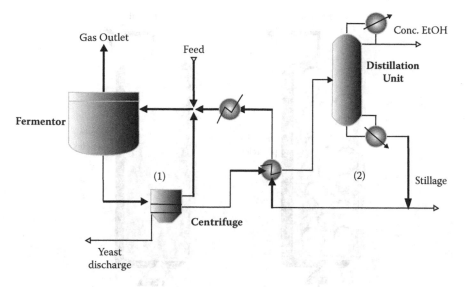

FIGURE 7.4 Biostill process for continuous ethanol production: (1) yeast recycling loop, (2) stillage recycling loop.

need of heat exchangers and increase the productivity (Costa et al., 2001). These types of configurations involving the application of reaction–separation integration are discussed in Chapter 9.

7.1.2.4 Fermentation of Sugar Solutions Using Immobilized Cells

One of the strategies employed to improve the ethanolic fermentation is the utilization of immobilized cells. Cell immobilization consists of the attachment of cells into a support or a location in a defined space in order to utilize, in a controlled way, their capacity to accomplish biological transformation. Thus, the cells do not leave the bioreactor, and continuous fermentation processes can be implemented. In this case, the substrates contained in the feed stream are transformed into products in the biocatalyst (cells + support) bed. These products abandon the system in the cell-free effluent stream. This leads to an easier product recovery as well as avoiding the risk of cell washout. A better control of the fermentation process is achieved compared to suspended cell cultivation for which microbial cells are continuously removed from the fermenter. On the other hand, the biocatalyst can be readily recovered if the process is carried out in batch regime. All of these advantages make reactors with immobilized cells exhibit higher productivities allowing the utilization of smaller bioreactors (lower capital costs).

However, employing immobilized cells implies that they do not reproduce during reactor operation. The growth means that cell layers are accumulated on the support surface until the moment they start to deattach from the solid phase leading to the system destabilization. To avoid this, the necessary conditions for the cells not to grow (nonviable cells) are ensured. In addition, if aeration is needed, a constant air supply has to be available, which can be difficult when a fixed bed

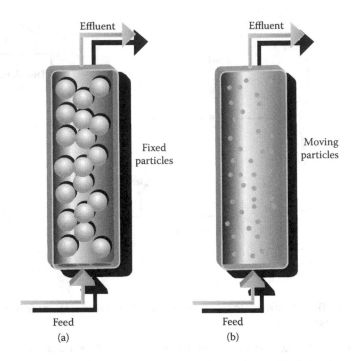

FIGURE 7.5 Most employed configurations for bioreactors with immobilized cells: (a) fixed-bed reactor, (b) fluidized-bed reactor.

of biocatalysts is used (Figure 7.5a). For this reason, an auxiliary tank is used to supply air to part of the effluent stream, which is then recirculated to the reactor to ensure the aerobic conditions of the culture broth within the bed. One alternative configuration is the fluidized-bed reactor where the liquid feed stream flows up inside the reactor containing mobile biocatalyst particles. In this way, the bed is expanded, as shown in Figure 7.5b. If aerobic conditions are required, the air can be directly injected into the bioreactor. Despite these advantages, processes using immobilized cells are not widespread in industrial microbiology today due to the complexity of the systems involved.

In the case of ethanolic fermentation, the implementation of continuous cultivation with immobilized cells can make possible processes with higher yields, greater productivities, and increased cell concentrations at the same time (Claassen et al., 1999), as presented in Table 7.2. Nevertheless, ethanol concentrations in the effluent tend to be lower than in other variants of continuous processes (see Table 7.1). Microbial cells for ethanol production are immobilized by entrapping within them porous, solid supports, such as calcium or sodium alginate, carrageenan or polyacrylamide. In addition, they can be adsorbed on the surface of materials, such as wood chips, bricks, synthetic polymers, or other materials with a large surface area (Gong et al., 1999). It is remarkable that support particles have influence on cellular metabolism, as has been shown in the case of solid-state fermentation, biofilm reactors, and immobilized cell reactors. Prakasham et al.

TABLE 7.2

Some Continuous Processes for Bioethanol Production from Sugarcane and Related Media Using Immobilized Cells

Microorganism	Carrier	Medium	Ethanol Conc. in Effluent/g/L	Productivity, in g/(L.h)	Yield, % of Theor. Max.	References
Saccharomyces cerevisiae	Sodium alginate and zeolitic base	Cane molasses	54.48	1.835	88.2	Caicedo et al., 2003
	Chrysotile	Cane syrup	25–75	16–25	80.4–97.3	Wendhausen et al., 2001
		Cane molasses		3.5–10		Monte Alegre et al., 2003
	Rice straw	Glucose[a]	45.8	17.84	93	Das et al., 1993
	Bagasse	Glucose	45	15.50	93	
	Alumino-silicate composite	Sucrose[b]	82.4–103.3	10.3–20.6	98–99	Gil et al., 1991
		Molasses[c]	96.9		98	
	Calcium alginate	Sucrose	50.6–60.0	10.2–12.1	66–79	Sheoran et al., 1998
		Molasses[d]	47.4–55.3	7.3–10.4	62–74	
		Glucose	30.6–41.0	2.98[e]	83.1	Gilson and Thomas, 1995
		Sucrose	41.4–69.2	15.7–31.5	87.4	Melzoch et al., 1994
S. carlbergensis	Calcium alginate[f]	Glucose	66.8–93.3	14.9–17.41	89	Tzeng et al., 1991
S. uvarum	Calcium alginate	Cane molasses	25–76	7.6–12.5		Grote and Rogers, 1985
Z. mobilis	Calcium alginate	Cane syrup	40–55	5–25	53–80	Grote and Rogers, 1985
		Cane molasses	63	6.3–12.5		

[a] 120 g/L of reducing sugars.
[b] 200 g/L + nutrients.
[c] High test molasses supplemented with ammonium sulfate.
[d] Higher values correspond to acid treated and clarified molasses.
[e] Measured in g EtOH/(10^{11} cells.h).
[f] Multistage fluidized-bed reactor.

(1999) claim that the simple addition of a small fraction of solids in submerged cultures facilitates cell anchorage. This kind of adhesion enhances the metabolic activity and is an easier and more economical method than the immobilization of cells. In batch flasks cultures, these authors showed that materials, such as river sand, delignified sawdust, chitin, and chitosan, make possible the adhesion of *S. cerevisiae* cells leading to higher ethanol production in comparison to free cells. Thus, the application of these techniques of "passive immobilization" to continuous cultures should be experimentally tested. Nowadays, most of the configurations using immobilized cells are, so far, used in commercial operations. Hence, preliminary design and simulation of this type of process could become a very useful tool for defining new research lines at pilot and semi-industrial levels considering the overall bioethanol production process (Sánchez and Cardona, 2008).

7.1.3 FERMENTATION OF MEDIA BASED ON STARCHY MATERIALS

7.1.3.1 Conversion of Saccharified Corn Starch into Ethanol

Yeasts are not able to metabolize the starch, so this polymer should be hydrolyzed before fermentation, as was discussed in Chapter 5. Most cereal ethanol is produced from corn.

To start the fermentation, the corn mash is cooled until reaching the cultivation temperature, usually 30° to 32°C, and then the cells of the process microorganism (generally *S. cerevisiae*) are inoculated. The corn mash is obtained by dry-milling technology and contains, besides the starch, all the other components of the grain (proteins, oil, fiber, etc.) that remain unmodified during the fermentation process. As in the case of molasses, the culture medium is also supplemented with urea or even with ammonium sulfate. Recently, some proteases are being added to the mash to provide an additional nitrogen source to the yeast resulting from the hydrolysis of corn proteins (Bothast and Schlicher, 2005). In most corn dry-milling plants producing ethanol, fermentation is accomplished in batch regime though milling, liquefaction, and saccharification as well as the subsequent distillation and dehydration, which are performed in a continuous regime. To ensure a continuous flow of materials, three or more fermenters are used in such a way that at any given moment one apparatus is an filling-up with the mash, the starch is fermented in another, and a third one is being unloaded and prepared for the next batch. Fermentation time is about 48 h. At the end of fermentation, the culture broth (fermented mash or beer) is sent to a storage tank from which the constant supply of beer to the distillation step is guaranteed.

In the case of ethanol production plants employing the corn wet milling, the fermentation is mostly carried out in a continuous regime. The technology is called cascade fermentation and is applied in large-scale facilities. Thus, a continuous cascade saccharification process is also employed. The effluent of this system feeds not only the bioreactors for yeast propagation, but also the prefermentation and fermentation trains. The fermentation train consists of a cascade of four continuous fermenters. The major infection source is concentrated in the saccharification step. To avoid the development of contaminating bacteria, the cultivation of yeasts is carried out at a pH of 3.5, which requires the construction of stainless steel bioreactors (Madson and Monceaux, 1995). The temperature is maintained below 34°C by recirculating the culture medium through an external heat exchanger. Airlift bioreactors are utilized in these technological configurations. The total residence time of the fermenters system is about 48 h and obtains an ethanol concentration in the effluent of the last tank of about 9% by weight (12% by volume; McAloon et al., 2000). Some ethanol-producing plants using the dry-milling technology also employ cascade fermentation; in those cases, higher yields are obtained by simultaneous saccharification and fermentation of corn mash.

7.1.3.2 Very High Gravity Fermentation

One of the proposed technologies for the development of high-performance processes using starchy materials consists of the fermentation of high and very high gravity mashes. During high gravity fermentation, the solids concentration in the medium exceeds 200 g/L, which implies a high substrate load and, consequently, high ethanol concentrations at the end of fermentation. Furthermore, a lower amount of process water is required as well as lower energy demands. However, this process implies more prolonged cultivation times and, sometimes, incomplete fermentation due to end-product inhibition, high osmotic pressure, and inadequate nutrition (Barber et al., 2002). To accelerate high gravity fermentations, the controlled addition of small amounts of acetaldehyde during the fermentation allows the reduction in cultivation time from 790 h to 585 h for initial glucose concentration of 300 g/L without effect on ethanol yield. It is believed that this positive effect may be caused by the ability of acetaldehyde to replenish the intracellular acetaldehyde pool and restore the cellular redox balance (Barber et al., 2002). Continuous operation can improve the performance of high gravity fermentations. In particular, continuous fermentation can reach almost the same ethanol production as batch fermentation, although it is likely that there is a threshold concentration of initial glucose above which an ethanol yield no longer increases (Zhao and Lin, 2003).

For industrial ethanol production, fermentation of wheat mashes of very high gravity (VHG) has been proposed. These mashes consist of wheat starch hydrolyzates containing 300 g or more of dissolved solids per liter of mash. VHG fermentation technology enables high ethanol concentrations to be obtained from very concentrated sugar solutions. To this aim, very low levels of dissolved oxygen are required as well as nitrogen sources that do not limit the cell growth, such as urea or ammonium salts (Jones and Ingledew, 1994a). In this way, 21.1% (by volume) ethanol concentrations are obtained in only four days of fermentation from VHG wheat mash. One of the strategies employed is the addition of commercial proteases during the VHG fermentation in order to release amino acids from soluble proteins contained in the wheat mash compensating, in this way, the addition of exogenous nitrogen sources (Jones and Ingledew, 1994b). Thomas et al. (1996) emphasize that considerable amounts of water can be saved by applying this technology to fuel ethanol production. Moreover, the implementation of VHG fermentation increases the throughput rate of an ethanol plant without the need of increasing the plant capacity. These authors provide a theoretical method for predicting the maximum concentration of ethanol in fermented mash that takes into account changes in the weight and volume of mash during fermentation. Bayrock and Ingledew (2001) designed and tested a system that combines the multistage continuous culture fermentation and the VHG cultivation for feed containing 150 to 320 g/L of glucose using *S. cerevisiae*. The maximum ethanol concentration obtained in the process was 132.1 g/L indicating the feasibility of implementing this technology in the industry, particularly in the continuous production of ethanol from wheat starch (Sánchez and Cardona, 2008).

VHG technology has been tested with successful results for oats, barley, rye, and triticale, as cited by Wang et al. (1999). The pretreatment of feedstock can play an important role when process integration is analyzed during this type of fermentation processes. Wang et al. (1999) propose the integration of a pretreatment process, pearling by abrasion of cereal grains such as rye or triticale, with VHG fermentation technology. The pearling of cereals removes approximately 12% of grain dry matter, which increases its starch content to 7 to 8%. This increase in starch content combined with the employment of high concentrations of sugars, which is the main feature of VHG fermentation, allows the increase in the final ethanol concentration of 64% in comparison to the use of nonpearled grains in conventional fermentations.

7.1.4 FERMENTATION OF MEDIA BASED ON LIGNOCELLULOSIC BIOMASS

7.1.4.1 Fermentation of Cellulose Hydrolyzates

The classic configuration employed for fermenting biomass hydrolyzates involves a sequential process where the hydrolysis of cellulose and the fermentation are carried out in different units (Sánchez and Cardona, 2008). This configuration is known as separate hydrolysis and fermentation (SHF). When this sequential process is employed, the solid fraction of pretreated lignocellulosic material undergoes hydrolysis (saccharification). This fraction contains the cellulose in a form accessible to acids or enzymes. Once hydrolysis is completed, the resulting cellulose hydrolyzate is fermented and converted into ethanol. *S. cerevisiae* is the most employed microorganism for fermenting the hydrolyzates of lignocellulosic biomass. This yeast ferments the hexoses contained in the hydrolyzate, but not the pentoses. One of the main features of the SHF process is that each step can be performed at its optimal operating conditions (especially temperature and pH). The overall scheme of this process is presented in Figure 7.6. Depending on the type of biomass pretreatment, the lignin can be separated in this step (as in the case of the organosolv process [see Chapter 4]) or remain in the stillage. This scheme involves the fermentation of hemicellulose hydrolyzate by pentose-assimilating yeasts in a way parallel to the fermentation of glucose using *S. cerevisiae*.

BC International Corporation (Dedham, MA, USA) has operated a pilot plant in Louisiana for producing ethanol from biomass by SHF. This plant has the

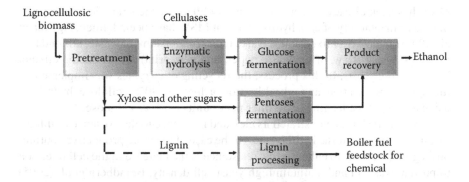

FIGURE 7.6 Block diagram of fuel ethanol production from lignocellulosic materials by separate hydrolysis and fermentation (SHF).

capacity to process about 500 ton/year of lignocellulosic materials. The University of Florida has granted the license to this company for worldwide commercialization of this technology, including the patent of the microorganism employed in the process, a recombinant strain of *Escherichia coli* to which *Zymomonas mobilis* genes, ensuring the biosynthesis of ethanol, has been introduced (Ingram et al., 1991). The company has been seeking investors for a $90 million project for an ethanol production facility using cane bagasse with a capacity of 23.2 mill gal EtOH/year, but the construction has not yet started. In fact, SHF is the technology with the most possibilities of being implemented at a commercial scale. For instance, the company Abengoa Bioenergía (Spain) is constructing a demonstration facility for ethanol production from lignocellulosic biomass with a capacity of 5 million L/year (Abengoa, 2008). The conversion of wheat and barley straw into ethanol will be done by an SHF scheme, as that shown in Figure 7.7, although this company plans to implement the pentose fermentation step at the midterm.

Some attempts to produce ethanol from municipal solid waste (MSW) have been done considering the cellulosic fraction of this type of materials (Sánchez and Cardona, 2008). Park et al. (2001) have studied the hydrolysis of waste paper contained in MSW obtaining significant sugars yield and evaluating the viscosity as an operating parameter. Bioethanol production from the cellulosic portion of

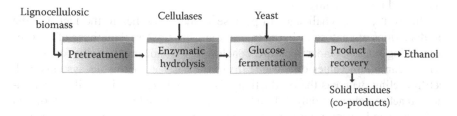

FIGURE 7.7 Block diagram of separate hydrolysis and fermentation (SHF) process for fuel ethanol production from lignocellulosic biomass without parallel fermentation of pentoses.

MSW has been already patented (Titmas, 1999) and some strategies for improving the fermentability of acid hydrolyzates of MSW have been defined. Nguyen et al. (1999) employed a mixed solids waste (construction lumber waste, almond tree prunings, wheat straw, office waste paper, and newsprint) for producing ethanol by SHF using yeasts. In this process, the recycling of enzymes was implemented through microfiltration and ultrafiltration achieving 90% cellulose hydrolysis using a net enzyme loading of 10 filter paper units (FPU)/g cellulose.

S. cerevisiae has demonstrated its elevated resistance to the presence of inhibitors in the lignocellulosic hydrolyzate. In the case of the more productive continuous regime, one way to enhance this resistance is the increase in the cell retention to prevent washout and maintain high yeast cell density. Brandberg et al. (2005) employed a microfiltration unit to recirculate the cells under microaerobic conditions achieving sugar conversion up to 99% for undetoxified dilute-acid pretreated hydrolyzates of softwood (spruce) supplemented with mineral nutrients, although the productivity was low.

7.1.4.2 Pentose Fermentation

When a technological flowsheet involving an SHF process is employed, the detoxified hemicellulose hydrolyzate can be unified with the cellulose hydrolyzate coming from the enzymatic reactor. The resulting stream contains mostly glucose, but also xylose and other hexoses released during biomass pretreatment. The simplest scheme includes the cultivation of S. cerevisiae that converts the glucose present in the medium into ethanol remaining in the xylose and the other hexoses. This implies a reduction in the amount of ethanol that could be obtained if the xylose and remaining sugars were utilized. To increase the amount of sugars converted into ethanol, yeast assimilating the xylose besides glucose can be used, but in this case the biomass utilization rates are lower relative to microorganisms that assimilate only hexoses. This is explained by the diauxic growth of this type of yeast (see Chapter 6). To offset this effect, sequential fermentations are employed in which S. cerevisiae utilizes the hexoses during the first days of cultivation and later xylose-utilizing yeast is added in order to complete the conversion to ethanol (Chandrakant and Bisaria, 1998), but achieved ethanol yields are not very high. Therefore, the most suitable configuration corresponds to the scheme of Figure 7.6 where both fermentations are performed independently.

One of the main challenges in pentose fermentation lies in the fact that the productivities of pentose-utilizing microorganisms are less than those of hexose-fermenting ones. Comparisons are conclusive: ethanol productivity for S. cerevisiae can attain values of 170 g/(L/h) in the case of continuous systems with cell recycling, whereas the productivity for C. shehatae at high cell concentrations reaches values of only 4.4 g/(L/h) (Olsson and Hahn-Hägerdal, 1996). On the other hand, there are a few cases where the immobilization of these yeasts increases the ethanol productivity (Chandrakant and Bisaria, 1998), unlike the case of hexose-fermenting yeasts or Z. mobilis. In spite of these drawbacks, pentose-utilizing microorganisms are important for the design of processes involving separate fermentation of hexoses and pentoses during the processing of resulting streams from biomass pretreatment. As occurs with *S. cerevisiae* or *Z. mobilis*, most pentose-fermenting yeasts are mesophiles (Sánchez and Cardona, 2008). On the other hand, although thermotolerant yeasts such as *Kluyveromyces marxianus* have demonstrated their capability to ferment glucose at 45°C (Singh et al., 1998), there are no data about the assimilation of xylose by this yeast, according to Ryabova et al. (2003). These authors described the separate fermentation of glucose and xylose by native strains of methylotrophic yeast *Hansenula polymorpha* at 37°C in flask cultures during 60 h achieving ethanol concentrations of 13.2 g/L and 2.98 g/L from glucose and xylose, respectively.

Lynd et al. (2001) report that for *Thermoanaerobacterium thermosaccharolyticum* cultivated in xylose-based media during batch and continuous cultures, the ethanol concentrations obtained are low (in the order of 25 g/L). These authors studied the influence of different factors limiting the substrate utilization for continuous cultures at progressively higher feed xylose concentrations. Their results indicate that the salt accumulation due to the utilization of bases for pH-control of fermentation limits the growth of this bacterium at elevated values of xylose content in the feed. These outcomes can explain the differences between the tolerance to added ethanol and the maximum concentration of produced ethanol for these microorganisms. Xylose-fermenting termophilic bacteria are prospective organisms to be co-cultured with cellulose hydrolyzing bacteria such as *Clostridium thermocellum* in order to directly convert pretreated lignocellulosic biomass into ethanol, a process called consolidated bioprocessing (CBP). Another promising microorganism capable of fermenting a great variety of sugars (including hexoses and pentoses) is the fungus *Mucor indicus* reaching ethanol yields of 0.46 g/g glucose when it is cultivated under anaerobic conditions. In addition, this fungus assimilates the inhibitors present in dilute-acid hydrolyzates (Sues et al., 2005). Other pentose-fermenting zygomycetes were also evaluated (Millati et al., 2005). The use of the fungus *Chalara parvispora* for ethanol production from pentose-containing materials has been patented (Holmgren and Sellstedt, 2006). Ogier et al. (1999) have compiled information about the main fermentative indexes for the pentose-assimilating yeasts *Candida shehatae*, *Pichia stipitis,* and *Pachysolen tannophilus*, and for the xylose-assimilating thermophilic bacteria *T. thermosaccharolyticum*, *T. ethanolicus*, and *Bacillus stearothermophilus*.

7.1.4.3   Co-Fermentation of Lignocellulosic Hydrolyzates

The co-fermentation of lignocellulosic hydrolyzates represents another technological option for utilizing all the sugars released during biomass pretreatment and cellulose hydrolysis. This kind of cultivation process is aimed at the complete assimilation of all the sugars resulting from lignocellulosic degradation by the microbial cells and consists of the employment of a mixture of two or more compatible microorganisms that assimilate both the hexoses and pentoses present in the medium. This means that the fermentation is carried out by a mixed culture. Some examples of mixed culture are summarized in Table 7.3. However, the use of mixed cultures presents the problem that microorganisms

TABLE 7.3

Some Examples of Co-Fermentation of Lignocellulosic Hydrolyzates Using Nonrecombinant Microorganisms

Technology	Bioagent	Feedstock/ Medium	Remarks	References
Co-fermentation (mixed culture)	*Saccharomyces cerevisiae* mutant+ *Pichia stipitis*	Glucose and xylose	Batch and continuous cultures; 100% glucose conversion and 69% xylose conversion	Laplace et al., 1993
	Respiratory deficient *S. diastaticus* + *P. stipitis*	Steam—exploded and enzymatically hydrolyzed aspen wood	Continuous culture: EtOH conc. 13.5 g/L, yield 0.25 g/g, productivity 1.6 g/(Lh); 100% conversion of glucose and xilose	Delgenes et al., 1996
Isomerization of xilose and fermentation	*S. cerevisiae* + xilose (glucosa)– isomerasa	Nonpretreated spent sulfite liquors, acid-hydrolyzed wheat straw	Batch process; yield 0.41 g/g; 51–84% xilose utilization	Chandrakant and Bisaria, 1998; Lindén and Hahn-Hägerdal, 1989

Source: Extracted from Cardona, C.A., and Ó.J. Sánchez. 2007. *Bioresource Technology* 98:2415–2457. Elsevier Ltd.

utilizing only hexoses grow faster than pentose-utilizing microorganisms leading to a more elevated conversion of hexoses into ethanol (Cardona and Sánchez, 2007). To solve this problem, the utilization of respiratory-deficient mutants of the hexose-fermenting microorganisms has been proposed. In this way, the fermentation and growth activities of the pentose-fermenting microorganisms are increased as they grow very slowly when cultivated along with rapid hexose-fermenting yeasts. In addition, the presence of hexose-assimilating microorganisms allows the reduction of the catabolic repression exerted by glucose on the pentose consumption in pentose-assimilating microorganisms (Laplace et al., 1993). Considering the indicators for the process using only the glucose-assimilating bacterium Z. *mobilis* grown on the biomass hydrolyzate, the productivities of the mixed culture are less than those of the bacterium, but the yields are comparable, which offers a space for further research (Delgenes et al., 1996). One of the additional problems in this kind of configuration is that pentose-fermenting yeasts present a greater inhibition by ethanol, which limits the use of concentrated substrates in the system.

Another variant of co-fermentation consists of the utilization of a single microorganism capable of assimilating both hexoses and pentoses in an optimal way allowing high conversion and ethanol yield. Although these microorganisms exist in nature (see previous section), their efficiency and ethanol conversion rates are reduced for the implementation of an industrial process. Hence, the addition to the culture medium of an enzyme transforming the xylose into xylulose (xylose-isomerase) has been proposed (see Table 7.3.). In this way, microorganisms exhibiting high rates of conversion to ethanol and elevated yields (like *S. cerevisiae*) can assimilate the xylulose, involving it in the metabolic pathways leading to the ethanol biosynthesis (see Chapter 6, Figure 6.2). On the other hand, a high efficiency in the conversion to ethanol can be reached through the genetic modification of yeasts or bacteria already adapted to the ethanolic fermentation—a topic that was discussed in Chapter 6. The microorganisms most commonly modified for this purpose are *S. cerevisiae* and *Z. mobilis* to which genes encoding the assimilation of pentoses have been introduced (see Chapter 6, Table 6.3). The other approach for genetic modification is the introduction of genes encoding the metabolic pathways for ethanol production to microorganisms that are capable of fermenting both hexoses and pentoses in their native form. The "design" of ethanologenic bacteria like *E. coli* or *Klebsiella oxytoca* is an example of such type of modification (see Chapter 6, Table 6.3). Using these recombinant microorganisms allows implementing the co-fermentation process intended to the more complete utilization of the sugars contained in the hydrolyzates of lignocellulosic biomass (Cardona and Sánchez, 2007).

7.2 MODELING OF ETHANOLIC FERMENTATION FOR PROCESS DESIGN PURPOSES

The procedures to design fuel ethanol production processes with an enhanced performance according to technoeconomic and environmental criteria require the development and utilization of an appropriate, sufficient, and accurate description of ethanolic fermentations. In particular, process analysis involves the employment of simulation tools to assess the performance of such cultivation processes, but the quality of simulation strongly depends on the quality of mathematical models describing the fermentation. Thus, the selection of already developed models, their modification, or the development of new ones has become a crucial aspect of the methodologies for the design of bioethanol production processes.

As many chemical and biochemical processes, fermentations can be modeled using different approaches to its description. The types of models employed depend on the nature of the phenomenon to be described (Figure 7.8). Thus, the cell population can be described through stochastic models involving a probabilistic function considering the probability of birth and death of microbial cells

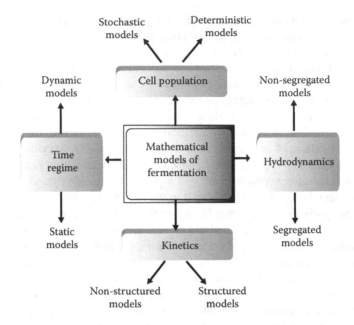

FIGURE 7.8 Types of mathematical models used for description of fermentation processes.

depending on cell age. If probabilistic functions are not considered in the model, the description becomes deterministic and other types of distribution can be used, such as the distribution of the average age of cell population as a function of time. In the simplest case, the age of the entire population corresponds to the current time in a batch fermentation process. When the time regime of a cultivation process is considered, the corresponding models can be either dynamic or static. Dynamic models are used to describe discontinuous and semicontinuous processes as well as to assess the dynamic behavior of continuous processes that are the bases for developing control schemes for fermentation. In these models, the time derivatives are important. If those derivatives are equal to zero, the model becomes static and the process is analyzed under steady-state conditions.

When the hydrodynamic features of the cultivation process are taken into account, it is necessary to employ segregated models considering, for example, the existence of cell aggregates or zones inside the fermenter with poor agitation. In these cases, the models can consider derivatives with respect to one or more of the three space coordinates (models with distributed parameters). If the hydrodynamic transport phenomena are not considered, the mentioned derivatives are null (models with nondistributed or concentrated parameters) and the models are nonsegregated. Actually, this type of models is used for description of the bioreactor rather than the cultivation process itself. Finally, to study the kinetic behavior of the cell culture, the models can be either structured (sometime, also called deterministic) or nonstructured (unstructured). In the former case, the internal structure of the biological system is considered, i.e., its constituent components and the interactions between them. Some fermentation processes have been described through structured models by considering several key intermediate metabolites (like RNA, ATP, NAD^+, etc.) besides the substrates and products. This approach has been applied to the analysis of metabolic pathways (metabolic flux analysis) leading to the biosynthesis of ethanol (for instance, see the work of Çakır et al., 2004). However, these models are complex and the determination of their parameters is quite difficult. In contrast, the nonstructured models simplify the description of fermentation behavior by considering the cells as black boxes with some "average" composition that is characterized through the overall cell biomass concentration.

For process design of technological configurations involving fermentation processes, especially when process synthesis procedures are applied, the utilization of complex structured segregated models are not desirable due to the considerable increase in computational effort and mathematical complexity of the calculations required to define the performance of the overall process at a plant scale. For this reason, nonstructured, nonsegregated models without consideration of cell distribution by age are preferable. However, these simplified models should reflect the main phenomena inherent to the description level defined. Thus, to evaluate ethanolic fermentation kinetics, such mathematical models should accurately describe at least the cell growth, consumption of the main substrate, and ethanol production. In the simplest case, ethanologenic fermentation can be described using a stoichiometric approach, which is useful for calculation of mass and energy balances of the cultivation process during both batch and continuous processes.

Lin and Tanaka (2006) reviewed the classical generic models for the description of ethanolic fermentation. These authors point out that the nonstructured models are frequently used during the routine control of fermentation processes, whereas structured models should be used for optimization and control of ethanol fermentation. The main limitation of these models is that they do not consider simultaneously the four factors affecting ethanol concentration (substrate limitation, substrate inhibition, product inhibition, and cell death). Most of the models reviewed by these authors are related to the simple ethanol fermentation, but it is also necessary to analyze other models, which apply to more complex processes, such as co-fermentation and simultaneous saccharification and fermentation (SSF).

Other specific models have been developed for special configurations of bioreactors or for specific fermentation conditions (Cardona and Sánchez, 2007). Gilson and Thomas (1995) developed a model for a fluidized-bed reactor with yeast cells immobilized on alginate beads. It was shown that the observed reduction in ethanol yield compared to free yeast cells was caused by substrate restrictions inside the beads and not by changes in the metabolic rate of the immobilized cells. Borzani (1987) derived a Monod-based model for evaluating the maximum value of the mash-feeding rate to be used in order to have a completely fermented medium during the fed-batch fermentation of molasses for ethanol production. Converti et al. (2003) provide a simplified modeling of the kinetics of fed-batch fermentation of sugarcane molasses that allows the prediction and control of the performance of this regime of cultivation. Certainly, this tool is very useful for the simulation of the entire process. Tsuji et al. (1986) evaluated the performance of continuous alcoholic fermentation using a vector-valued objective function. This analysis considered the trade-off among three criteria (ethanol productivity, the ethanol concentration in the broth, and the substrate conversion) on the basis of the noninferior set defined in a vector-valued function space. Costa et al. (2001) used an intrinsic model that took into account cell volume fraction and the dependence of kinetic constants on temperature for continuous vacuum fermentation using yeasts. With the help of this model, the process was optimized using response surface analysis that allowed the determination of operational conditions maximizing high yield and productivity. In the same way, dynamic simulation was performed using the concepts of factorial design in order to determine the best control structures for the process. Maia and Nelson (1993) presented a model of gravitational sedimentation intended to describe the recycling of cells to the bioreactor using a parallel plate sedimenter during continuous fermentation. This model allowed the optimization of the operating conditions in order to efficiently recycle yeasts at high cell density.

7.2.1 MODELING OF ETHANOLIC FERMENTATION FROM SUGARS

During the process synthesis procedures aimed at identifying those technological configurations for fuel ethanol production with better technoeconomic and environmental performance, the authors of this book carried out a comprehensive review of the available models for describing ethanolic fermentation kinetics (Sánchez, 2008). Considering that the three main types of feedstocks employed for fuel ethanol production can be broken down into glucose, the first stage of the analysis included the selection of the most appropriate kinetic models that take into account the growth of *S. cerevisiae* or *Z. mobilis* on glucose-rich media. As a kind of "sample" that evidences the great diversity of kinetic models studied, some kinetic relationships for cell growth rate are shown in Table 7.4. In this table, substrate consumption and product formation rates are not shown by space limitations, but these expressions depend, in turn, on either cell growth rate or cell concentration. All the presented models consider the substrate limitation, which is mostly expressed through Monod-type equations (Monod, 1949). Similarly, all the models include expressions describing the growth rate inhibition by the ethanol formed. One of the most useful models corresponds to number 4 in Table 7.4, which describes the alcoholic fermentation using yeasts from glucose (Lee et al., 1983). In particular, this model has been employed for continuous processes, as reported by Tsuji et al. (1986). It was one of the models employed for the subsequent stages of process synthesis procedures to be discussed in this book. On the other hand, number 5 proposed by Garro et al. (1995) was selected for fermentation using *Z. mobilis*.

Cell biomass recirculation is a practice employed to increase cell concentration inside the fermenter in order to diminish the cultivation time in batch fermentation or the residence time in continuous fermentation. Considering that this type of configuration will be analyzed during process synthesis, it is necessary to include

TABLE 7.4
Some Kinetic Models Describing Specific Cell Growth Rate during Ethanologenic Fermentations Employing Glucose-Based Media

No.	Model of Growth[a]	Observations	Ref.
1.	$\mu = \mu_{max} \dfrac{S}{K_S + S} e^{-k_1 P}$	*Saccharomyces cerevisiae*, k_1 shift for continuous culture	Aiba et al., 1968
2.	$\mu = \mu_{max} \dfrac{S}{K_S + S}(1 - K_{PX} P)$	*S. cerevisiae* alginate inmovilization, ≥ 10% sugars	Birol et al., 1998
3.	$\mu = \mu_{max} \dfrac{S}{K_S + S}\left[1 - \left(\dfrac{P}{P_m}\right)^{\alpha}\right]$	*S. cerevisiae*, some constants taken of de Birol et al. (1998), ≥ 10% sugars	Luong, 1985
4.	$\mu = \mu_{max} \dfrac{S}{K_S + S}\left[1 - \left(\dfrac{P}{P_m}\right)\right]$	*S. cerevisiae*, model taken of Lee et al. (1983)	Tsuji et al., 1986
5.	$\mu = \mu_{max} \dfrac{S}{K_S + S}\left[1 - \left(\dfrac{P}{P_m}\right)^{n}\right]$	*Z. mobilis*, model taken of Garro et al. (1995)	da Silva Henriques et al., 1999
6.	$\mu = \dfrac{\mu_{max}\left[1 - \left(\dfrac{P}{P_{ma}}\right)^{a}\right]\left[1 - \left(\dfrac{P - P_{0b}}{P_{mb} - P_{0b}}\right)^{b}\right] S}{K_m + S + \dfrac{S(S - S_i)}{K_i - S_i}}$	*Z. mobilis*, modification of the model of Veeramallu and Agrawal (1990)	da Silva Henriques et al., 1999

[a] X = Biomass concentration, P = ethanol concentration, μ = cell growth rate.

mathematical expressions taking into account such recirculation in the models to avoid a sham growth of cell biomass concentration within the fermenter. In addition, microorganisms undergo stress when cell concentration increases above 100 g/L in the case of yeasts. Another issue to be considered is the substrate inhibition that limits the utilization of very concentrated media due to factors like catabolic repression and osmotic pressure exerting a negative influence on cell growth rate. Therefore, one of the expressions chosen for yeast growth rate (r_X) is as follows:

$$r_X = \mu_{max} \frac{S}{K_S + S}\left(\frac{K_i}{K_i + S}\right)\left(1 - \frac{P}{P_m}\right)\left(1 - \frac{X}{X_m}\right) X \qquad (7.1)$$

where S, X, and P are the concentrations (in g/L) of substrate (glucose), cell biomass (*S. cerevisiae*), and product (ethanol), respectively, μ_{max} is the maximum specific growth rate (in h^{-1}), K_S and K_i are the semisaturation and substrate inhibition constants, P_m is the maximum product concentration, and X_m is the maximum cell biomass concentration in the medium. The models describing the substrate consumption rate (r_S) and product formation rate (r_P) are represented by the following equations:

$$r_S = -\left(\frac{1}{Y_{X/S}}\right) r_X \tag{7.2}$$

$$r_P = \left(\frac{1}{Y_P}\right) r_X \tag{7.3}$$

where $Y_{X/S}$ is the cell biomass yield from substrate (in g cell/g substrate) and $Y_P = Y_{X/S}/Y_{P/S}$ (in g cell/g EtOH); $Y_{P/S}$ is the product yield from substrate (in g EtOH/g substrate).

Substrate consumption rate and product formation rate for the bacterium Z. *mobilis* were selected according to the model 5 of Table 7.4, which was slightly modified to consider cell biomass recirculation as follows:

$$r_X = \mu_{max} \frac{S}{K_S + S}\left(1-\frac{P}{P_m}\right)^n\left(1-\frac{X}{X_m}\right) X \tag{7.4}$$

$$r_S = -\left(\frac{1}{Y_{X/S}}\right) r_X - mX \tag{7.5}$$

$$r_P = \alpha r_X + \beta X \tag{7.6}$$

where α and β are kinetic constants.

For sucrose-based media, the Equations (7.1) through (7.6) were also applied using not only glucose as a substrate, but also fructose. In this case, it was considered that the cell growth rate is equal to the sum of the growth rates from both substrates. For starch-based media, the same kinetic expressions for glucose content were used considering the content of this sugar in the saccharified starch.

The stoichiometric approach can be useful as an approximation method to describe the ethanologenic fermentation from media containing glucose, fructose, or sucrose. In this case, the overall biological transformations are considered through specific key chemical reactions including the global reaction for biomass cell biosynthesis. Thus, the fermentation employing sucrose-based media may be described by the following reactions:

$$C_{12}H_{22}O_{11} + H_2O \rightarrow C_6H_{12}O_6 + C_6H_{12}O_6$$

$$\text{Sucrose} \quad \text{Water} \quad \text{Glucose} \quad \text{Fructose}$$

$$C_6H_{12}O_6 + 1.14\ NH_3 \rightarrow 5.71\ CH_{1.8}O_{0.5}N_{0.2} + 2.57\ H_2O + 0.29\ CO_2$$

Glucose Ammonium   Yeast   Water   Carbon dioxide

$$C_6H_{12}O_6 + 1.14\ NH_3 \rightarrow 5.71\ CH_{1.8}O_{0.5}N_{0.2} + 2.57\ H_2O + 0.29\ CO_2$$

Fructose Ammonium   Ethanol   Water   Carbon dioxide

$$C_6H_{12}O_6 \rightarrow 2\ C_2H_5OH + 2\ CO_2$$

Glucose   Ethanol   Carbon dioxide

$$C_6H_{12}O_6 \rightarrow 2\ C_2H_5OH + 2\ CO_2$$

Fructose   Ethanol   Carbon dioxide

The yeast cells are assumed to have a "molecular formula" that is derived from the elemental analysis of cell biomass. Despite its simplicity, this approach has the advantage of providing suitable relationships to perform mass and energy balances required to simulate the alcoholic fermentation process. Moreover, other transformations can be considered by adding more stoichiometric reactions describing, for example, the formation of fermentation by-products like acetaldehyde and glycerol:

$$C_6H_{12}O_6 \rightarrow C_3H_5(OH)_3 + C_2H_4O + CO_2$$

Glucose   Glycerol Acetaldehyde Carbon dioxide

The biological transformations can be directed by adjusting the corresponding conversions of substrate into products. This approach is particularly useful when modular-sequential process simulators are used, such as Aspen Plus (Aspen Technology, Inc., USA) or SuperPro Designer (Intelligen, Inc., USA). These commercial simulators have modules for calculating chemical and biochemical transformations according to this approach, though a kinetic approximation can be applied through other different modules.

7.2.2 MODELING OF CO-FERMENTATION OF HEXOSES AND PENTOSES

As mentioned above, the liquid fraction resulting from lignocellulosic biomass pretreatment can be unified with the cellulose hydrolyzate obtained after the enzymatic treatment of the cellulose contained in the solid fraction coming from the mentioned pretreatment process. The liquid stream produced contains all the soluble sugars derived from the biomass, mostly glucose and xylose. This stream can be fermented by microorganisms able to convert these two sugars into ethanol. For this, recombinant bacteria can be used. Leksawasdi et al. (2001) employed an engineered strain of Z. mobilis able to co-ferment hexoses and pentoses to process a solution containing a mixture of glucose and pentose. In addition, they developed and experimentally tested an accurate mathematical description that considers substrate limitation by the two sugars, substrate inhibition by both sugars, and ethanol inhibition. For this, it employs the concepts of threshold ethanol concentration for which inhibition of growth begins, and maximum ethanol concentration for which biomass growth becomes zero. Inhibition constants for the substrates are taken into account for considering glucose and xylose concentrations inhibiting both cell growth and ethanol biosynthesis as a result of catabolic repression. Kinetic expressions can be found in the work of Leksawasdi et al. (2001). This description was used by the authors of this book to consider cofermentation processes during process synthesis procedures. The equations of the model are as follows:

$$r_x = [\alpha r_{x,1} + (1-\alpha)r_{x,2}]X$$

$$r_{x,1} = \mu_{max,1}\left(\frac{S_1}{K_{sx,1}+S_1}\right)\left(1-\frac{P-P_{ix,1}}{P_{mx,1}-P_{ix,1}}\right)\left(\frac{K_{ix,1}}{K_{ix,1}+S_1}\right)$$

$$r_{x,2} = \mu_{max,2}\left(\frac{S_2}{K_{sx,2}+S_2}\right)\left(1-\frac{P-P_{ix,2}}{P_{mx,2}-P_{ix,2}}\right)\left(\frac{K_{ix,2}}{K_{ix,2}+S_2}\right)$$

$$r_{s,1} = -\alpha q_{s,max,1}\left(\frac{S_1}{K_{ss,1}+S_1}\right)\left(1-\frac{P-P_{is,1}}{P_{ms,1}-P_{is,1}}\right)\left(\frac{K_{is,1}}{K_{is,1}+S_1}\right)X$$

$$r_{s,2} = -(1-\alpha)q_{s,max,2}\left(\frac{S_2}{K_{ss,2}+S_2}\right)\left(1-\frac{P-P_{is,2}}{P_{ms,2}-P_{is,2}}\right)\left(\frac{K_{is,2}}{K_{is,2}+S_2}\right)X$$

$$r_p = [\alpha r_{p,1} + (1-\alpha)r_{p,2}]X$$

$$r_{p,1} = q_{p,max,1}\left(\frac{S_1}{K_{sp,1}+S_1}\right)\left(1-\frac{P-P_{ip,1}}{P_{mp,1}-P_{ip,1}}\right)\left(\frac{K_{ip,1}}{K_{ip,1}+S_1}\right)$$

$$r_{p,2} = q_{p,max,2}\left(\frac{S_2}{K_{sp,2}}\right)\left(1-\frac{P-P_{ip,2}}{P_{mp,2}-P_{ip,2}}\right)\left(\frac{K_{ip,2}}{K_{ip,2}+S_2}\right)$$

(7.7)

where r_x is the overall cell growth rate, $r_{x,1}$ and $r_{x,1}$ are the cell growth rate from glucose and xylose, respectively; $r_{s,1}$ and $r_{s,1}$ are the glucose and xylose consumption rates, respectively; r_p is the overall ethanol production rate; $r_{p,1}$ and $r_{p,1}$ are the ethanol production rates from glucose and xylose, respectively; X, S_1, S_2, and P are the concentrations (in g/L) of cell biomass, glucose, xylose, and ethanol, respectively; α, $\mu_{max,1}$, $\mu_{max,2}$, $q_{s,max,1}$, $q_{s,max,2}$, $q_{p,max,1}$, $q_{p,max,2}$, $K_{sx,1}$, $K_{sx,2}$, $K_{ss,1}$, $K_{ss,2}$, $K_{sp,1}$, $K_{sp,2}$, $P_{mx,1}$, $P_{mx,2}$, $P_{ms,1}$, $P_{ms,2}$, $P_{mp,1}$, $P_{mp,2}$, $K_{ix,1}$, $K_{ix,2}$, $K_{is,1}$, $K_{is,2}$, $K_{ip,1}$, $K_{ip,2}$, $P_{ix,1}$, P_{ix}, $P_{is,1}$, $P_{is,2}$, $P_{ip,1}$, $P_{ip,2}$ are the kinetic parameters.

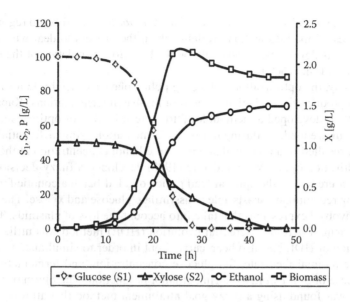

FIGURE 7.9 Batch co-fermentation of a hydrolyzate of lignocellulosic biomass. Process behavior was calculated by using the model of Leksawasdi et al. (2001). Initial sugar concentrations: glucose, 100 g/L, xylose, 50 g/L, cell biomass, 0.003 g/L.

CASE STUDY 7.1 MODELING OF CO-FERMENTATION FERMENTATION

Based on the model of Leksawasdi et al. (2001), the simulation of alcoholic fermentation from biomass was performed with initial glucose concentration of 100 g/L and initial xylose concentration of 50 g/L. These concentrations approximately correspond to those of lignocellulosic hydrolyzates. Inhibition of growth rate can be observed from Figure 7.9 due to relatively high amounts of ethanol in the broth, as can be seen after 25 h. The use of more concentrated culture media leads to the underutilization of expensive feedstocks, which cannot be transformed into ethanol despite their availability in the broth. According to the model, when a medium containing up to 400 g/L of fermentable sugars is employed, an ethanol concentration of about 71.2 g/L is reached only after 80.5 h of cultivation. An ethanol concentration of 24.7 g/L is attained at 48 h remaining more than 347 g/L of substrate (compared with data of Figure 7.9 for a medium with a lower substrate concentration). Therefore, this model proved its suitability for describing the complex inherent phenomena of the co-fermentation of lignocellulosic hydrolyzates.

7.3 ANALYSIS OF FED-BATCH ETHANOLIC FERMENTATION

In Section 7.1.2.2., the principle and applications of fed-batch fermentation technologies for ethanol production were discussed. The operation of fermenters under fed-batch conditions is very difficult to model due to the fact that microbial cells grow under permanently changing conditions including volume changes explained by feeding of fresh medium into the bioreactor (Cardona and Sánchez, 2007). To tackle this problem, da Silva Henriques et al. (1999) developed a hybrid neural model for alcoholic fermentation by *Z. mobilis* in a fed-batch regime. The model uses all the information available about the process to deal with the difficulties in its development and could be the basis for formulation of the optimal feed policy of the reactor.

Precisely, the optimization of feeding policy plays a crucial role for increasing both productivity and ethanol yield of fed-batch fermentations. Kapadi and Gudi (2004) developed a methodology for determination of optimal feed rates of fresh culture medium during fed-batch fermentation using differential evolution that resulted in a predicted augment of ethanol concentration at the end of each cultivation cycle. Wang and Jing (1998) developed a fuzzy-decision-making procedure to find the optimal feed policy of a fed-batch alcoholic fermentation using recombinant yeasts able to assimilate glucose and xylose. The kinetic model involved expressions that take into account the loss of plasmids. To solve this problem, a hybrid differential evolution (HDE) method was utilized. The application of HDE has also been carried out in order to simultaneously determine the optimal feed rate, fed glucose concentration, and fermentation time for the case of *S. diastaticus* during ethanol production. The optimal trade-off solution was found using a fuzzy goal attainment method that allowed obtaining a good agreement between experimental and computed results (Chen and Wang, 2003). HDE also has been used for estimation of kinetic parameters during batch culture of the mentioned yeast (Wang et al., 2001). Other strategies of optimization have demonstrated their usefulness for evaluation of hybrid configurations involving reaction–separation integration (Cardona and Sánchez, 2007). For instance, vacuum fermentation technology has been modeled, simulated, and optimized by means of factorial design and response surface analysis (da Silva et al., 1999).

7.4 DYNAMICS OF CONTINUOUS FERMENTATION SYSTEMS

Besides design procedures, process systems engineering applied to fuel ethanol production processes consists of operation and control, especially when different technologies are to be implemented at industrial level. This is particularly important for continuous ethanologenic fermentation. If such a system is operated under conditions corresponding to a stable steady state, any small perturbation in input parameters (like dilution rate, temperature, or substrate concentration of the feed) will be compensated for by the same system. If the system is operated near an unstable steady-state, any small perturbation could not be offset by the culture and the system function can result in conditions of lower productivity or oscillate with the time.

The problem of multiple steady-states is related to the fact that for a same single value of process operating parameters, typically the dilution rate and inlet substrate concentration, the system can attain different steady-states, each with different performance indicators (yield, productivity, conversion). The analysis becomes much more complicated if considering that these steady-states can be stable or unstable. Often, just the unstable steady-state exhibits higher

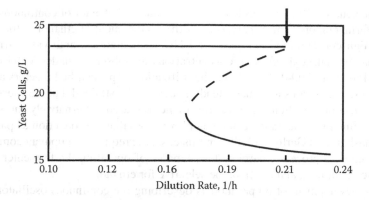

FIGURE 7.10 Operational diagram for continuous cultivation of yeasts. Continuous lines correspond to stable states while the dashed line corresponds to unstable states.

productivities or ethanol yields. This makes the industrial operation of these processes very complex because small variations in dilution rate or composition of culture medium can make the system migrate to the steady-state with a lower, though more stable, performance indicator. This situation is schematically illustrated in Figure 7.10 for continuous yeast cultivation. Just under the conditions corresponding to the point marked with the arrow, the system presents its highest productivity. In this way, an optimum operation of the bioreactor is obtained when the system is near the point in which the fermentation destabilizes. If a perturbation occurs, the system can fall down to its stable state with lower productivity. The goal of control is to keep the system precisely in its optimum operating point.

This indicates the importance of conducting studies about stable states in continuous bioreactors. These studies could provide the optimal values of operation variables in order to design highly effective processes. Perego et al. (1985) showed that instability during the operation of continuous fermentation from sugarcane molasses depends at a high degree on the temperature of cultivation. However, these authors did not report any mathematical description of the process in order to explain this characteristic behavior. Laluce et al. (2002) constructed a special five-stage continuous fermentation system with cell recycling and different temperatures in each stage. With this system, they experimentally assessed the effect of fluctuations in operating temperature that occur under industrial conditions on fermentation performance. These fluctuations produced variations in the cell concentration and cell viability. Hojo et al. (1999) showed that microaeration plays an important role in the stabilization of concentrations of ethanol, substrate, and cells during continuous cultivation of sugarcane syrup with cell recirculation of *S. cerevisiae*. Without air addition at low rates (0.05 vvm), these concentrations had significant fluctuations. These authors adjusted the obtained data to one simple model, but no dynamic simulation of the studied process was performed.

One source of fluctuations leading to oscillatory behavior of continuous ethanolic fermentation using *S. cerevisiae* is the high content of ethanol in the broth. This high concentration is typical of VHG fermentations, and wide variations in ethanol, cell, and substrate concentrations are observed under VHG conditions. Bai et al. (2004) showed that the utilization of packed-bed reactors attenuates these oscillations and quasi-steady-states can be attained, but the causes and mechanism of this attenuation require further research. Alternatively, an oscillatory regime of fermentation can be employed for ethanol production as patented by Elnashaie and Garhyan (2005). In this case, the required equipment comprises a fermenter, a process control system capable of operating the fermenter under chaotic conditions, and a membrane selective for ethanol.

The development of proper models describing the continuous oscillatory fermentation allows the deep stability analysis of cultures presenting this behavior that is characteristic of continuous cultures of *Z. mobilis* and *S. cerevisiae* under certain conditions, such as specific dilution rates or ethanol concentrations in the broth. Tools like dynamic simulation and, especially, bifurcation analysis can provide valuable information for design of more effective continuous fermentation processes. Dynamic simulation is required for control of fermentation processes including those carried out in batch, fed-batch, and continuous regimes. For instance, through a nonstructured mathematical model that considers four state variables (concentrations of cells, substrate, product, and CO_2 evolution rate), Thatipamala et al. (1996) developed an algorithm for the prediction of nonmeasurable state variables and critic parameters varying with time, which allowed the online estimation of these variables and the adaptive optimization of a continuous bioreactor for ethanol production.

Oscillatory behavior of fermentations imposes great challenges for bioprocess design. Several experimental runs with forced oscillations of *Z. mobilis* culture were carried out in order to formulate and test a model describing the oscillatory behavior (Daugulis et al., 1997; McLellan et al., 1999). The model makes use of the concept of "dynamic specific growth rate," which considers inhibitory culture conditions in the recent past affecting subsequent cell behavior. Through dynamic simulation, it was shown that the lag in the cells response was the major factor contributing to the oscillations. Moreover, the change in morphology to a more filamentous form may explain the change in specific growth rate and product formation characteristics. However, Zhang and Henson (2001) point out that dynamic simulation has several limitations for analyzing the dynamic behavior of fermentation processes only a limited number of simulations tests can be performed and that it does not easily reveal the model characteristics leading to certain dynamic behaviors. In contrast, nonlinear analysis allows a deeper insight into this type of processes. Nonlinear analysis provides tools for studying the appearance of multiple steady states with changes in parameter values of the model. These authors performed the bifurcation analysis for models describing continuous alcoholic fermentation of *Z. mobilis* and *S. cerevisiae* and concluded that employed tools allowed revealing an important characteristic of the employed models as the lack of model robustness to small parameter variations and the coexistence of multiple stable solutions under the same operating conditions. An experimentally verified, unsegregated, two-compartment model of ethanol fermentation was utilized to assess the dynamic behavior of a stirred-tank bioreactor with a membrane for the *in situ* removal of ethanol (Garhyan and Elnashaie, 2004; Mahecha-Botero et al., 2006). Through bifurcation analysis, it was shown that the operation of the reactor under periodic/chaotic attractor's conditions gives higher substrate conversions, yields, and production rates than the corresponding steady-states. It also has been shown that the membrane acts as a stabilizer of the process eliminating the oscillations (Cardona and Sánchez, 2007).

CASE STUDY 7.2 STABILITY ANALYSIS OF CONTINUOUS ETHANOLIC FERMENTATION

To describe the oscillatory behavior of continuous ethanolic fermentation using yeasts or bacteria, the mathematical model employed should contain the necessary expressions showing the variation of the state variables with time. Some models have been proposed with this aim and contrasted with the experimental data in order to validate their suitability. Among these models, the mathematical description of Jarzębski (1992) should be highlighted. This model is based on stochastic assumptions taking into account the distribution of cell population and corresponds to the continuous fermentation in a stirred tank using the yeast *Saccharomyces cerevisiae*. This model divides the cell biomass into three groups: viable biomass (cells able to reproduce and biosynthesize ethanol), nonviable biomass (cells able to biosynthesize ethanol, but not to reproduce), and dead biomass (cells able neither to reproduce nor biosynthesize ethanol). Moreover, the kinetic model includes terms describing the maintenance of viable and nonviable cells. The results derived from this model were compared to the experimental data obtained by Perego et al. (1985). The equations of the model are as follows:

$$\frac{dX_v}{dt} = (\mu_v - \mu_{nv} - \mu_d - D)X_v$$

$$\frac{dX_{nv}}{dt} = \mu_{nv}X_v(\mu_d - D)X_{nv}$$

$$\frac{dX_d}{dt} = \mu_d(X_v - X_{nv}) - DX_d \tag{7.8}$$

$$\frac{dP}{dt} = \frac{\mu_v X_v}{Y_{x/p}} - m_p X_{nv} - DP$$

$$\frac{dS}{dt} = \frac{\mu_v X_v}{Y_{x/s}} - m_S X_{nv} - D(S_0 - S)$$

where X, S, and P are the concentrations of cell biomass in the effluent stream, S_0 is the concentration of substrate in the feed stream (in kg/m^3), D is the dilution rate (in h^{-1}), μ is the specific growth rate (in h^{-1}). The subindexes v, nv, and d refer to viable, nonviable, and dead cells, respectively. The biomass growth (death) rate expressions are as follows:

$$\mu_v = \mu_{max} \frac{S}{K_1 + S}\left(1 - \frac{P}{Pc}\frac{S}{K_2 + S}\right)$$

$$\mu_{nv} = \mu_{max} \frac{S}{K_1 + S}\left(1 - \frac{P}{Pc}\frac{S}{K_2 + S}\right) - \mu_v \qquad (7.9)$$

$$\mu_d = \mu_v$$

where μ_{max}, K_1, K_2, and Pc are kinetic constants, which can be found in the original work of Jarzębski (1992).

In a previous work (Restrepo et al., 2007), the simulation of the continuous system using the available information and parameters reported by Jarzębski (1992) was performed, but the results were not satisfactory because there was no appropriate correspondence with the experimental data reported by Perego et al. (1985) for continuous fermentation using sugarcane molasses as a feedstock. Using the software Matlab™ (Mathworks, Inc., USA), the curves corresponding to the dynamic behavior of such fermentation system were obtained (Figure 7.11). Nondynamic analysis (bifurcation analysis) is a powerful tool to study the oscillatory behavior of continuous ethanolic fermentations. This tool was used in the previous mentioned work in order to assess whether the model can describe the dynamic behavior of the system. For the analysis of the fermentation system studied, the software MatCont developed in the University of Gent (Belgium) was employed as a toolbox in Matlab package. The results of bifurcation analysis for the Jarzębski's model are shown in Figure 7.12. From this diagram in the zone having a physical sense (positive dilution rates), it can be observed that the dynamic system behaves in an ordinary way and presents no sustained oscillations (Hopf bifurcations). In the zone of negative dilution rates, limit point (LP or node-saddle bifurcation) and Hopf bifurcation are present indicating that the model does have the possibility to represent the oscillations.

For representing the sustained oscillations and the instability of the continuous fermentation, the Jarzębski's model was modified and adjusted in such a way that the model represents not only the oscillatory fermentation reported by Perego, Jr. et al. (1985) appropriately, but also the data for continuous nonoscillatory fermentation obtained by these same authors. The modification consisted in the simplification of the nonviable biomass growth rate (μ_{nv}) considering that the inhibition of cell biomass is mainly due to the high ethanol concentration. In addition, the expression describing the biomass death rate (μ_{nv}) was changed in such a way that this rate was proportional to a fraction d of the viable biomass growth rate (μ_{nv}). Therefore, the model modified comprises the system (7.8) and the following rate expressions:

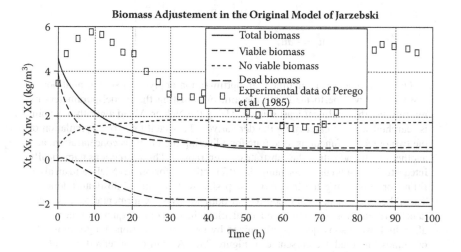

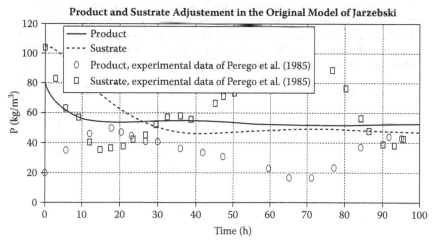

FIGURE 7.11 Dynamic behavior of continuous ethanolic fermentation of sugarcane molasses using *Saccharomyces cerevisiae*. The continuous curves were obtained by the dynamic simulation of the model proposed by Jarzębski (1992). The experimental data were taken from the work of Perego et al. (1985). (a) Cell biomass behavior: total biomass (———), viable biomass (‑‑‑), nonviable biomass (☐), dead biomass (‑‑‑‑), experimental data for cell biomass (☐). (b) Substrate (glucose) and product (ethanol) behavior: substrate (———), ethanol (‑‑‑), substrate experimental data (☐), ethanol experimental data (○).

$$\left. \begin{aligned} \mu_v &= \mu_{max} \frac{S}{K_1 + S}\left(1 - \frac{P}{Pc}\frac{S}{K_2 + S} \right) \\ \mu_{nv} &= \mu_{max} \frac{S}{K_1 + S}\left(1 - \frac{P}{Pc}\frac{S}{K_2 + S} \right) - \mu_v \\ \mu_d &= -d\mu_v \end{aligned} \right\} \tag{7.10}$$

Once the model was modified, an optimization applying a simple multiobjective strategy was performed with the aim of adjusting the model parameters by minimizing the nonlinear least squares. Using the Nelder–Mead algorithm, a point is searched without calculating the derivatives. This facilitates the formulation of the two objective functions used (one for adjusting the biomass concentration, and another one for adjusting the substrate concentration). These functions numerically integrate the system using the numeric differentiation formulae (NDF) and an algorithm for controlling the integration step size with a defined absolute tolerance. Then, the experimental data reported for both steady-state fermentation and oscillatory fermentation were employed to calculate the sum of the squares of the residuals, which were subsequently minimized by using the mentioned algorithm. The outcomes obtained are presented in Figure 7.13. As can be observed, the adjustment of the model to the experimental data was appropriate and allowed describing the behavior of the two types of fermentations studied. Unlike the nonmodified Jarzębski model, all the concentrations had physical sense. In the previous work reported (Restrepo et al., 2007), the values of the kinetic parameters were calculated. These values are presented in Table 7.5.

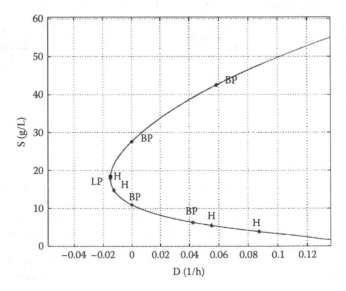

FIGURE 7.12 Bifurcation diagram of the continuous ethanolic fermentation using the model of Jarzębski (1992) for an inlet substrate concentration (S_0) of 137.5 kg/m³. H = Hopf bifurcation, BP = bifurcation point, LP = limit point.

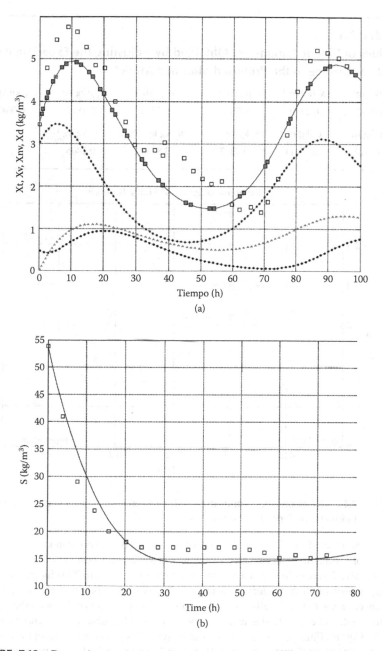

FIGURE 7.13 Dynamic simulation using the proposed modified model (continuous curves). The continuous curves were calculated using the model proposed. The experimental data were taken from the work of Perego et al. (1985). (a) Cell biomass behavior: total biomass (——), viable biomass (---), nonviable biomass (---), dead biomass (···), experimental data for cell biomass (□). (b) Substrate behavior: substrate (——), experimental data for substrate (□).

TABLE 7.5

Values of Kinetic Parameters Obtained by Adjusting the Experimental Data according to the Proposed Modified Model

K_1/kg m^{-3}	K_2/kg m^{-3}	μ_{max}/h^{-1}	μ'_{max}/h^{-1}	$Y_{x/p}$/kg kg^{-1}	$Y_{x/s}$/kg kg^{-1}
0.0842	6.2479	0.2623	0.2218	0.1817	0.0647

Pc/kg m^{-3}	Pc'/kg m^{-3}	m_s/kg kg^{-1} h^{-1}	m_p kg kg^{-1} h^{-1}	d/kg kg^{-1}
76.3222	202.6611	5.7277	3.8419	0.2405

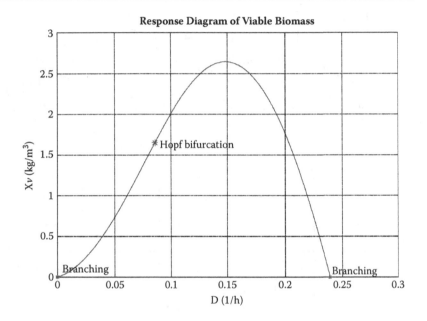

FIGURE 7.14 Response diagram of the continuous ethanolic fermentation process for viable cell biomass in dependence on the dilution rate D.

When the bifurcation analysis has been performed based on the modified model, a Hopf bifurcation can be noted in the response diagram near $D = 0.089$ h^{-1} for the viable biomass in dependence on the dilution rate (Figure 7.14) indicating the appearance of the oscillatory fermentation. Similarly, the point with the maximum concentration of viable cells was also established. This point matches with the maximum ethanol concentration and corresponds to an inlet substrate concentration of 137.5 kg/m^3. Finally, through bifurcation analysis, it is possible to delimit the zones of the operating diagram S_0 versus D (Figure 7.15). In particular, the parameter S_0 has demonstrated a strong influence on the fermentation behavior. Small changes in the nutrient composition of the culture broth entering the continuous fermenter can provoke significant changes in the response variables (concentrations of biomass, ethanol, and residual substrate). Knowing these zones, the task of operating the fermenter in the stable zones becomes easier. Moreover, the design of the control

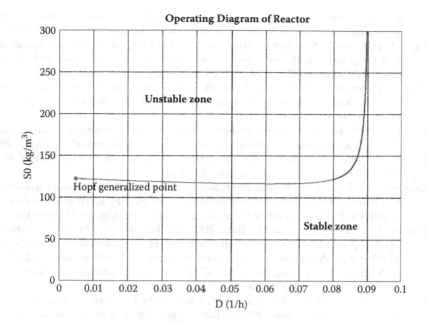

FIGURE 7.15 Operating diagram of the continuous ethanolic fermentation for inlet substrate concentration (S_0) and dilution rate (D).

structure can be based on these nonlinear analysis tools. Therefore, the modified model can predict when the instability of the fermentation will occur. At this point, the instability of the system can be considered during the early step of conceptual design, which permits a better design of the automatic control system. This issue is of great importance during the operation of industrial ethanolic fermentations.

REFERENCES

Abengoa. 2008. *Biocarburantes de Castilla y León informs.* Abengoa. http://www. abengoa.com/sites/abengoa/en/noticias_y_publicaciones/noticias/historico/noticias/2008/07_julio/20080715_noticias.html (accessed February 2009).

Aiba, S., M. Shoda, and M. Nagatani. 1968. Kinetics of product inhibition in alcohol fermentation. *Biotechnology and Bioengineering* 10:845–864.

Aldiguier, A.S., S. Alfenore, X. Cameleyre, G. Goma, J.L. Uribelarrea, S.E. Guillouet, and C. Molina-Jouve. 2004. Synergistic temperature and ethanol effect on *Saccharomyces cerevisiae* dynamic behaviour in ethanol bio-fuel production. *Bioprocess and Biosystems Engineering* 26:217–222.

Alfenore, S., X. Cameleyre, L. Benbadis, C. Bideaux, J.-L. Uribelarrea, G. Goma, C. Molina-Jouve, and S.E. Guillouet. 2004. Aeration strategy: A need for very high ethanol performance in *Saccharomyces cerevisiae* fed-batch process. *Applied Microbiology and Biotechnology* 63:537–542.

Bai, F.W., L.J. Chen, W.A. Anderson, and M. Moo-Young. 2004. Parameter oscillations in a very high gravity medium continuous ethanol fermentation and their attenuation on a multistage packed column bioreactor system. *Biotechnology and Bioengineering* 88 (5):558–566.

Barber, A.R., M. Henningsson, and N.B. Pamment. 2002. Acceleration of high gravity yeast fermentations by acetaldehyde addition. *Biotechnology Letters* 24:891–895.

Bayrock, D.P., and W.M. Ingledew. 2001. Application of multistage continuous fermentation for production of fuel alcohol by very-high-gravity fermentation technology. *Journal of Industrial Microbiology & Biotechnology* 27:87–93.

Birol, G., P. Doruker, B. Kirdar, Z.I. Önsan, and K.Ö. Ülgen. 1998. Mathematical description of ethanol fermentation by immobilised *Saccharomyces cerevisiae*. *Process Biochemistry* 33 (7):763–771.

Borzani, W. 1987. Kinetics of ethanol production during the reactor feeding phase in constant fed-batch fermentation of molasses. *Biotechnology and Bioengineering* 29:844–849.

Bothast, R.J., and M.A. Schlicher. 2005. Biotechnological processes for conversion of corn into ethanol. *Applied Microbiology and Biotechnology* 67 (19–25).

Brandberg, T., N. Sanandaji, L. Gustafsson, and C.J. Franzén. 2005. Continuous fermentation of undetoxified dilute acid lignocellulose hydrolysate by *Saccharomyces cerevisiae* ATCC 96581 using cell recirculation. *Biotechnology Progress* 21:1093–1101.

Caicedo, L.A, M.M. Cuenca, and M. Díaz. 2003. Escalado de la producción de etanol a nivel de planta piloto empleando un reactor con células inmovilizadas. (Scale-up of ethanol production at pilot scale level using a reactor with immobilized cells, in Spanish.) Paper presented at the XXII Congreso Colombiano de Ingeniería Química, Bucaramanga, Colombia.

Çakır, T., K.Y. Arga, M.M. Altıntaş, and K.Ö. Ülgen. 2004. Flux analysis of recombinant *Saccharomyces cerevisiae* YPB-G utilizing starch for optimal ethanol production. *Process Biochemistry* 39 (12):2097–2108.

Cardona, C.A., and Ó.J. Sánchez. 2007. Fuel ethanol production: Process design trends and integration opportunities. *Bioresource Technology* 98:2415–2457.

Chandrakant, P., and V.S. Bisaria. 1998. Simultaneous bioconversion of cellulose and hemicellulose to ethanol. *Critical Reviews in Biotechnology* 18 (4):295–331.

Chen, S.L. 1981. Optimization of batch alcoholic fermentation of glucose syrup substrate. *Biotechnology and Bioengineering* 23:1827–1836.

Chen, Y.-F., and F.-S. Wang. 2003. Crisp and fuzzy optimization of a fed-batch fermentation for ethanol production. *Industrial & Engineering Chemistry Research* 42:6843–6850.

Claassen, P.A.M., J.B. van Lier, A.M. López Contreras, E.W.J. van Niel, L. Sijtsma, A.J.M. Stams, S.S. de Vries, and R.A. Weusthuis. 1999. Utilisation of biomass for the supply of energy carriers. *Applied Microbiology and Biotechnology* 52:741–755.

Converti, A., S. Arni, S. Sato, J.C.M. Carvalho, and E. Aquarone. 2003. Simplified modeling of fed-batch alcoholic fermentation of sugarcane blackstrap molasses. *Biotechnology and Bioengineering* 84 (1):88–95.

Costa, A.C., D.I.P. Atala, F. Maugeri, and R. Maciel. 2001. Factorial design and simulation for the optimization and determination of control structures for an extractive alcoholic fermentation. *Process Biochemistry* 37:125–137.

Cysewski, G.R., and C.R. Wilke. 1977. Rapid ethanol fermentations using vacuum and cell recycle. *Biotechnology and Bioengineering* 19:1125–1143.

Da Silva, A.C.S., and C.M. Vaz. 1989. Continuous process of optimized fermentation for the production of alcohol. U.S. Patent US4889805.

da Silva, F.L.H., M.A. Rodrigues, and F. Maugeri. 1999. Dynamic modelling, simulation and optimization of an extractive continuous alcoholic fermentation process. *Journal of Chemical Technology and Biotechnology* 74:176–182.

da Silva Henriques, A.W., A.C. da Costa, T.L.M. Alves, and E.L. Lima. 1999. A hybrid neural model of ethanol production by *Zymomonas mobilis*. *Applied Biochemistry and Biotechnology* 77–79:277–291.

Das, D., N.R. Gaidhani, K. Murari, and P.S. Gupta. 1993. Ethanol production by whole cell immobilization using lignocellulosic materials as solid matrix. *Journal of Fermentation and Bioengineering* 75 (2):132–137.

Daugulis, A.J., P.J. McLellan, and L. Jinghong. 1997. Experimental investigation and modeling of oscillatory behavior in the continuous culture of *Zymomonas mobilis*. *Biotechnology and Bioengineering* 56 (1):99–105.

Delgenes, J.P., J.M. Laplace, R. Moletta, and J.M. Navarro. 1996. Comparative study of separated fermentations and cofermentation processes to produce ethanol from hardwood derives hydrolysates. *Biomass and Bioenergy* 11 (4):353–360.

Echegaray, O., J. Carvalho, A. Fernandes, S. Sato, E. Aquarone, and M. Vitolo. 2000. Fed-batch culture of *Saccharomyces cerevisiae* in sugarcane blackstrap molasses: Invertase activity of intact cells in ethanol fermentation. *Biomass and Bioenergy* 19:39–50.

Ehnstroem, L.K.J. 1984. Fermentation method. U.S. Patent US4460687.

Elnashaie, S.S.E.H, and P. Garhyan. 2005. Chaotic fermentation of ethanol. U.S. Patent US2005170483.

Garhyan, P., and S.S.E.H. Elnashaie. 2004. Static/dynamic bifurcation and chaotic behavior of an ethanol fermentor. *Industrial & Engineering Chemistry Research* 43:1260–1273.

Garro, O.A., E. Rodriguez, P.U. Roberto, and D.A.S. Callieri. 1995. Mathematical modelling of the alcoholic fermentation of glucose by *Zymomonas mobilis*. *Journal of Chemical Technology and Biotechnology* 63 (4):367–373.

Gil, G.H., W.J. Jones, and T.G. Tornabene. 1991. Continuous ethanol production in a two-stage, immobilized/suspended-cell bioreactor. *Enzyme and Microbial Technology* 13:390–399.

Gilson, C.D., and A. Thomas. 1995. Ethanol production by alginate immobilised yeast in a fluidized bed bioreactor. *Journal of Chemical Technology and Biotechnology* 62:38–45.

Gong, C.S., N.J. Cao, J. Du, and G.T. Tsao. 1999. Ethanol production from renewable resources. *Advances in Biochemical Engineering/Biotechnology* 65:207–241.

Grote, W., and P.L. Rogers. 1985. Ethanol production from sucrose-based raw materials using immobilized cells of *Zymomonas mobilis*. *Biomass* 8:169–184.

Hawgood, N., S. Evans, and P.F. Greenfield. 1985. Enhanced ethanol production in multiple batch fermentations with an auto-flocculating yeast strain. *Biomass* 7:261–278.

Hojo, O., C.O. Hokka, and A.M. Souto Maior. 1999. Ethanol production by a flocculant yeast strain in a CSTR type fermentor with cell recycling. *Applied Biochemistry and Biotechnology* 77–79:535–545.

Holmgren, M., and A. Sellstedt. 2006. Producing ethanol through fermentation of organic starting materials, involves using fungus, e.g., *Chalara parvispora*, capable of metabolizing pentose compounds. Sweden Patent SE527184.

Ingram, L.O., T. Conway, and F. Alterthum. 1991. Ethanol production by *Escherichia coli* strains co-expressing *Zymomonas* PDC and ADH genes. U.S. Patent US5000000.

Jarzębski, A.B. 1992. Modeling of oscillatory behavior in continuous ethanol fermentation. *Biotechnology Letters* 14 (2):137–142.

Jones, A.M., and W.M. Ingledew. 1994a. Fuel alcohol production: Appraisal of nitrogenous yeast foods for very high gravity wheat mash fermentation. *Process Biochemistry* 29:483–488.

Jones, A.M., and W.M. Ingledew. 1994b. Fuel alcohol production: Assessment of selected commercial proteases for very high gravity wheat mash fermentation. *Enzyme and Microbial Technology* 16:683–687.

Kapadi, M.D., and R.D. Gudi. 2004. Optimal control of fed-batch fermentation involving multiple feeds using differential evolution. *Process Biochemistry* 39:1709–1721.

Kida, K., S.I. Asano, M. Yamadaki, K. Iwasaki, T. Yamaguchi, and Y. Sonoda. 1990. Continuous high-ethanol fermentation from cane molasses by a flocculating yeast. *Journal of Fermentation and Bioengineering* 69 (1):39–45.

Kida, K., S. Morimura, K. Kume, K. Suruga, and Y. Sonoda. 1991. Repeated-batch ethanol fermentation by a flocculating yeast, *Saccharomyces cerevisiae* IR-2. *Journal of Fermentation and Bioengineering* 71 (5):340–344.

Kosaric, N., and J. Velikonja. 1995. Liquid and gaseous fuels from biotechnology: Challenge and opportunities. *FEMS Microbiology Reviews* 16:111–142.

Kuriyama, H., H. Ishibashi, I. Umeda, T. Murakami, and H. Kobayashi. 1993. Control of yeast flocculation activity in continuous ethanol fermentation. *Journal of Chemical Engineering of Japan* 26 (4):429–431.

Laluce, C., C.S. Souza, C.L. Abud, E.A.L. Gattas, and G.M. Walker. 2002. Continuous ethanol production in a nonconventional five-stage system operating with yeast cell recycling at elevated temperatures. *Journal of Industrial Microbiology & Biotechnology* 29:140–144.

Laplace, J.M., J.P. Delgenes, R. Moletta, and J.M. Navarro. 1993. Cofermentation of glucose and xylose to ethanol by a respiratory-deficient mutant of *Saccharomyces cerevisiae* co-cultivated with a xylose-fermenting yeast. *Journal of Fermentation and Bioengineering* 75 (3):207–212.

Lee, J.M., J.F. Pollard, and G.A. Couman. 1983. Ethanol fermentation with cell recycling: Computer simulation. *Biotechnology and Bioengineering* 25 (2):497–511.

Leksawasdi, N., E.L. Joachimsthal, and P.L. Rogers. 2001. Mathematical modeling of ethanol production from glucose/xylose mixtures by recombinant *Zymomonas mobilis*. *Biotechnology Letters* 23:1087–1093.

Lin, Y., and S. Tanaka. 2006. Ethanol fermentation from biomass resources: Current state and prospects. *Applied Microbiology and Biotechnology* 69:627–642.

Lindén, T., and B. Hahn-Hägerdal. 1989. Fermentation of lignocellulose hydrolysates with yeasts and xylose isomerase. *Enzyme and Microbial Technology* 11:583–589.

Luong, J. H. T. 1985. Kinetics of ethanol inhibition in alcohol fermentation. *Biotechnology and Bioengineering* 27:280–285.

Lynd, L.R., K. Lyford, C.R. South, G.P. van Walsum, and K. Levenson. 2001. Evaluation of paper sludges for amenability to enzymatic hydrolysis and conversion to ethanol. *TAPPI Journal* 84:50–69.

Madson, P.W., and D.A. Monceaux. 1995. Fuel ethanol production. In *The Alcohol Textbook*, ed. T.P. Lyons, D.R. Kelsall and J.E. Murtagh. Nottingham, U.K.: University Press.

Mahecha-Botero, A., P. Garhyan, and S.S.E.H. Elnashaie. 2006. Non-linear characteristics of a membrane fermentor for ethanol production and their implications. *Nonlinear Analysis: Real World Applications* 7:432–457.

Maia, A.B.R.A., and D.L. Nelson. 1993. Application of gravitational sedimentation to efficient cellular recycling in continuous alcoholic fermentation. *Biotechnology and Bioengineering* 41:361-369.

Maiorella, B.L., H.W. Blanch, and C.R. Wilke. 1984. Economic evaluation of alternative ethanol fermentation processes. *Biotechnology and Bioengineering* 26:1003–1025.

McAloon, A., F. Taylor, W. Yee, K. Ibsen, and R. Wooley. 2000. *Determining the cost of producing ethanol from cornstarch and lignocellulosic feedstocks*. Technical Report NREL/TP-580-28893, National Renewable Energy Laboratory, Washington, D.C.

McLellan, P.J., A.J. Daugulis, and L. Jinghong. 1999. The incidence of oscillatory behavior in the continuous fermentation of *Zymomonas mobilis*. *Biotechnology Progress* 15:667–680.

Melzoch, K., M. Rychtera, and V. Hábová. 1994. Effect of immobilization upon the properties and behaviour of *Saccharomyces cerevisiae* cells. *Journal of Biotechnology* 32:59–65.

Millati, R., L. Edebo, and M.J. Taherzadeh. 2005. Performance of *Rhizopus, Rhizomucor*, and *Mucor* in ethanol production from glucose, xylose, and wood hydrolyzates. *Enzyme and Microbial Technology* 36:294–300.

Monod, J. 1949. The growth of bacterial cultures. *Annual Review of Microbiology* 3:371–394.

Monte Alegre, R., M. Rigo, and I. Joekes. 2003. Ethanol fermentation of a diluted molasses medium by *Saccharomyces cerevisiae* immobilized on chrysotile. *Brazilian Archives of Biology and Technology* 46 (4):751–757.

Morimura, S., Y.L. Zhong, and K. Kida. 1997. Ethanol production by repeated-batch fermentation at high temperature in a molasses medium containing a high concentration of total sugar by a thermotolerant flocculating yeast with improved salt-tolerance. *Journal of Fermentation and Bioengineering,* 83 (3):271–274.

Navarro, A.R, H.R. Marangoni, A. de Cabada, and D.A.S. Callieri. 1986. Producción de etanol por fermentación con alta concentración de levaduras: su aplicación en destilerías ya instaladas. *Revista Argentina de Microbiología* 18 (1):7–11.

Nguyen, Q.A., F.A. Keller, M.P. Tucker, C.K. Lombard, B.M. Jenkins, D.E. Yomogida, and V.M. Tiangco. 1999. Bioconversion of mixed solids waste to ethanol. *Applied Biochemistry and Biotechnology* 77–79:455–472.

Ogier, J.-C., D. Ballerini, J.-P. Leygue, L. Rigal, and J. Pourquié. 1999. Production d'éthanol à partir de biomasse lignocellulosique (Ethanol production from lignocellulosic biomass, in French). *Oil & Gas Science and Technology— Revue de l'IFP* 54 (1):67–94.

Olsson, L., and B. Hahn-Hägerdal. 1996. Fermentation of lignocellulosic hydrolysates for ethanol production. *Enzyme and Microbial Technology* 18:312–331.

Park, E.Y., A. Michinaka, and N. Okuda. 2001. Enzymatic hydrolysis of waste office paper using viscosity as operating parameter. *Biotechnology Progress* 17:379–382.

Perego, Jr., L., J.M. Cabral de S. Dias, L.H. Koshimizu, M.R. de Melo Cruz, W. Borzani, and M.L.R. Vairo. 1985. Influence of temperature, dilution rate and sugar concentration on the establishment of steady-state in continuous ethanol fermentation of molasses. *Biomass* 6:247–256.

Prakasham, R.S., B. Kuriakose, and S.V. Ramakrishna. 1999. The influence of inert solids on ethanol production by *Saccharomyces cerevisiae*. *Applied Biochemistry and Biotechnology* 82:127–134.

Restrepo, J.B., Ó.J. Sánchez, G. Olivar, and C.A. Cardona. 2007. Análisis de estabilidad de la fermentación continua para la producción de alcohol carburante (Stability analysis of continuous fermentation for fuel ethanol production, in Spanish). Paper presented at the XXIV Congreso Nacional de Ingeniería Química, Cali, Colombia.

Roukas, T. 1996. Ethanol production from non-sterilized beet molasses by free and immobilized *Saccharomyces cerevisiae* cells using fed-batch culture. *Journal of Food Engineering* 21:87–96.

Ryabova, O.B., O.M. Chmil, and A.A. Sibirny. 2003. Xylose and cellobiose fermentation to ethanol by the thermotolerant methylotrophic yeast *Hansenula polymorpha*. *FEMS Yeast Research* 4:157-164.

Sánchez, Ó.J. 2008. Síntesis de Esquemas Tecnológicos Integrados para la Producción Biotecnológica de Alcohol Carburante a partir de tres Materias Primas Colombianas (Synthesis of Integrated Flowsheets for Biotechnological Production of Fuel Ethanol from Three Colombian Feedstocks, Ph.D. thesis, in Spanish), Departamento de Ingeniería Química, Universidad Nacional de Colombia sede Manizales, Manizales.

Sánchez, Ó.J., and C.A. Cardona. 2008. Trends in biotechnological production of fuel ethanol from different feedstocks. *Bioresource Technology* 99:5270–5295.

Shabtai, Y., S. Chaimovitz, A. Freeman, E. Katchalski-Katzir, C. Linder, M. Nemas, M. Perry, and O. Kedem. 1991. Continuous ethanol production by immobilized yeast reactor coupled with membrane pervaporation unit. *Biotechnology and Bioengineering.* 38:869–876.

Sheoran, A., B.S. Yadav, P. Nigam, and D. Singh. 1998. Continuous ethanol production from sugarcane molasses using a column reactor of immobilized *Saccharomyces cerevisiae* HAU-1. *Journal of Basic Microbiology* 38 (2):123–128.

Shojaosadati, S.A. 1996. The use of biomass and stillage recycle in conventional ethanol fermentation. *Journal of Chemical Technology and Biotechnology* 67:362–366.

Singh, D., P. Nigam, I.M. Banat, R. Marchant, and A.P. McHale. 1998. Ethanol production at elevated temperatures and alcohol concentrations. Part II: Use of *Kluyveromyces marxianus* IMB3. *World Journal of Microbiology and Biotechnology* 14:823–834.

Sues, A., R. Millati, L. Edebo, and M.J. Taherzadeh. 2005. Ethanol production from hexoses, pentoses, and dilute-acid hydrolyzate by *Mucor indicus. FEMS Yeast Research* 5:669–676.

Thatipamala, R., G.A. Hill, and S. Rohani. 1996. On-line state estimation and adaptive optimization using state equations for continuous production of bioethanol. *Journal of Biotechnology* 48:179–190.

Thomas, K.C., S.H. Hynes, and W.M. Ingledew. 1996. Practical and theoretical considerations in the production of high concentrations of alcohol by fermentation. *Process Biochemistry* 31 (4):321–331.

Titmas, J.A. 1999. Apparatus for hydrolyzing cellulosic material. U.S. Patent US5879637.

Tsuji, S., K. Shimizu, and M. Matsubara. 1986. Performance evaluation of ethanol fermentor systems using a vector-valued objective function. *Biotechnology and Bioengineering* 30:420–426.

Tzeng, J.-W., L.-S. Fan, Y.-R. Gan, and T.-T. Hu. 1991. Ethanol fermentation using immobilized cells in a multistage fluidized bed bioreactor. *Biotechnology and Bioengineering* 38:1253–1258.

Veeramallu, U., and P. Agrawal. 1990. A structured kinetic model for *Zymomonas mobilis* ATCC10988. *Biotechnology and Bioengineering* 36 (7):694–704.

Vriesekoop, F., and N.B. Pamment. 2005. Acetaldehyde addition and pre-adaptation to the stressor together virtually eliminate the ethanol-induced lag phase in *Saccharomyces cerevisiae. Letters in Applied Microbiology* 41:424–427.

Wang, F.-S., and C.-H. Jing. 1998. Fuzzy-decision-making problems of fuel ethanol production using a genetically engineered yeast. *Industrial & Engineering Chemistry Research* 37:3434–3443.

Wang, F.-S., T.-L. Su, and H.-J. Jang. 2001. Hybrid differential evolution for problems of kinetic parameter estimation and dynamic optimization of an ethanol fermentation process. *Industrial & Engineering Chemistry Research* 40:2876–2885.

Wang, S., K.C. Thomas, K. Sosulski, W.M. Ingledew, and F.W. Sosulski. 1999. Grain pearling and very high gravity (VHG) fermentation technologies for fuel alcohol production from rye and triticale. *Process Biochemistry* 34:421–428.

Wendhausen, R., A. Fregonesi, P. Moran, I. Joekes, J. Augusto, R. Rodrigues, E. Tonella, and K. Althoff. 2001. Continuous fermentation of sugarcane syrup using immobilized yeast cells. *Journal of Bioscience and Bioengineering* 91 (1):48–52.

Wieczorek, A., and H. Michalski. 1994. Continuous ethanol production by flocculating yeast in the fluidized bed bioreactor. *FEMS Microbiology Reviews* 14:69–74.

Zhang, Y., and M.A. Henson. 2001. Bifurcation analysis of continuous biochemical reactor models. *Biotechnology Progress* 17:647–660.

Zhao, Y., and Y.-H. Lin. 2003. Growth of *Saccharomyces cerevisiae* in a chemostat under high glucose conditions. *Biotechnology Letters* 25:1151–1154.

8 Analysis of Ethanol Recovery and Dehydration

In this chapter, main technologies for concentration of dilute aqueous solutions of ethanol obtained during the fermentation step are discussed. The main ethanol dehydration technologies, including the most promising, are disclosed as well. The main advantages and drawbacks of the discussed dehydration schemes are included as well. The utilization of process simulation tools is illustrated through the analysis of different technological schemes for ethanol separation and dehydration. The role of thermodynamic–topological analysis during the conceptual design of separation and dehydration technological configurations is also highlighted.

The recovery of ethanol produced by different technological configurations and from diverse types of feedstocks is accomplished in a very similar way. The ethanol content in the culture broth resulting from fermentation processes oscillates between 2.5 and 10% (by weight). The utilization of fuel ethanol as a gasoline oxygenate requires a high-purity ethanol, so it is necessary to concentrate the ethyl alcohol up to 99% to obtain the anhydrous ethanol, which is the suitable form used for ethanol-gasoline blends. Because the presence of water in fuel ethanol can lead to failures in the engine during the combustion of such blends (Sánchez and Cardona, 2005). The first step of the ethanol recovery scheme is the concentration of ethanol contained in culture broths. This process is carried out in distillation columns achieving ethanol content of about 50%. The next step is the rectification of this concentrated stream to obtain a product with a composition near to the azeotropic mixture of ethanol and water. In the following section, these two steps are discussed.

8.1 CONCENTRATION AND RECTIFICATION OF ETHANOL CONTAINED IN CULTURE BROTHS

The culture broth can contain or not cells of yeast or other ethanol-producing microorganisms. This stream is called *fermented wort* (especially when sugarcane is the feedstock employed), *fermented mash* (when cereal grains are used), or simply *wine* or *beer*. Besides ethanol and depending on the raw material employed, other substances can be found in a culture broth, such as:

- Nonfermented sugars
- Oligosaccharides resulting from the incomplete saccharification of starch or cellulose
- Ground and spent cereal grains
- Lignin (in dependence on the biomass pretreatment method)
- Other fermentation products like glycerol
- Lactic acid produced from contaminant bacteria
- Small amounts of acetic acid released during hemicellulose hydrolysis
- Dissolved carbon dioxide, salts, and excretion products from microbial cells metabolism
- Other compounds and materials

However, the two main components are water (80 to 90%) and ethyl alcohol. Making use of the higher volatility and lower boiling point of ethanol, the unit operation used as a rule for its separation is the conventional distillation at a pressure equal or higher than atmospheric pressure.

Usually, two distillation columns are utilized to elevate the ethanol concentration up to 90 to 92%. Concentrations higher than 95.6% (or 89.4 molar %) are impossible to obtain by conventional distillation due to the similar composition of both the saturated vapor and saturated liquid achieved in the top of the distillation column. This composition is named azeotropic and is one of the main thermodynamic limits imposed on ethanol purification process. To produce anhydrous ethanol (99.5% or more), nonconventional separation technologies are required (see Section 8.2).

The technological scheme of the concentration and rectification steps of the ethanol contained in the culture broth is shown in Figure 8.1. During fermentation, carbon dioxide is generated as a result of the microbial metabolism. In the gas outlet stream and, along with CO_2, small amounts of volatilized ethanol, as well as even much smaller amounts of water and other volatile substances, can be found. To avoid the ethanol loss, this gaseous stream is fed to a scrubber where a counter-current stream of water absorbs more than 98% ethanol. The scrubber is filled with a plastic-packed bed favoring the contact between the rising gaseous stream and the downward liquid stream. The gaseous ethanol-stripped stream is released into the atmosphere while the liquid stream containing about 2.5% ethanol is unified with the culture broth coming from fermenter in order to be fed to the distillation column (Wooley et al., 1999).

The first distillation column is called the concentration or beer column. This column has a determined number of plates that can be of various types. It has been suggested that the fixed-valve Nutter-type plates are the most suitable to handle streams containing solids exhibiting a relatively good efficiency (about 48%). The concentrated ethanol stream (35 to 50%) is removed from the column by a side stream. The overhead vapors contain mostly CO_2 (approximately 84%), a significant amount of ethanol (12%), and a small amount of water.

The bottoms of this column, called stillage or vinasses, concentrate the nonvolatile substances and suspended solids that enter along with the culture broth.

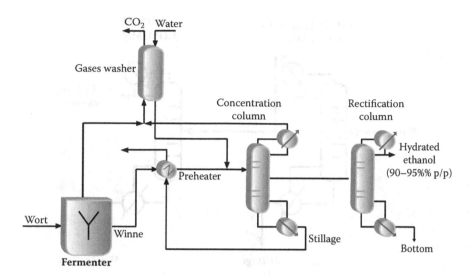

FIGURE 8.1 Technological scheme of the concentration and rectification steps during fuel ethanol production.

Stillage composition depends on the feedstock employed for fuel ethanol production. The heat of this bottom's stream is utilized to preheat the stream feeding the same distillation column. As an example of the location along this column of these streams, the design information for a concentration column corresponding to the biomass-to-ethanol process is provided (Wooley et al., 1999). This column has 32 plates, the feed stream is supplied to the fourth plate from the top, the side ethanol stream is removed from the eighth plate, and the reflux ratio required is 6.1.

The second column or rectification column is fed with the side ethanol stream from the concentration column. The operation of this column allows for a distillate with 90 to 92% ethanol concentration, which is sent to the dehydration step. The bottoms of this column have a very low ethanol content (less than 1%), being mostly water. For illustration purposes, the design data of the rectification column for the above-mentioned lignocellulosic ethanol process are provided as well (Wooley et al., 1999). The column has 69 plates and an additional feed stream corresponding to the recycle stream from the dehydration step, which is fed into the nineteenth plate from the top. The feed stream from the concentration column is supplied in the forty-fourth plate; fixed-valve Nutter-type plates are employed and the reflux ratio required is 3.2.

8.2 ETHANOL DEHYDRATION

The distillate product from the rectification column represents the stream entering the ethanol dehydration scheme, where the fuel ethanol with ethanol content greater than 99.5% should be produced. To achieve this high purity from streams containing 90 to 92% ethanol, it is necessary to employ nonconventional separation

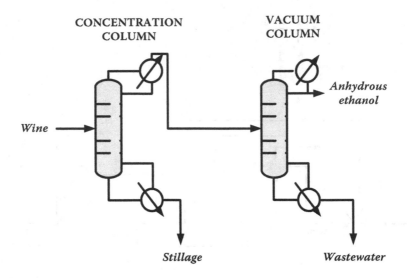

FIGURE 8.2 Diagram of ethanol dehydration by pressure-swing distillation.

operations like pressure-swing distillation, azeotropic distillation, extractive distillation, adsorption, and pervaporation (Sánchez and Cardona, 2005). All of these operations have found industrial application in the fuel ethanol industry.

8.2.1 PRESSURE-SWING DISTILLATION

To achieve the separation of an azeotropic mixture by using pressure-swing distillation, the manipulation of the column pressure is required, e.g., by utilizing a second distillation column working under vacuum conditions (Figure 8.2). This type of distillation makes use of the change of the vapor–liquid phase equilibrium at lower pressures than atmospheric (vacuum) leading to the disappearance of the azeotrope. The pressure required to eliminate the azeotrope in an ethanol–water mixtures is less than 6 kPa. But to obtain a high purity product, distillation columns with a large number of plates (above 40) and a high reflux ratio are needed. These conditions imply significant capital costs (large column diameters) and increased energy costs due to the maintenance of vacuum in distillation towers with many plates. This configuration has no fluxes or refluxes connecting the two columns. In general, pressure-swing distillation cannot always be employed; its utilization is limited to mixtures with azeotropes susceptible to be displaced with small changes of pressure, which is not exactly the case in ethanol–water systems.

8.2.2 AZEOTROPIC DISTILLATION

Most methods involving distillation for ethanol dehydration utilized in the industry comprise at least three steps: (1) distillation of dilute ethanol until it reaches a concentration near the azeotropic point, (2) distillation using a third component

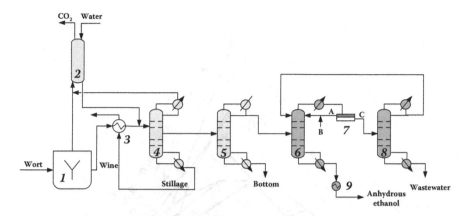

FIGURE 8.3 Technological scheme for ethanol separation and dehydration by azeotropic distillation using benzene as the entrainer: (1) fermenter, (2) scrubber, (3) preheater, (4) concentration column, (5) rectification column, (6) azeotropic column, (7) decanter, (8) column for entrainer recovery, (9) product cooler. A = benzene-enriched stream, B = benzene make-up stream, C = water-enriched stream.

added that allows the ethanol removal, and (3) distillation to recover the third component and reutilize it in the process (Montoya et al., 2005). The azeotropic distillation corresponds to this scheme. This technology consists of the addition of an entrainer to the ethanol–water mixture to form a new azeotrope. The azeotrope formed is ternary (involves three components) and allows a much easier separation in schemes involving two or three distillation columns. Among the substances most used as entrainers for separation of ethanol–water mixtures are benzene, toluene, n-pentane, and cyclohexane.

In the case of benzene, the process comprises one dehydration (azeotropic) column, which is fed with the mixture containing 90 to 92% ethanol from rectification column (Figure 8.3). The benzene is added in the upper plate. From the lower part of the azeotropic column, ethanol is removed with water content below 1%, while the overhead vapors in the column top, which correspond to a mixture with a composition equal or near to the composition of the ethanol–water–benzene ternary azeotrope, are condensed and sent to a liquid–liquid separator (decanter). Due to the mixture properties, the ternary azeotrope is located in the immiscibility zone of the ethanol–water–benzene system (Figure 8.4), so once condensed, it is separated into two liquid phases: one phase with high benzene content that is recirculated as a reflux to the azeotropic column, and the another phase with higher water content that is fed to a smaller column for entrainer recovery (stripping column). The distillate from the stripping column has a significant benzene concentration and, for this reason, this stream is recycled back to the azeotropic column or to the decanter. The bottoms of the stripping column contain mostly water. If these bottoms have an important amount of ethanol, they are recirculated to the concentration column; in this way, the separation of water and ethanol is attained and the entrainer is recovered. As the process is operated in continuous

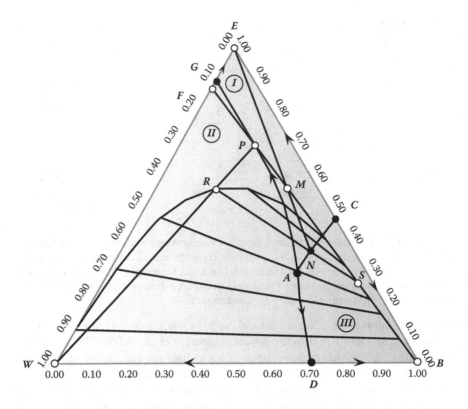

FIGURE 8.4 Ternary diagram of vapor–liquid–liquid equilibrium of ethanol–water–benzene system at 1 atm. The compositions are given in molar fractions. E = ethanol (T_b = 78.3°C), W = water (T_b = 100.0°C), B = benzene (T_b = 80.13°C), A = ternary azeotrope (T_b = 63.9°C), G = ethanol–water binary azeotrope (T_b = 78.1°C), C = ethanol–benzene binary azeotrope (T_b = 67.8°C), D = water–benzene binary azeotrope (T_b = 69.2°C). T_b refers to the boiling point. Distillation regions are indicated by Roman numerals.

regime, the benzene is permanently recirculating within the system. Nevertheless, small amounts of this compound leave the scheme along with the ethanol or water streams, thus a make-up stream is required. This latter stream is fed to the first plate (from the top) of the azeotropic column or is mixed with the reflux stream coming from the decanter to this same column.

The phase equilibrium properties are crucial for the design of an azeotropic distillation scheme. This equilibrium can be represented in the ternary diagram shown in Figure 8.4 for the case of benzene. The principles of topologic thermodynamics can be applied to the analysis of this diagram (Pisarenko et al., 2001). To provide more clarity, molar fractions are employed through the analysis are identical for compositions expressed in mass fractions. The feeding of the starting ethanol–water mixture is indicated by the straight line FB and is accomplished in such a way that the point M representing the pseudo-starting state of the system is located inside the distillation region I. This region is delimited by the two distillation boundaries, which coincide in the ternary azeotrope with the minimum boiling point. The distillation boundaries define the process constraints because any distillation operation (indicated by straight lines of mass balances) cannot have distillates and bottoms whose compositions are in different regions. When drawing a balance line corresponding to the indirect distillation for the point M (the prolongation of the straight line EM until the distillation boundary represented by the curve AC), bottoms with a composition corresponding to pure ethanol E and distillate with a composition near to the ternary azeotrope represented by the point N are obtained. The composition of this distillate corresponds to the immiscibility zone of the system so it is separated into two liquid phases indicated by the points R and S that are determined following the tie lines of the liquid–liquid equilibrium plot (bimodal plot). The point R represents the liquid phase with higher water content (the raffinate) and the point S represents the liquid phase with higher benzene content (the extract) that is evidenced by its higher proximity to the vertex B (pure benzene) compared to point R. The stream with the composition of the point S is recirculated as the reflux to the azeotropic column. The raffinate stream, in turn, undergoes distillation in the stripping column, which is represented by the balance line WRP that is located in the distillation region II. The composition of point P corresponds to the composition of the distillate stream from the stripping column that is recycled back to either the azeotropic column or the decanter. This type of analysis allows one to predict the behavior of the system without carrying out a rigorous assessment (short-cut method). These short-cut methods allow one to obtain valuable information for the subsequent rigorous modeling of the system. In particular, the application of these methods facilitates the specification of the operating conditions in the distillation columns when commercial process simulators employing rigorous methods are used.

The above-described distillation receives the name *hetero-azeotropic distillation* considering that the entrainers form azeotropes located within the immiscibility zone of the system. This implies its separation into two liquid phases. The utilization of *n*-octane as a co-entrainer along with benzene has been proposed in order to decrease the energy costs of the traditional process (Chianese and Zinnamosca, 1990). The simulation and optimization accomplished based on a mass-transfer model (nonequilibrium model) for this process show that if the values of operating parameters of the column are adjusted to minimize the amount of plates in the azeotropic column, it is possible to reduce the capital costs, but increasing the heat flow rates required implies an increase of the energy costs. In terms of energy costs, the most influencing process parameters are the reflux ratio and flow rate of the stream recirculated from the stripping column to the azeotropic column. For these parameters, their optimum values have been obtained according to economic considerations (Mortaheb and Kosuge, 2004). However, the utilization of benzene as an entrainer is not desirable due to its carcinogenic properties. In addition, the azeotropic distillation using this compound leads to the appearance of multiple steady states and the occurrence of a parametric sensibility related to small changes in column pressure (Wolf Maciel and Brito, 1995). Taking into account these drawbacks, the use of less contaminant entrainers has been attempted. In particular, some new ethanol-producing facilities in Brazil employ cyclohexane as the entrainer for ethanol dehydration by hetero-azeotropic distillation.

Significant efforts to reduce the elevated energy consumption of the azeotropic distillation using such entrainers, such as benzene, cyclohexane, diethyl ether, and *n*-pentane, have been made. Under real conditions, distillation columns are configured in such a way that the dehydration process is operated using the heat recovered from the primary distillation system (concentration and rectification columns). Alternatively, rectification and stripping columns can be operated using the heat released in the azeotropic column. For fuel ethanol industry in the United States, the consumption of thermal energy during the separation and dehydration steps employing the azeotropic distillation is about 4.73 MJ/L ethanol on average (Madson and Monceaux, 1995).

8.2.3 EXTRACTIVE DISTILLATION

In this type of distillation, a third substance called an extractive agent or solvent is added to the ethanol–water system. The extractive agent modifies the relative volatility of the mixture components without forming new azeotropes and allowing, in this manner, the separation.

The solvent added should have a low volatility such that its separation in the second distillation column where it is recovered would be much easier. In addition, the solvent should have a high boiling point. It has been suggested that the high relative viscosity of the solvents employed in extractive distillation could decrease the mass transfer efficiency of the process (Meirelles and Telis, 1994). Ethylene glycol has been traditionally used for these purposes, but the energy costs are higher than in the case of the azeotropic distillation using benzene. Nevertheless, Meirelles et al. (1992) point out that extractive distillation can become competitive under specific operating conditions. In fact, in a previous work that implied the simulation of ethanol–water mixtures resulting from fermentation broths by extractive and azeotropic distillation (Montoya et al., 2005), the possibility of achieving lower energy costs during the extractive distillation using ethylene glycol was demonstrated (see Section 8.3).

The technological scheme for ethanol dehydration by extractive distillation using ethylene glycol is shown in Figure 8.5. The solvent is fed into a tray located above the ethanol feed stream coming from the rectification column. Unlike the azeotropic distillation, anhydrous ethanol is recovered in the distillate stream of the extractive column, while a stream with a ternary composition containing almost the total amount of ethylene glycol is removed from the bottoms. This stream is sent to the column for solvent recovery where the ethylene glycol is collected in the bottoms thanks to its low volatility. These bottoms are recycled back to the extractive column. Water contained in the starting mixture is recovered in the distillate of the recovery column. When ethylene glycol is recirculated, the reduction of stream temperature down to 80°C is required in order to feed it to the extractive column ensuring, in this way, a better separation. This is achieved by exchanging heat with the feed stream of this column (Montoya et al., 2005).

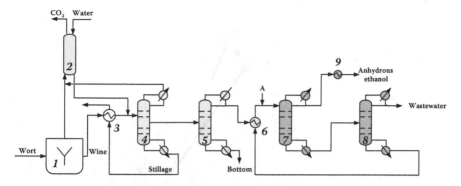

FIGURE 8.5 Technological scheme for ethanol separation and dehydration by extractive distillation using ethylene glycol as the solvent: (1) fermenter, (2) scrubber, (3) preheater, (4) concentration column, (5) rectification column, (6) heat exchanger, (7) extractive column, (8) column for solvent recovery, (9) product cooler. A = ethylene glycol make-up stream.

To avoid a possible thermal decomposition of ethylene glycol due to its high boiling point, the recovery column operates at 0.2 atm in such a way that the bottoms temperature does not reach 150°C.

The effect of the solvent can be considered as the breaking of the binary azeotrope between the ethanol and water. The ethylene glycol modifies the relative volatility of ethanol in such a way that it can then be recovered through direct distillation. The ternary diagram corresponding to the extractive distillation scheme is illustrated in Figure 8.6. The solvent is added to the starting mixture F. The straight line FG represents this. The point M indicates the starting state of the system within the extractive column. The dashed curve AI' represents those compositions with a relative volatility equal unity (iso-volatility curve). As the point M is located in the region where the relative volatility of the ternary mixtures are greater than unity, their direct distillation is possible. This is represented by the balance line EM, which is prolonged until point N in the base of the ternary diagram. This point corresponds to the binary water–ethylene glycol mixtures. In this way, pure ethanol distillate (E) and bottoms containing water and ethylene glycol are obtained. These bottoms are fed to the recovery column. In this column, the distillation starting from point N implies the production of pure water distillate (point W) and pure ethylene glycol bottoms (point G), which are recycled back to the extractive column.

For small- and medium-scale ethanol producing facilities, batch extractive distillation is more suitable than the continuous process due to its higher flexibility to handle the separation of small volumes and feed streams with great variability in their compositions under limited capital costs conditions. In particular, the task of finding the best conditions to operate a batch extractive distillation process that includes a single tank located between two columns (rectification and stripping columns) using ethylene glycol as the solvent has been undertaken (Ruiz Ahón and Luiz de Medeiros, 2001). Varying the feed state and the purity degree and employing ideal thermodynamic models, they defined the best policy

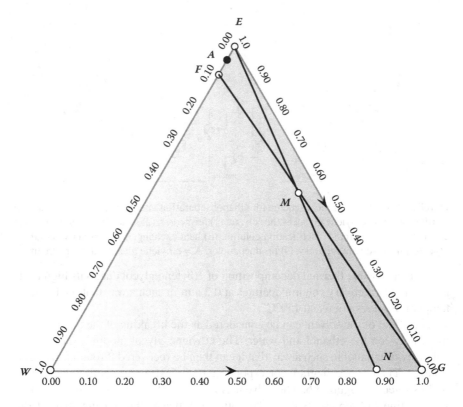

FIGURE 8.6 Ternary diagram of the vapor–liquid equilibrium for ethanol–water–ethylene glycol system at 1 atm. The compositions are given in mass fractions E = ethanol (T_b = 78.3°C), W = water (T_b = 100.0°C), G = ethylene glycol (T_b = 197.1°C), A = ethanol–water binary azeotrope (T_b = 78.1°C). T_b refers to the boiling point.

for removing the streams in the top of the rectification column and in the bottoms of the stripping column in such a way that anhydrous ethanol could be obtained.

In early works, gasoline was proposed as a solvent for ethanol dehydration by extractive distillation (Chianese and Zinnamosca, 1990). In this case, gasoline (C_7 and C_8 oil fractions) causes the inversion of the water–ethanol volatility allowing the removal of water with small amounts of residual ethanol and light hydrocarbon fractions from the column top while ethanol mixed with the solvent is withdrawn from the bottoms (unlike the ethylene glycol). No water is present in these bottoms so this stream can be used as a gasohol (a blend of gasoline and ethanol as the oxygenate). Models contemplating the dynamics of extractive distillation for ethanol dehydration have been developed (Wolf Maciel and Brito, 1995). These models take into account the simultaneous mass and heat transfer, hydraulic behavior of the trays, and rigorous representation of the phase equilibrium. Based on these studies, it is noted that the composition of the distillate stream that contains the anhydrous ethanol varies slightly with changes in the solvent flowrate, though it does vary with changes in reflux and feed flowrates.

8.2.4 SALINE EXTRACTIVE DISTILLATION

The use of saline extractive agents in extractive distillation processes has been studied in the past few years. Among these agents, potassium acetate should be highlighted (Sánchez and Cardona, 2005). As the salts are nonvolatile components, the distillate obtained in distillation columns is much easier to separate. Therefore, the energy costs are lower compared to traditional extractive distillation. The salt effect consists of the preferential solvation of the ions formed from the salt dissociation with the less volatile component that leads to the increase of the relative volatility of the most volatile component (ethanol).

Two possible configurations have been simulated for ethanol dehydration using potassium acetate (Ligero and Ravagnani, 2003). In the first case, the starting point is a diluted ethanol solution that is fed to the saline extractive distillation column followed by a multistage evaporation and spray drying for salt recovery. In the column distillate, the anhydrous ethanol is obtained due to the breaking of the azeotrope induced by the salt. The second case has a previous distillation column to concentrate the ethanol solution, which is fed to the saline extractive distillation column with the subsequent salt recovery through spray drying. This second option presents best results from the viewpoint of energy costs. In addition, the salt used is not toxic, which indicates that this process can replace the traditional azeotropic distillation that uses such toxic entrainers as benzene.

Rigorous models for saline extractive distillation for ethanol–water–$CaCl_2$ system have been developed. These models take into account the contribution of the salt to the liquid phase enthalpy (Llano-Restrepo and Aguilar-Arias, 2003). This type of model allows simulating the phase equilibrium and thermodynamic properties of solvent–electrolyte mixed systems having a strong nonlinear character that becomes an obstacle for their study. In this way and starting from theoretical considerations, the production of water-free ethanol using $CaCl_2$ as the separation agent was predicted. These rigorous models can make possible the utilization of this type of process again by the industry considering that saline extractive distillation was discarded in the past due to, among other reasons, the difficulty in its design, which was caused by the complexity of the nonideal behavior of the phase equilibrium and the intricate modeling of the thermodynamic properties of these systems. This process was also discarded because of the technical problems derived from the dissolution and subsequent crystallization of the salt, as well as the need to use materials resistant to the corrosion (Pinto et al., 2000).

Besides inorganic salts, the possibility of employing hyperbranched polymers like polyesteramides and hyperbranched polyesters during the extractive distillation of ethanol–water mixtures has also been studied. These polymers exhibit a remarkable separation efficiency and selectivity, and their physical–chemical properties can be tailored depending on the required application. The recovery of these polymers can be carried out by washing, evaporation, drying, or crystallization (Sánchez and Cardona, 2005; Seiler et al., 2003).

8.2.5 ADSORPTION

Adsorption is another unit operation widely employed in the industry for ethanol dehydration. In this operation, the ethanol–water mixture passes through an
apparatus (usually cylindrical) that contains a bed with an adsorbent material.
Due to the difference in the affinity of molecules of water and ethanol with
respect to the adsorbent, the water remains entrapped in the bed while the ethyl
alcohol passes though this same bed increasing its concentration in the stream
leaving the apparatus.

In early work carried out in the 1980s, the utilization of different biomaterials
as adsorbents for ethanol dehydration, such as corn, xylan, pure cellulose, corn stover, corn flour, wheat straw, and sugarcane bagasse, was proposed. Such materials
whose structure is based on polysaccharides demonstrated their ability to separate
water and ethanol (Hong et al., 1982; Ladisch and Dyck, 1979; Ladisch et al., 1984;
Westgate and Ladisch, 1993). In particular, the broken corn grains showed good
properties to absorb water in aqueous solutions of ethanol. In pilot plant studies, the possibility of concentrating ethanol from 91% to more than 99% using a
bed of broken corn grains was demonstrated. The bed can be used during several
adsorption–regeneration cycles (up to more than 30; Tanaka and Otten, 1987).

The bed can then be utilized as a starch source for ethanol production. In
principle, the energy costs were lower than those of azeotropic distillation. More
recently, some works have shown the great adsorption ability of different starchy
materials that have adsorptive properties similar to those of the inorganic adsorbents when the mixture to be separated contains about 10% water. In addition, the
enzymatic modification using α-amylase contributes to enhance the adsorptive
properties of starch (Beery and Ladisch, 2001a, b).

However, the adsorption of water employing the so-called molecular sieves to
dehydrate ethanol has been the technology that has acquired more development in
past years in the fuel ethanol industry. In fact, this technology has been replacing
the azeotropic distillation. The molecular sieves are granular rigid materials with
a spherical or cylindrical shape manufactured from potassium aluminosilicates.
They are classified according to the nominal diameter of the large amount of
internal pores that provide access to the interstitial free volume found in their
microcrystalline structure (Madson and Monceaux, 1995). For ethanol dehydration, sieves with an average diameter of the interstitial passageways of 3 ang-
stroms (Type 3Å sieves) are commonly used. The water molecule has a diameter
lower than that of the interstitial passageways of this type of sieves, while the
ethanol molecule does not. In addition, the water can be adsorbed onto the internal surface of the passageways in the molecular sieve structure. Thus, ethanol
molecules pass out of the apparatus without adsorbing onto the bed.

The adsorption operation requires that, once the adsorbent bed is saturated
with the substance to be separated, the desorption of this substance should be
accomplished to make possible the reutilization of the adsorbent material (regeneration cycle). For regeneration of the sieves, hot gas is needed that rapidly dete-
riorates them especially if the bed is fed with a liquid stream during the previous

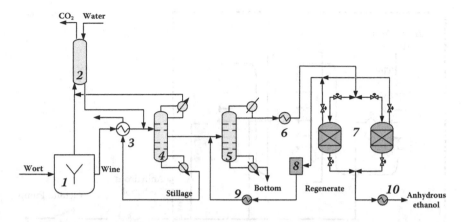

FIGURE 8.7 Technological scheme for ethanol separation and dehydration by adsorption using molecular sieves: (1), fermenter, (2) scrubber, (3) preheater, (4) concentration column, (5) rectification column, (6) heat exchanger, (7) molecular sieves, (8) regenerate tank, (9) head exchanger, (10) product cooler.

water adsorption cycle. To counter this deterioration, the pressure-swing adsorption (PSA) technology was developed. This technology involves the use of two adsorption beds. While one bed produces vapors of anhydrous ethanol superheated under pressure, the other one is regenerated under vacuum conditions by recirculating a small portion of superheated ethanol vapors through the saturated sieves (Figure 8.7). The system feeding is carried out using the overhead vapors from the rectification column. The ethanolic vapors obtained in the regeneration cycle and that can contain 28% water are recirculated to the rectification column (Montoya et al., 2005; Wooley et al. 1999). In this way, the molecular sieves life can be prolonged for several years, which, in turn, leads to very low costs related to the replacement of adsorbent material and, therefore, reduced operating costs (Guan and Hu, 2003; Madson and Monceaux, 1995).

8.2.6 PERVAPORATION

Different applications of the membranes for both concentration of ethanol solutions and ethanol dehydration have been developed recently. Although the reverse osmosis was first proposed (Leeper and Tsao, 1987), the pervaporation definitively boosted the introduction of membranes into the fuel ethanol industry (Sánchez and Cardona, 2005). The pervaporation (evaporation through membranes) is an operation based on the separation of two components by a selective membrane under a pressure gradient in which the component passing across the membrane is removed as a gaseous stream (permeate), while the other component remains in the liquid phase and is removed as a more concentrated stream (retentate), as shown in Figure 8.8. This operation began to exhibit industrial feasibility when polyvinyl alcohol (PVA) composite membranes were developed at the end of 1980s. These membranes present a high selectivity by favoring the water passing across them

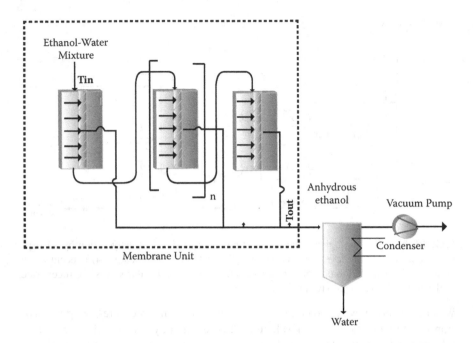

FIGURE 8.8   Schematic diagram of pervaporation for ethanol dehydration. $T_{in} > T_{out}$.

and a high retention of several organic solvents. Just the PVA membranes allow the change of the phase equilibrium of ethanol–water solutions in such a way that the mass transfer between the liquid and vapor phases is determined by the semi-permeable membrane and not by the free interphase (Sander and Soukup, 1988). Thus, it is possible to obtain a concentrated ethanol stream with a composition above the azeotropic. The driving force of pervaporation is maintained thanks to the application of a vacuum in the permeate side of the membrane. This driving force is evidenced by the difference of partial pressures or activities of the component passing across the membrane. This difference of partial pressures can be increased making the feed temperature as high as possible. The permeate is later condensed to generate a liquid stream that can be recycled back to the rectification column. Diverse chemical modifications of PVA membranes have been proposed to improve the hydrophilic/hydrophobic ratio in order to enhance the selectivity and flux of these membranes (Chiang and Chen, 1998).

The pervaporation offers a series of advantages compared to azeotropic or extractive distillations because the product contains no entrainer or solvent traces. In addition, this technology is easily adjustable and very flexible with respect to the changes of the feed concentration. The start-up and stop of the process require minimum labor and supervision. Finally, the pervaporation units are compact and need no large areas compared to the big towers of azeotropic distillation schemes (Sánchez and Cardona, 2005; Sander and Soukup, 1988). As this operation consumes less energy than the conventional operations based on distillation, the use of pervaporation has become very promising considering the permanent development of membrane manufacturing technologies, which have allowed the increase of membrane selectivity and permeate flux.

Besides the flat membranes, hollow-fiber membranes have been manufactured to dehydrate ethyl alcohol. Tsuyomoto et al. (1991) produced this type of membrane by the partial hydrolysis of polyacrylonitrile in order to introduce carboxylic groups and obtain a polyionic complex. These membranes were successfully tested to concentrate 95% ethanol solutions at 60°C showing separation factors higher than 5,000. Currently, there exist about 100 plants worldwide employing ethanol and isopropanol dehydration (Baker, 2004), some of them located in facilities for ethanol production from corn. In practice, the chemical process industries use only plate-and-frame modules for pervaporation. These units are sealed with graphite compression gaskets that universally resist organic liquids. Spiral-wound and hollow-fiber modules require adhesives that are not resistant to all solvents (Wynn, 2001).

Other separation technology based on membranes has been proposed (the vapor permeation), which has been much less studied than pervaporation (Sánchez and Cardona, 2005). This process has the same principle of pervaporation, but the feed stream is also gaseous. Vapor permeation offers certain advantages over pervaporation (Jansen et al., 1992). First of all, the construction of vapor permeation modules is simpler because evaporation heat is not required, i.e., there is no need to heat the fractions of the retentate feeding the subsequent modules. In practice, the flux of vapor permeation is higher than the pervaporation flux because the driving force is less affected by the polarization of concentration and by the reduction of partial pressures. Second, increasing the pressure or decreasing the superheating can significantly increase the flux. Moreover, the membrane can be impregnated with different substances (especially salts) to enhance or regulate its selectivity and flux, which is favored by the fact that the membrane is not in contact with the liquid phase. For instance, the impregnation of composite PVA membranes with cesium fluoride (CsF) provoked almost a two-fold increase of the flux during the vapor permeation of 95% ethanol mixtures.

8.3 EVALUATION OF SEPARATION AND DEHYDRATION SCHEMES

For process synthesis procedures intended for the design of fuel ethanol production processes, the selection of the technological scheme for ethanol separation and dehydration has paramount importance. Thus, the utilization of salts as extractive agents (saline extractive distillation) has demonstrated certain energetic advantages compared to other dehydration schemes, according to some reports (Barba et al., 1985; Llano-Restrepo and Aguilar-Arias, 2003). Cited results indicate that energy costs of saline distillation were lower than in the case of azeotropic distillation (using benzene, pentane, or diethyl ether), extractive distillation (using ethylene glycol or gasoline), or solvent extraction, being almost the same as the costs of pervaporation. Pinto et al. (2000) used Aspen Plus® for simulation and optimization of the saline extractive distillation for several substances (NaCl, KCl, KI, and CaCl$_2$). This configuration was compared to the simulated scheme of conventional extractive distillation with ethylene glycol and to data for azeotropic distillation. Obtained results showed considerably lower energy consumption for the process with salts. However, for this latter case, the recovery of salts was not simulated. Thus, if evaporation and recrystallization of salts is contemplated, energy requirements could significantly increase, taking into account the energy expenditures. In this way, the utilization of commercial simulators shows the viability for predicting the behavior of a given process configuration providing the appropriate thermodynamic models of studied systems, as illustrated in the following case study.

CASE STUDY 8.1   EVALUATION OF DEHYDRATION STEP FOR ETHANOL PRODUCTION

In a previous work (Quintero et al., 2007), a simulation of the ethanol dehydration process in addition to an evaluation of energy and capital costs was performed. For this, several representative configurations were compared under the same input conditions as the case of a fermented mash with a given ethanol concentration (11.4%) and considering the physical–chemical properties of the components of the culture broth. To this end, commercial process simulators were utilized for solving mass and energy balances and performing the economic evaluation.

The simulation of mass and energy balances was accomplished by using the commercial package Aspen Plus v11.1 (Aspen Technology, Inc., Burlington, MA). The composition of the feed stream to the separation and dehydration steps corresponded to the culture broth resulting from the fermentation step after the cell biomass removal by centrifugation. The composition of this stream for the particular case of fermentation of starchy materials is as follows (in % by weight): water 63.70, ethanol 11.40, CO$_2$ 10.80, protein 7.44, linoleic acid 1.71, hemicellulose 1.50, cellulose 1.12, oleic acid 1.40, ash 0.38, glucose 0.22, dextrins 0.20, lignin 0.05, starch 0.01, and others. The flowrate of this feed stream was set to 152,153 kg/h for all the separation and dehydration schemes analyzed. Part of the data for simulation of physical properties was obtained from Wooley and Putsche (1996) who compiled information from the literature, estimated the properties when necessary, and determined a consistent set of physical properties for some key components of fuel ethanol production process. The remaining data were obtained from secondary sources of information (handbooks, monographs, papers, presentations, etc.). During the simulation, the nonrandom two-liquid (NRTL) thermodynamic model was used to calculate the activity coefficients in the liquid phase and the Hayden–O'Conell equation of state was used to model the vapor phase. The biological transformations were simulated based on a stoichiometric approach as shown in the Chapter 7.

Before the definitive simulation of the schemes, a preliminary simulation of the processes involving distillation columns was performed. The preliminary simulation of such columns was done by using the DSTWU module of Aspen Plus, which employs a short-cut method based on the equations and correlations of Win–Underwood–Gilliland. This module was chosen taking into account the presence of the binary ethanol–water azeotrope. This module also provides an initial estimation of the minimum amount of theoretical stages, minimum reflux ratio, location of the feed stage, and component distribution.

Once performed, the preliminary simulation distillation processes were analyzed using rigorous thermodynamic models. Thus, the rigorous calculation of operating conditions in distillation columns was carried out using the module RadFrac of Aspen Plus based on the method inside-out that employs the MESH equations. This method implies the simultaneous solving of mass balance equations (M), phase equilibrium equations (E), expressions for summation of the compositions (S), and heat balance equations (H) for all the components in all the stages of the given distillation column. Other than the results obtained by the DSTWU module, the information required by the simulator for the specification of input data of distillation columns was obtained with the help of short-cut methods based on the principles of the topological thermodynamics, especially the analysis of the statics (Pisarenko et al., 2001). Sensitivity analyses were performed to study the effect of main operating variables (reflux ratio, temperature of feed streams, ratio between the solvent and feed, etc.) on the purity of ethanol obtained and the energy consumption. The final result was the establishment of the operating conditions that allow accomplishing more efficient ethanol dehydration processes. The estimation of the energy consumption was carried out based on the results of the simulation, taking into account the thermal energy required by the heat exchangers and reboilers. Capital costs, overall operating costs, and other related expenditures were calculated by using Aspen Icarus Process Evaluator™ v11.1 (Aspen Technology, Inc.).

Pressure-swing adsorption was the studied technology for the analysis of ethanol dehydration by adsorption with molecular sieves. The simulation of this process considered that the adsorption was carried out in vapor phase so the distillate from the rectification column was not condensed and its temperature was raised to 116°C before sending it to the adsorption unit. The technology simulated corresponded to the PSA. The desorption cycle considered vacuum conditions at 0.14 atm. The vapors from desorption were recirculated to the rectification column where the used ethanol was recovered (see Figure 8.7). Vacuum distillation, azeotropic distillation using benzene as the entrainer and extractive distillation using ethylene glycol as the solvent were also analyzed as ethanol dehydration configurations. Each one of the schemes included a scrubber for recovery of 98% of ethanol volatilized and a preheater that brought the stream to be fed to the first distillation column until its saturation point. To concentrate the wine, two distillation columns were considered. In the first (concentration) column, ethanol content was raised up to 50%, while in the second one (rectification column), the ethanol concentration reached 91%. Both columns, the scrubber, and the pre-heater were included in all the dehydration schemes simulated (see Figure 8.1).

Obtained results showed the inconvenience of using pressure-swing (vacuum) distillation for dehydrating ethanol. The simulation indicated that large distillation columns with many stages (above 40) are required to obtain a high purity product. In addition, high reflux ratios are needed. These conditions imply high energy costs due to the high heat duty of the column reboilers and to the maintenance of vacuum conditions in the second column having a large amount of trays. Thus, the energy consumption of this dehydration scheme reaches 12.17 MJ/L of ethanol. According to the performed calculations, the distillate of the vacuum column has an ethanol content of 99.3%. The capital costs of this scheme along with the corresponding costs of the other dehydration schemes are presented in Table 8.1.

TABLE 8.1

Capital and Operating Costs (in US$) for Different Separation and Dehydration Technologies Used for Fuel Ethanol Production

Item	Units	Vacuum Distillation	Azeotropic Distillation	Extractive Distillation	Molecular Sieves	References
Ethanol produced	kg/yr	141,560,084	142,609,349	141,897,940	142,726,998	Montoya et al. (2005);
Total capital costs	US$	14,156,063	9,547,963	9,525,920	12,809,706	Quintero et al. (2007);
Total operation costs	US$/yr	11,539,808	8,943,642	8,023,714	7,730,563	Sánchez (2008)
Utilities	US$/yr	9,063,508	7,113,850	6,266,715	53,821,429	
Labor	US$/yr	600,000	600,000	600,000	600,000	
Maintenance costs	US$/yr	381,000	78,000	75,100	191,000	
Other	US$/yr	1,495,300	1,151,592	1,081,899	1,118,134	
Unit capital costs	US$/kg	0.1000	0.0670	0.0671	0.0897	
Unit operation costs	US$/kg	0.0815	0.0627	0.0565	0.0542	

In contrast, adsorption with molecular sieves showed the best results regarding operation costs, i.e., this technology presents lower energy costs (7.68 MJ/L). Elevated capital costs for the configuration involving adsorption are related to the complexity of the automation and control system inherent to the pressure swing adsorption technology. The higher energy consumption for azeotropic (9.77 MJ/L) and extractive (8.44 MJ/L) distillation is explained by the presence of two additional distillation towers that increase the energy costs. The product for these latter schemes contains traces of the entrainer or the solvent unlike the dehydration by adsorption where these third components are not utilized. Simulation results indicate that the extractive distillation can be competitive compared to azeotropic distillation from an energy point of view, as pointed out by Meirelles et al. (1992). According to the simulations performed, the amount of ethylene glycol required to attain the desired dehydration of ethanol was 17,900 kg/h, but it is necessary to create only 50 kg/h of this solvent thanks to the recirculation stream from the recovery column. The amount of benzene required for azeotropic distillation was 19,980 kg/h, but the amount of fresh benzene in the make-up stream was only 17 kg/h. It is worth noting that the convergence of the simulation of this dehydration scheme was a difficult task that required numerous successive simulation runs. This behavior can be explained by the appearance of multiple steady states and the presence of a parametric sensitivity with small changes in the column pressure when benzene is the entrainer used, as indicated by Wolf and Brito (1995). The obtained results represented a suitable approximation of the results published in different sources (Chianese and Zinnamosca, 1990; Luyben, 2006) as well as the predictions of the thermodynamic-topological analysis. It should be emphasized that the azeotropic distillation using benzene is not an environmentally friendly process and the operation of such dehydration schemes imply the utilization of a carcinogenic substance that can involve potential risks for the operating staff.

From outcomes achieved, it is evident that process simulation represents a powerful tool to design the downstream processes of such biotechnological processes as fuel ethanol production. Similarly, the importance of applying a suitable thermodynamic approach to study the separation operations is another crucial factor influencing the success of the simulation procedures.

REFERENCES

Baker, R.W. 2004. *Membrane technology and applications*. Chichester, U.K.: John Wiley & Sons.

Barba, D., V. Brandani, and G. di Giacomo. 1985. Hyperazeotropic ethanol salted-out by extractive distillation. Theoretical evaluation and experimental check. *Chemical Engineering Science* 40:2287–2292.

Beery, K.E., and M.R. Ladisch. 2001a. Adsorption of water from liquid-phase ethanol–water mixtures at room temperature using starch-based adsorbents. *Industrial & Engineering Chemistry Research* 40:2112–2115.

Beery, K.E., and M.R. Ladisch. 2001b. Chemistry and properties of starch based dessicants. *Enzyme and Microbial Technology* 28:573–581.

Chianese, A., and F. Zinnamosca. 1990. Ethanol dehydration by azeotropic distillation with a mixed-solvent entrainer. *The Chemical Engineering Journal* 43:59–65.

Chiang, W.-Y., and C.-L. Chen. 1998. Separation of water-alcohol mixture by using polymer membranes - 6. Water-alcohol pervaporation through terpolymer of PVA grafted with hydrazine reacted SMA. *Polymer* 39 (11):2227–2233.

Guan, J., and X. Hu. 2003. Simulation and analysis of pressure swing adsorption: Ethanol drying process by the electric analogue. *Separation and Purification Technology* 31:31–35.

Hong, J., M. Voloch, M.R. Ladisch, and G.T. Tsao. 1982. Adsorption of ethanol–water mixtures by biomass materials. *Biotechnology and Bioengineering* 26:725–730.

Jansen, A.E., W.F. Versteeg, B. van Engelenburg, J.H. Hanemaaijer, and B.Ph. ter Meulen. 1992. Methods to improve flux during alcohol/water azeotrope separation by vapor permeation. *Journal of Membrane Science* 68:229–239.

Ladisch, M.R., and K. Dyck. 1979. Dehydration of ethanol: New approach gives positive energy balance. *Science* 205 (4409):898–900.

Ladisch, M.R., M. Voloch, J. Hong, P. Bienkowski, and G.T. Tsao. 1984. Cornmeal adsorber for dehydrating ethanol vapors. *Industrial & Engineering Chemistry Process Design and Development* 23 (3):437–443.

Leeper, S.A., and G.T. Tsao. 1987. Membrane separations in ethanol recovery: An analysis of two applications of hyperfiltration. *Journal of Membrane Science* 30:289–312.

Ligero, E., and T. Ravagnani. 2003. Dehydration of ethanol with salt extractive distillation—A comparative analysis between processes with salt recovery. *Chemical Engineering and Processing* 42:543–552.

Llano-Restrepo, M., and J. Aguilar-Arias. 2003. Modeling and simulation of saline extractive distillation columns for the production of absolute ethanol. *Computers and Chemical Engineering* 27 (4):527–549.

Luyben, W. 2006. Control of a multiunit heterogeneous azeotropic distillation process. *AIChE Journal* 52 (2):623–637.

Madson, P.W., and D.A. Monceaux. 1995. Fuel ethanol production. In *The Alcohol Textbook*, ed. T.P. Lyons, D.R. Kelsall, and J.E. Murtagh. Nottingham, U.K.: University Press.

Meirelles, A., and V. Telis. 1994. Mass transfer in extractive distillation of ethanol/water by packed columns. *Journal of Chemical Engineering of Japan* 27 (6):824–827.

Meirelles, A., S. Weiss, and H. Herfurth. 1992. Ethanol dehydration by extractive distillation. *Journal of Chemical Tecnology and Biotecnology* 53:181–188.

Montoya, M.I., J.A. Quintero, Ó.J. Sánchez, and C.A. Cardona. 2005. Efecto del esquema de separación de producto en la producción biotecnológica de alcohol carburante (Effect of the product separation scheme in the biotechnological production of fuel ethanol, in Spanish). Paper presented at the II Simposio Sobre Biofábricas, Medellín, Colombia.

Mortaheb, H.R., and H. Kosuge. 2004. Simulation and optimization of heterogeneous azeotropic distillation process with a rate-based model. *Chemical Engineering and Processing*. 43:317–326.

Pinto, R.T.P., Wolf-Maciel. M.R., and L. Lintomen. 2000. Saline extractive distillation process for ethanol purification. *Computers and Chemical Engineering* 24:1689–1694.

Pisarenko, Yu.A., L.A. Serafimov, C.A. Cardona, D.L. Efremov, and A.S. Shuwalov. 2001. Reactive distillation design: Analysis of the process statics. *Reviews in Chemical Engineering* 17 (4):253–325.

Quintero, J.A., M.I. Montoya, Ó.J. Sánchez, and C.A. Cardona. 2007. Evaluación de la deshidratación de alcohol carburante mediante simulación de procesos (Evaluation of fuel ethanol dehydration through process simulation, in Spanish). *Biotecnología en el Sector Agropecuario y Agroindustrial* 5 (2):72–83.

Ruiz Ahón, V., and J. Luiz de Medeiros. 2001. Optimal programming of ideal and extractive batch distillation: Single vessel operations. *Computers and Chemical Engineering* 25:1115–1140.

Sánchez, Ó.J. 2008. Síntesis de Esquemas Tecnológicos Integrados para la Producción Biotecnológica de Alcohol Carburante a partir de tres Materias Primas Colombianas (Synthesis of Integrated Flowsheets for Biotechnological Production of Fuel Ethanol from Three Colombian Feedstocks, Ph.D. thesis, in Spanish), Departamento de Ingeniería Química, Universidad Nacional de Colombia sede Manizales, Manizales.

Sánchez, Ó.J., and C.A. Cardona. 2005. Producción biotecnológica de alcohol carburante II: Integración de procesos (Biotechnological production of fuel ethanol II: Process integration, in Spanish). *Interciencia* 30 (11):679–686.

Sander, U., and P. Soukup. 1988. Design and operation of a pervaporation plant for ethanol dehydration. *Journal of Membrane Science* 36:463–475.

Seiler, M., D. Köhler, and W. Arlt. 2003. Hyperbranched polymers: New selective solvents for extractive distillation and solvent extraction. *Separation and Purification Technology* 30:179–197.

Tanaka, B., and L. Otten. 1987. Dehydration of aqueous ethanol. *Energy in Agriculture* 6:63–76.

Tsuyomoto, M., H. Karakane, Y. Maeda, and H. Tsugaya. 1991. Development of poly-ion complex hollow fiber membrane for separation of water-ethanol mixtures. *Desalination* 80:139–158.

Westgate, P.J., and M.R. Ladisch. 1993. Sorption of organics and water on starch. *Industrial & Engineering Chemistry Research* 32:1676–1680.

Wolf Maciel, M.R., and R.P. Brito. 1995. Evaluation of the dynamic behavior of an extractive distillation column for dehydration of aqueous ethanol mixtures. *Computers and Chemical Engineering* 19 (Suppl.):S405–S408.

Wooley, R., and V. Putsche. 1996. Development of an ASPEN PLUS physical property database for biofuels components. Report NREL/MP-425-20685, National Renewable Energy Laboratory, Washington, D.C.

Wooley, R., M. Ruth, J. Sheehan, K. Ibsen, H. Majdeski, and A. Galvez. 1999. Lignocellulosic biomass to ethanol process design and economics utilizing co-current dilute acid prehydrolysis and enzymatic hydrolysis. Current and futuristic scenarios. Technical Report NREL/TP-580-26157, National Renewable Energy Laboratory, Washington, D.C.

Wynn, N. 2001. Pervaporation comes of age. *CEP* 97 (10):66–72.

9 Integrated Processes for Fuel Ethanol Production

The intensification of fuel ethanol production processes has become a priority during the design of technological configurations with enhanced technical, economic, and environmental performance. Process integration achieves this intensification through the combination of several unit operations and processes in a same single unit or through the better utilization of the energy flows. In this chapter, main aspects of the intensification of ethanol production by different ways of process integration (reaction–reaction, reaction–separation, separation–separation, heat integration) are presented. Examples of the application of the integration principle to bioethanol production are provided emphasizing their efficiency from an energy point of view. The potential offered by the implementation of technologies with a high degree of integration, such as the simultaneous saccharification and co-fermentation and consolidated bioprocessing, are discussed highlighting their advantages and limitations as well as their possibilities for further development. In addition, some configurations involving the ethanol removal from culture broth are analyzed because one of the most important challenges in ethanol production is the reduction of end product inhibition on the growth rate of ethanol-producing microorganisms.

9.1 PROCESS INTEGRATION

Process efficiency plays a crucial role when the performance of different technological configurations are to be considered. To reach this improved efficiency, several conventional approaches may be used, which make possible to a certain degree the intensification of processes, a necessary condition for the design of technologies with enhanced performance. But for attaining a higher degree of process intensification, the application of new concepts within the framework of a new paradigm is required—a product and process engineering paradigm, according to Stankiewicz and Moulijn (2002). These authors define *process intensification* as the development of new equipment and procedures leading to a "dramatic improvement" in chemical processes through the reduction of the ratio between the equipment size and the production capacity, energy consumption, and waste production, resulting in cheaper and sustainable technologies, i.e., any chemical engineering development leading to a significantly smaller, cleaner, and more energy-efficient technology. Process intensification can be carried out by using new types of equipment and unconventional processing methods, such as integrated processes and processes using alternative energy sources such as light, ultrasound, and the like. It also can be done by implementing new process control methods, such as intentional unsteady-state operation. In this way, process intensification can be considered as a major headway toward the design of essentially more efficient technologies with much better performance comparing them to processes based on individual unit operations preferentially connected in a sequential mode (Cardona et al., 2008).

One of the main ideas of process intensification is to combine different process functions (separation, mixing, chemical reaction, biological transformation, fluids transport; Li and Kraslawski, 2004) and to utilize the energy flows of the same process in order to achieve better process performance. The combination of functions implies the physical combination of unit operation and processes through their simultaneous accomplishment in the same single unit or by their coupling (conjugation). Similarly, the combined utilization of energy flows allows a better exploitation of available energy sources. This physical combination of material and energy flows leads to the integration of processes oriented to their intensification. In this way, the possibilities of improving the performance of the overall process in terms of saving energy and reducing capital costs are greater when the integration of several operations into one single unit is carried out (Cardona et al., 2008).

Process integration offers many advantages in comparison to nonintegrated processes. Particularly in the case of reaction–reaction and reaction–separation processes, integration allows increasing the conversion of reactants and, consequently, the volumetric productivity. This increased conversion is explained by the fact that some key components formed during the chemical or biochemical transformation are removed from the reaction zone leading to the acceleration of the direct reaction in reversible reactions, or to the reduction of the inhibition effects in the case of some biological processes (Cardona et al., 2008). For integrated processes, the increased conversion makes possible a better utilization of the feedstocks, and the increased selectivity allows the reduction in the amount of nondesired products, which implies the reduction of the waste streams. In this way, the process integration approach contributes to the design of environmentally friendly technologies. From the viewpoint of production costs, the integration allows the development of more compact processes due to the reduction in the amount and size of processing units. Therefore, capital costs may be reduced as well as energy consumption. The reduction of energy costs is related to the decrease in the size of processing units. Smaller units have lower steam and cooling water requirements. Moreover, the integration approach allows achieving a synergetic effect in the heat transfer leading to the reduction of energy consumption. The reduction of the energy needs leads to the decrease in the size of the heat exchangers, which contributes to the compactness of the technological configuration, one of the most important features of integrated processes. Furthermore, the compactness of some integrated schemes allows the reduction in the amount of external recycling streams, which are substituted by internal recycles.

However, integrated processes exhibit some disadvantages when compared to nonintegrated processes. First of all, the controllability of integrated processes is much more complex. Often, the integration leads to the existence of multiple steady-states in the system. For this reason, integrated processes require robust control loops, which are expensive and difficult to design. In addition, the use of third substances in some integrated schemes, such as extractive reaction where the addition of an extractive agent is necessary, indicates the need for using recovery units in order to decrease the operating costs of the process (Cardona et al., 2008). One of the most difficult issues during the design of integrated processes is related to the lack of appropriate models for describing this type of configuration. Most of the developed models correspond to short-cut methods where main phenomena taking place in the system are quite simplified. These methods are mostly based on equilibrium models. However, this kind of method has allowed the preliminary and conceptual design of many integrated processes as well as the assessment of the viability of their implementation.

In a previous work (Rivera and Cardona, 2004), the classification of integrated processes was provided. Such processes can be divided into two main classes depending on whether unit operation or unit process is being combined. The integrated process is homogenous when two or more unit operations or two or more reactions (unit processes) are combined and heterogeneous when the combination is carried out between one unit operation (physical process) and one chemical reaction. Each case can be accomplished through either simultaneous or conjugated configuration. In the first case, the physical and/or chemical processes are simultaneously carried out in a single unit. In the second case, the processes are carried out in different apparatuses connecting them by fluxes or refluxes, i.e., by coupling two or more units (Cardona et al., 2008). In relation to the process steps that can be combined, integrated processes can be of the following types: reaction–reaction, reaction–separation, or separation–separation.

In this context, the design of technologies with improved performance according to technical, economic, and environmental criteria for producing fuel ethanol is required. The reduction of energy consumption along with the decrease in the capital costs through process integration offers promising opportunities for the improvement of the overall process for bioethanol production. Thus, this reduction can contribute to the worldwide development of the biofuels industry with its inherent economic, social, and environmental benefits. The aim of this chapter is to study and recognize the vast possibilities of process integration during the conceptual design and development of high-performance technologies for production of fuel ethanol from different feedstocks.

Process integration, as a mean for process intensification, is a successful approach for designing improved technological configurations for fuel ethanol production in which energy consumption, production costs, and negative environmental impacts can be reduced. This fact is remarkably important taking into account that the main objective of using liquid biofuels, like bioethanol, is the progressive displacement of fossil fuels. This implies the sustainable exploitation of the huge biomass resources of our planet and the use of clean and renewable energy sources. Solutions provided by the process integration approach have to be proved at an industrial level in order to develop energy efficient, environmentally friendly, and even "politically correct" processes for fuel ethanol production. In fact, some technologies directly involving the principle of integration for bioethanol production have already been successfully implemented.

9.2   REACTION–REACTION INTEGRATION FOR BIOETHANOL PRODUCTION

Process integration is gaining more and more interest due to the advantages related to its application in the case of bioethanol production: reduction of energy costs, decrease in the size and number of process units, and intensification of the biological and downstream processes, among others. For instance, the combination in a same unit of the enzymatic hydrolysis and the microbial transformation leads to the reduction of the negative effect due to the inhibition of the enzymes by the product of the reaction catalyzed by them. This corresponds to an integration of the reaction–reaction type.

In general, reaction–reaction integration has been proposed for the integration of different biological transformations taking place during ethanol production (Cardona and Sánchez, 2007). This type of integration mainly includes the combination of the enzymatic reactions for hydrolysis of starch or cellulose with the microbial conversion of formed sugars into ethyl alcohol. There exist different possibilities for reaction–reaction integration during production of ethanol from starch (Figure 9.1) and lignocellulosic biomass (Figure 9.2).

Considering as the starting point the nonintegrated separate hydrolysis and fermentation (SHF) process, several cases of reaction–reaction integration may be analyzed for ethanol production from both starchy and lignocellulosic materials.

9.2.1   PROCESS INTEGRATION BY CO-FERMENTATION

In Chapter 7, Section 7.1.4.3, the co-fermentation was presented as a way for a more complete utilization of all the sugars present in the hydrolyzates of lignocellulosic biomass. This process can be considered an example of reaction–reaction integration since two biochemical processes (fermentation of glucose and fermentation of xylose) are combined and simultaneously accomplished in the same single vessel. For this, mixed cultures can be used as shown in Table 7.3. In addition, an enzyme transforming the xylose into another compound more assimilable by conventional yeasts can be added to the culture medium in order to allow the utilization not only of glucose, but also of xylose. The other approach consists of using recombinant microorganisms able to assimilate these two sugars as presented in Chapter 6, Table 6.3. From the viewpoint of process systems engineering, the modeling of this process plays a crucial role for simulation procedures intended to assess different technological configurations of biomass-to-ethanol conversion in the framework of process synthesis. The aspects concerning this topic were discussed in Chapter 7, Section 7.2.2. In general, through co-fermentation, it is possible to implement simultaneous fermentation processes with higher compactness and lower production costs since a unit for pentose fermentation is not required (Figure 9.3).

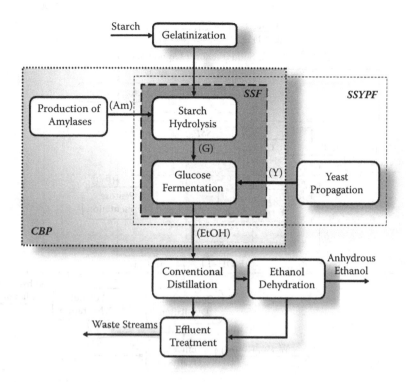

FIGURE 9.1 Possibilities for reaction–reaction integration during fuel ethanol production from starchy materials: SSF = simultaneous saccharification and fermentation; SSYPF = simultaneous saccharification, yeast propagation and fermentation; CBP = consolidated bioprocessing. Main streams components: Am = amylases, G = glucose, Y = yeasts, EtOH = ethanol. (From Cardona, C.A., and Ó.J. Sánchez. 2007. *Bioresource Technology* 98:2415–2457. Elsevier Ltd. With permission.)

Nevertheless, the process itself corresponds to an SHF scheme because the cellulose has to be hydrolyzed in a previous bioreactor using cellulases.

9.2.2 Process Integration by SSF

One of the most important advances in the bioethanol industry is the development and implementation of processes in which the hydrolysis of the glucan (starch, cellulose) and the conversion of sugars into ethanol are carried out simultaneously in the same single unit. This process is known as simultaneous saccharification and fermentation (SSF) and has been successfully implemented in the production of ethanol from corn, especially in dry-milling plants.

9.2.2.1 SSF of Starch

SSF technology born in the 1970s was assimilated by the starch-processing industry for ethanol production obtaining high and sustainable yields on the

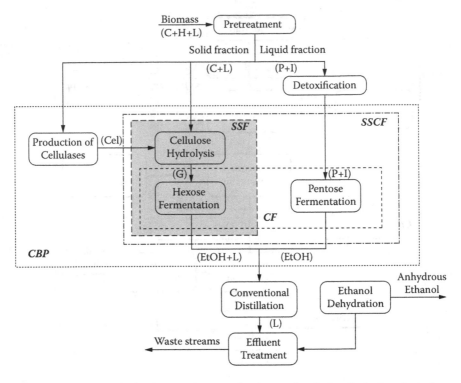

FIGURE 9.2 Possibilities for reaction–reaction integration during fuel ethanol production from lignocellulosic biomass: CF = co-fermentation; SSF = simultaneous saccharification and fermentation; SSCF = simultaneous saccharification and co-fermentation; CBP = consolidated bioprocessing. Main stream components: C = cellulose, H = hemicellulose, L = lignin, Cel = cellulases, G = glucose, P = pentoses, I = inhibitors, EtOH = ethanol. (From Cardona, C.A., and Ó.J. Sánchez. 2007. *Bioresource Technology* 98:2415–2457. Elsevier Ltd. With permission.)

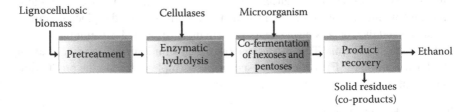

FIGURE 9.3 Block diagram of fuel ethanol production from lignocellulosic biomass involving the co-fermentation of hexoses and pentoses.

order of 0.410 L/kg of corn (Madson and Monceaux, 1995). In the case of the saccharification step when starch is used as feedstock, glucoamylase experiences the inhibitory effect caused by glucose released as the hydrolysis of this biopolymer advancement. This effect is more pronounced at high conversions of starch into ethanol. In contrast, integration by SSF makes it possible for yeasts to consume the glucose immediately as it forms under the action of the amylases on starch. In addition, the risk of bacterial contamination of the wort is drastically reduced because of the low level of glucose in the medium during the SSF process. The elimination of the external step of saccharification, the major source of infection by bacteria, also contributes to the reduction of contamination (Madson and Monceaux, 1995). In the same way, capital costs are reduced as a consequence of the increase in the compactness of the system (fewer numbers of units). Moreover, low glucose concentrations in the medium decrease the osmotic pressure over the yeasts because the use of concentrated solutions is avoided (Bothast and Schlicher, 2005). Energy costs can also be reduced considering that the SSF process is operated at temperatures less than those of the separate saccharification process; this implies the reduction in the steam consumption. All these synergic features have allowed gains of ethanol yields higher than those of the SHF process.

The main disadvantage of the SSF process is that the optimum temperature of glucoamylase (65°C) does not coincide with the optimum temperature for yeast growth (30°C). Fortunately, starch saccharification can be carried out at 30 to 35°C although at a slower rate. For this reason, higher enzyme dosages are required. Finally, processing times for batch SSF are longer than the corresponding times for batch SHF.

Most ethanol production facilities utilizing the corn dry-milling technology employ batch SSF processes. The duration of this process is 48 to 72 h achieving final ethanol concentrations in the medium of 10 to 12% by volume (Bothast and Schlicher, 2005). A number of modifications of the SSF of starchy materials have been proposed in order to decrease the production costs. Some of them are included in Table 9.1. Montesinos and Navarro (2000) have studied the possibility of utilizing raw wheat flour during the batch SSF process with the aim of reducing costs attaining a decrease in the process time. On the other hand, the SSF performed at a temperature above 34°C using a thermotolerant yeast, which enabled the reduction of cooling requirements and the improvement of the conversion process, as claimed in the patent of Otto and Escovar-Kousen (2004).

CASE STUDY 9.1   ENERGY EVALUATION OF SSF VERSUS SHF FOR CORN ETHANOL PRODUCTION

To evaluate the contribution of reaction–reaction integration to the improvement of energy efficiency of the corn-to-ethanol process, two technological configurations were simulated and their energy consumption was calculated. The first process corresponded to a nonintegrated configuration based on SHF of the dextrins produced during the liquefaction of corn starch. In addition, the considered configuration

TABLE 9.1

Integration of Reaction–Reaction Processes by Simultaneous Saccharification and Fermentation (SSF) for Fuel Ethanol Production from Different Feedstocks

Technology	Bioagent	Feedstock/Medium	Remarks	References
Batch SSF (mixed culture)	*Saccharomyces cerevisiae* + *Fusarium oxysporum*	Sweet sorghum stalks	Fungus produces cellulases and hemicellulases for hydrolysis process; formed sugars are converted into ethanol by concerted action of both microorganisms; 108–132% yield; EtOH conc. 35–49 g/L.	Mamma et al. (1995, 1996)
Batch SSF (mixed culture with co-product formation)	*S. cerevisiae* + *Candida tropicalis* + *Chaetomium thermophile* cellulases and xylanases	Alkali pretreated corn cobs	*C. tropicalis* produces xylitol and ethanol; EtOH conc. 21 g/L, xylitol conc. 20 g/L; EtOH yield 0.32 g/g, xylitol yield 0.69 g/g; 37°C	Latif and Rajoka (2001)
Batch SSF	*S. cerevisiae* + *Aspergillus niger* glucoamylase	Raw wheat flour	Previous liquefaction with α-amylase; 21–31 h cultivation; EtOH conc. 67 g/L	Montesinos and Navarro (2000)
Batch SSF	Yeasts + *Trichoderma reesei* cellulases supplemented with β-glucosidase	Pretreated lignocellulosic biomass	3–7 d of cultivation; EtOH conc. 40–50 g/L for *S. cerevisiae*, 16–19 g/L for *Kluyveromyces marxianus*; 90–96% substrate conversion	Ballesteros et al. (2004) De Bari et al. (2002) Hari Krishna et al. (1998) Lynd et al. (2001) South et al. (1993) Wyman (1994)

Semicontinuous SSF	S. cerevisiae + commercial cellulase supplemented with β-glucosidase	Paper sludge	Special design of solids-fed reactor; EtOH conc. 35–50 g/L; 0.466 g/g EtOH yield; 74–92% cellulose conversion; 1–4 months of operation	Fan et al. (2003)
Continuous SSF	S. cerevisiae + microbial amylases	Grains	Yield 2.75 gal/bushel; industrially implemented	Madson and Monceaux (1995)
	Co-immob. Zymomonas mobilis and glucoamylase on κ-carrageenan	Liquefied corn liquid	Fluidized-bed reactor; approx..100 g/L dextrin feed; conversion 53.6–89.3%; EtOH conc. 22.9–36.44 g/L; productivity 9.1–15.1 g/(L.h)	Krishnan et al. (1999)
	S. cerevisiae + commercial cellulase supplemented with β-glucosidase	Dilute-acid pretreated hardwood	CSTR; residence time 2–3 d; 83% conversion; EtOH conc. 20.6 g/L	South et al. (1993)
Continuous SSYPF	S. cerevisiae + microbial amylases	Corn, milo, wheat	Yield 2.75–2.8 gal/bushel; industrially implemented	Madson and Monceaux (1995)

Source: Adapted from Cardona, C.A., and Ó.J. Sánchez. 2007. *Bioresource Technology* 98:2415–2457. Elsevier Ltd.

TABLE 9.2

Energy Comparison of Two Processes for Fuel Ethanol Production from Corn Grains by Dry-Milling

Item	Integrated Process (SSF)	Non-Integrated Process (SHF)
Ethanol produced (kg EtOH/h)	17,837	17,838
Energy consumption (MJ/h)	270,487	290,856
Energy consumption (MJ/L EtOH)	11.67	12.55

comprised the dry-milling of corn grains, ethanol dehydration by adsorption with molecular sieves, and effluent treatment step, which allows the recovery of a co-product called dried distiller's grain with solubles (DDGS) that is commercialized as an animal feed. The flowsheet for this configuration is illustrated in Figure 11.6 of Chapter 11.

The simulation of the above-described process was performed using the simulator Aspen Plus™ using a capacity of 17,839 kg/h. The simulation procedure for distillation columns was the same of that described in Case Study 8.1 (see Chapter 8, Section 8.3).

The second studied process corresponded to the integrated configuration based on SSF of the dextrins produced during the liquefaction of corn starch. This process only differs from the SHF process in that the saccharification and fermentation are carried out simultaneously in the same unit at 33°C (see Figure 11.7 of Chapter 11). The remaining steps are similar for both processes. This configuration has already been simulated in previous works (Cardona et al., 2005a, 2005b; Quintero et al., 2008) applying the same simulation features mentioned above.

The obtained results are shown in Table 9.2. Calculated data allow observing that the effect of the reaction–reaction integration studied on the energy performance of the process is favorable. The integration of the enzymatic reaction (saccharification) and the microbial transformation (fermentation) allowed the reduction of energy costs by 7%, which is a significant energy savings.

9.2.2.2 SSYPF of Starchy Materials

Fuel ethanol industry has advanced in SSF technology by incorporating the yeast propagation (from active dry yeasts) in the fermenter during initial saccharification, a process called simultaneous saccharification, yeast propagation, and fermentation (SSYPF), as indicated in Figure 9.1. High sugar concentrations are not achieved in the fermenter avoiding the inhibition of the enzymatic hydrolysis that is characteristic for the amylases. Due to this, the bacterial growth is inhibited because of the lack of substrate caused by the immediate conversion of glucose into ethanol. Maintaining the pH, nutrients, and sterility in relation to bacteria, the complete conversion of available starch into ethanol is accomplished. SSYPF technology has been utilized in several plants in North America mainly employing corn, milo (a variety of sorghum), and wheat (Madson and Monceaux, 1995).

The objective of this configuration is to increase the contact time between the mash and yeasts in order to reduce the bacterial growth. This allows reaching higher yields, an earlier ethanol production (that also reduces the contamination), and reducing the need of handling large amounts of yeasts (Novozymes & BBI International, 2005). At the beginning of an SSYPF process, pH is adjusted to 5.2 for favoring the growth of microorganisms and, as the cultivation goes forward, pH is diminished to 4.5 at the end of fermentation (Madson and Monceaux, 1995).

9.2.2.3 SSF of Lignocellulosic Materials

Conversion of cellulose into ethanol can be carried out through SSF, as in the case of starch. For this conversion, several enzymes with cellulolytic activity (basically endolucanases, cellobiohydrolases, and β-glucosidase) are added to the suspension obtained by mixing water with the solid fraction resulting from the pretreatment step and that contains cellulose and lignin. In the same way, process microorganisms (yeasts) are added to this mixture in the bioreactor where SSF immediately converts the formed glucose into ethanol. Taking into account that sugars (glucose, cellobiose) are more inhibitive for the conversion process than ethanol, SSF can reach higher rates, yields, and ethanol concentrations in comparison with SHF (Wyman et al., 1992). The increased ethanol concentration in the culture broth allows the reduction of energy costs during distillation. In addition, SSF offers an easier operation and a lower equipment requirement than the sequential process since no hydrolysis reactors are needed. Moreover, the presence of ethanol in the broth makes the reaction mixture less vulnerable to the action of undesired microorganisms (Wyman, 1994). Nevertheless, SSF is inconvenient in that the optimal conditions for hydrolysis and fermentation are different, which leads to a difficult control and optimization of process parameters (Claassen et al., 1999). In addition, larger amounts of exogenous enzymes are required (Cardona and Sánchez, 2007).

Since the time in which the first introduction to SSF from biomass began, the duration of the batch process has decreased from 14 days required for conversion of 70% cellulose into ethanol with final concentrations of 20 g/L to 3 to 7 days needed for reaching 90 to 95% conversions with final ethanol concentrations of 40 to 50 g/L (Cardona and Sánchez, 2007; Wyman, 1994). The concept of the SSF process was first described by Takagi et al. (1977). Takagi, Suzuki, and Gauss (Gauss et al., 1976) had previously patented the SSF technology for bioethanol production by which the yeasts simultaneously metabolize the glucose into ethanol *in situ* during the enzymatic saccharification of the cellulose. This patent expired in 1993 and it has been utilized for small-scale demonstrations, according to Ingram and Doran (1995), but until now, no commercial plants have been built at an industrial level (Cardona and Sánchez, 2007).

The overall process for ethanol production from lignocellulosic biomass by SSF is depicted in Figure 9.4. This process allows the recovery of energy along with ethanol production. The lignin, one of the major polymers present in lignocellulosic biomass, is not practically changed during the process and remains in stillage generated in the product recovery step. This lignin can be separated by

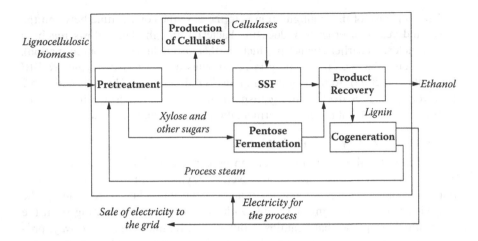

FIGURE 9.4 Simplified diagram of the integrated process for fuel ethanol production from lignocellulosic biomass by simultaneous saccharification and fermentation (SSF).

centrifugation from the liquid fraction of stillage and burnt in order to extract its high energy content. The obtained thermal energy is converted into process steam, which is employed within the same process. If co-generation is considered, part of the released energy is transformed into electricity that can supply all the needs of the plant, with a remaining surplus that can be sold to the grid. These energy recovery possibilities are an important feature of ethanol production processes when lignocellulosic biomass is utilized, and they allow the compensation of energy demands represented mostly by the steam requirements for the pretreatment reactor, distillation columns, and dehydration scheme.

Intensive research to carry out the SSF of lignocellulosic biomass has been carried out. Some examples of these efforts are presented in Table 9.1. Considering that enzymes account for an important part of production costs, it is necessary to find methods reducing the cellulases doses to be used. Thus, the integration of the cellulase production process using *Trichoderma reesei* with ethanolic fermentation has been proposed. As a small amount of enzymes remains entrapped in fungal cells producing cellulases, the addition of the whole culture broth of this process to the SSF reactor was proposed. Besides the fungal biomass and obtained cellulases, this broth contains residual cellulose and lignin. This allows the increase of the β-glucosidase activity, essential for the reduction of cellobiose levels, and a more complete utilization of sugars employed during the production of cellulases (Wyman, 1994). The addition of surfactants has been proposed for these purposes as well. Alkasrawi et al. (2003) showed that the addition of the nonionic surfactant Tween-20 to the steam exploded wood during the batch SSF using *Saccharomyces cerevisiae* has some effects: 8% increase in ethanol yield, 50% reduction in cellulases dosage (from 44 FPU/g cellulose to 22 FPU/g cellulose), an increase of enzyme activity at the end of the process, and a decrease in the time required for reaching the highest ethanol concentration. It is postulated that the surfactant avoids or diminishes the nonuseful adsorption of cellulases to the lignin. However, Saha et al. (2005) obtained marginal increases (3.5%) in saccharification of rice hulls using 2.5 g/L of Tween-20.

In early works, Wyman et al. (1992) carried out a systematic evaluation of the ethanol yields obtained from several lignocellulosic materials pretreated with sulfuric acid and fermented with *S. cerevisiae* in batch SHF and SSF processes. Different dosages of cellulase supplemented with β-glucosidase were used. Obtained results show that corn cobs, corn stover, wheat straw, and weeping love grass, in that order, were adequately pretreated with acid, achieved high rates of enzymatic hydrolysis, and demonstrated high ethanol yields during 7 to 8 d of SSF. Switchgrass, in turn, exhibited the worst results. The different types of analyzed pretreated woody crops showed similar cellulose-to-ethanol conversions though the yields were slightly less than corn stover, which was the material with the best values (92 to 94%). South et al. (1993) obtained 91% and 96% conversions for the batch SSF of dilute acid pretreated hardwood and poplar wood, respectively, using *S. cerevisiae* and cellulase supplemented with β-glucosidase. These values were attained during 3 d, but the authors did not report the concentration of produced ethanol. Hari Krishna et al. (1998) also evaluated the optimal conditions of the SSF of sugarcane leaves, as they did for the SHF process (see Chapter 5, Section 5.2.2). These authors defined the best conditions for the 3-d cultivation process as 40°C and a pH of 5.1, which allows for achieving 31 g/L of ethanol from an initial substrate load as high as 15%. Nevertheless, the enzyme dosage was quite high (100 FPU/g cellulose). Similarly, the leaves of *Antigonum leptopus*, a weedy creeper abundant in the areas not employed for conventional agriculture, was evaluated for ethanol production by SSF using yeasts showing higher yields than in the case of SHF (Hari Krishna et al. 1999).

Dale and Moelhman (2001) tested different commercial cellulases during batch SSF processes for ethanol production from biomass. Best results corresponded to Validase TR of Valley Research and CEP of Enzyme Development Corp. These authors state that Amano and Genencor cellulases were not adequate for this process. However, this information should be corroborated for each type of biomass since these authors do not clarify what material was utilized.

Softwood is more difficult to degrade by SSF than hardwood (Sánchez and Cardona, 2008). Stenberg et al. (2000a) employed the resulting slurry of the steam pretreatment of SO_2-impregnated spruce in SSF tests using *S. cerevisiae* and determined that the best initial load of substrate was 5% (w/w) obtaining 82% yield of the theoretical based on the cellulose and soluble hexoses present at the start of the SSF process. The productivity was doubled related to SHF. Higher dry weights negatively affected the fermentation process because of the inhibitors present in the slurry (the pretreated material was not washed). The load of cellulase was in the range of 5 to 32 FPU/g cellulose.

One of the most relevant factors in the alcoholic fermentation is the possibility of infection by acidolactic bacteria. Stenberg et al. (2000b) studied the influence of lactic acid formed during fermentation on the course of batch SSF of steam exploded spruce chips (softwood) using baker's yeast. It is interesting that the contamination was higher when washed substrate was used, in comparison with the utilization of the whole suspension resulting from pretreatment. This indicates that acidolactic bacteria are more sensitive to inhibitors' presence than yeasts.

As mentioned above, one of the main disadvantages of SSF processes using lignocellulosic biomass lies in the different optimum conditions of enzymatic hydrolysis of cellulose and fermentation. Cellulases work in an optimal way at 40° to 50°C and pH of 4 to 5, whereas the fermentation of hexoses with *S. cerevisiae* is carried out at 30°C and pH of 4 to 5, and fermentation of pentoses is optimally performed at 30° to 70°C and pH of 5 to 7 (Olsson and Hahn-Hägerdal, 1996; Sánchez and Cardona, 2008). Varga et al. (2004) proposed a nonisothermal regime for batch SSF process applied to wet oxidized corn stover. In the first step of SSF, small amounts of cellulases were added at 50°C in order to obtain better mixing conditions. In the second step, more cellulases were added along with the yeast *S. cerevisiae* at 30°C. In this way, the final solid concentration in the hydrolyzate could be increased up to 17% dry matter concentration achieving 78% ethanol yield.

In general, increased cultivation temperature accelerates metabolic processes and lowers the refrigeration requirements (Sánchez and Cardona, 2008). Yeasts such as *Kluyveromyces marxianus* have been tested as potential ethanol producers at temperatures higher than 40°C (Ballesteros et al., 2004; Hari Krishna et al., 2001). Kádár et al. (2004) compared the performance of thermotolerant *K. marxianus* and *S. cerevisiae* during batch SSF of waste cardboard and paper sludge not finding great differences between both microorganisms at 40°C, although cellulose conversions (55 to 60%) and ethanol yields (0.30 to 0.34 g/g cellulose) were relatively low. Ballesteros et al. (2001) carried out several fed-batch SSF tests at 42°C during 72 h using *K. marxianus* in the case of by-products of olive oil extraction (olive pulp and fragmented olive stones). Their results showed 76% ethanol yields of theoretical for olive pulp and 59% yield for acid-catalyzed steam-exploded olive stones. With the aim of increasing ethanol yields from olive pulp, Ballesteros et al. (2002) employed liquid hot water (LHW) pretreatment reaching an 80% yield of the theoretical value and recovering potentially valuable phenolic compounds.

If, when thermotolerant yeasts are used, the microbial cells can also assimilate pentoses, the SSF process can become more perspective (Cardona and Sánchez, 2007). Yeasts such as *Candida acidothermophilum*, *C. brassicae*, *S. uvarum,* and *Hansenula polymorpha* can be used for these purposes. In this case, the addition of a larger amount of nutrients to the medium is required. Alternatively, the utilization of higher cell concentrations could be implemented for obtaining better results (Olsson and Hahn-Hägerdal, 1996; Ryabova et al., 2003). The difficulty lies in the fact that higher temperatures enhance the inhibitory effect of ethanol. Therefore, the isolation and selection of microorganisms that could be adapted in a better way to these hard conditions should be continued. Kádár et al. (2004) make reference to several reports about the utilization of thermotolerant microorganisms for ethanol production.

One approach to accomplish the SSF of biomass without the addition of cellulases consists in the utilization of mixed cultures in such a way that the hydrolysis and fermentation of lignocellulosic biomass be carried out simultaneously (Cardona and Sánchez, 2007). This procedure was applied to sweet sorghum stalks employing cellulase- and hemicellulase-producing fungus *Fusarium oxysporum* along with *S. cerevisiae* (Mamma et al., 1995). Considering that sweet sorghum stalks contain several carbohydrates, such as sucrose, glucose, hemicellulose, and cellulose, the obtained yields in this process were higher than the theoretical yield from only glucose (0.51 g EtOH/g glucose); this is explained by the additional bioconversion of cellulose and hemicellulose into ethanol (Mamma et al., 1996). However, the final ethanol concentrations in this type of SSF process were quite low considering the separation process. Logically, the mixed culture presents a high complexity during its implementation at the industrial level and has the additional disadvantage that optimal growth conditions for two or more different microorganisms are not the same. Besides, part of the substrate is deviated for the growth of the enzyme-synthesizing microorganism. Panagiotou et al. (2005a) carried out the SSF of cellulose with *F. oxysporum* demonstrating the production of ethanol under anaerobic conditions. In addition, the metabolite profiling of the microorganisms cultivated in different media was achieved through the measurement of the intracellular concentration of key metabolites (Panagiotou et al., 2005a, 2005b, 2005c).

The SSF process also can be done in continuous regime. In the same work of South et al. (1993) cited above, the behavior of a continuous-stirred tank reactor (CSTR) for continuous SSF of hardwood using the same microorganism and the same enzymes was investigated. For a residence time of 3 d, 83% conversion was achieved, i.e., less than in the case of batch SSF, which is explained by the decrease in the reactivity when conversion in the biomass hydrolysis is increased. Ethanol concentration reached 20.6 g/L for a residence time of about 2 d. Obtained results showed that enzyme and substrate concentrations in the feed within studied ranges did not influence cellulose conversion. This indicates the existence of mass transfer restrictions related to cellulose, besides inhibitory effects caused by the substrate and products. The fact that the enzymatic hydrolysis rate decreases with the course of hydrolysis has often been reported in the literature. According to Zhang and Lynd (2004), this decreased reactivity of residual cellulose can be due to less surface area, fewer accessible chain ends, and/or adsorption of inactive cellulase on the surface of lignocellulosic particles. The addition of fresh substrate can stimulate the release of more soluble sugars indicating the loss of cellulose reactivity at the end of hydrolysis or the increase of reactivity for the "new" encounters enzyme-substrate compared with the "old" ones.

9.2.2.4 Modeling of SSF of Cellulose

In Chapter 5, Section 5.2.3.2, the importance of cellulose hydrolysis description was discusses, taking into account the development of suitable simulation tools to be employed during process synthesis procedures. The difficulties that arose due to the several factors affecting the kinetics of cellulose hydrolysis were also highlighted in that section. Evidently, these difficulties are also present when modeling of cellulose SSF is performed. One of the most comprehensive mathematical models for this process in both batch and continuous regimes corresponds to the description provided by South et al. (1995). As mentioned in Section 5.2.3.2, these authors developed a model based on experimental data obtained in their previous work using commercial fungal cellulase (South et al., 1993) employing pretreated wood as the feedstock. The kinetic model includes the cellulose conversion, the formation and disappearance of cellobiose and glucose, the formation of cells, and the biosynthesis of ethanol structure, i.e., the main enzymatic and microbial phenomena taking place during the SSF process. In addition, a Langmuir-type model taking into account the adsorption of cellulases on the solid particles of cellulose and lignin is also considered. Moreover, this mathematical description contemplates a population model for the residence time of solid particles entering the bioreactor in the case of the continuous regime. Thus, expressions describing the dependence of cellulose conversion on the residence time of nonsoluble, solid particles of biomass were derived. This last description confers great validity to the model because it provides a better approximation to real processes, which cannot be suitably explained by the traditional models for CSTR with soluble substances.

CASE STUDY 9.2   MODELING OF SSF OF BIOMASS IN BATCH AND CONTINUOUS REGIME

The importance of modeling SSF processes is invaluable considering the design of fuel ethanol production processes employing lignocellulosic materials as feedstocks. In a previous work (Sánchez et al., 2005), the analysis of SSF for conversion of cellulose into ethanol was performed in both batch and continuous regimes. The kinetic model of such a process was based on the mathematical description developed by South et al. (1995). However, considering that this model will be employed in subsequent procedures and algorithms for process synthesis of ethanol production, the expressions were simplified to not add more complexity to the calculations to be performed during process synthesis and optimization procedures. This is justified because process synthesis tools deal with many alternative process flowsheets. These flowsheets involve all the processing steps for conversion of feedstocks into products. As pointed out by Grossmann et al. (2000), there exist different levels of detail for the mathematical description of each unit processes and operations involved in each flowsheet (see Chapter 2). In fact, for the task of process synthesis, it is not desirable to consider models with a higher degree of detail especially if equation-oriented simulators are used, or optimization-based process synthesis procedures are applied. The simplification of the kinetic model mentioned above wasn't meant to consider its population and adsorption components, but to take into account the rigorous description of the kinetic processes involved.

For simulation of the batch SSF process, the rate equations were extracted from South et al. (1995). Equations (9.1) and (9.2) correspond to the enzymatic hydrolysis of cellulose and cellobiose, respectively. Equations (9.3) through (9.5) represent cell biomass production, glucose uptake and formation, and ethanol biosynthesis, respectively:

$$r_s = -(k \cdot (1-x)^n + c) \cdot \frac{ES}{\varsigma_s} \cdot \left[\frac{k_{S/G}}{C + k_{S/G}} \right] \left[\frac{k_{S/P}}{P + k_{S/P}} \right] \tag{9.1}$$

$$r_c = -1.056 \cdot r_s - \left[\frac{k_c \cdot C \cdot Bg}{K_m \cdot \left(1 + \dfrac{G}{k_{C/G}}\right) + C} \right] \tag{9.2}$$

$$r_x = \frac{(Xc \cdot \mu_{max} G)}{G + k_G} \cdot \left(1 - \frac{P}{k_{C/P}}\right) \tag{9.3}$$

$$r_G = (-1.056 \cdot r_s - r_c) \cdot 1.053 - \frac{r_x}{Y_{x/G}} \tag{9.4}$$

$$r_P = r_x \cdot \frac{Y_{P/G}}{Y_{X/G}} \tag{9.5}$$

The nomenclature of all the variables and kinetic parameters involved in the above equations are presented in Table 9.3. In Equation (9.1), the last two terms represent the inhibition by cellobiose and ethanol. These terms influence all the rate equation directly or indirectly. Similar expressions can be observed in Equation (9.2) for the inhibitory effect of glucose on the β-glucosidase activity. In Equation (9.3), the expression for biomass formation rate has a lowering term due to high ethanol concentrations present in the broth. For this case, a cellulose conversion (x) of 0.70 was preset. The general mass balance expression for each one of i components (cellulose S, cellobiuse C, cell biomass X, glucose G, and ethanol P) is:

$$\frac{d(C_i)}{dt} = r_i \tag{9.6}$$

In all cases, a lignocellulosic substrate with cellulose loading of 60 g/L and initial concentrations of cellobiose, biomass, glucose, and ethanol of 0, 1, 8.5, and 0 g/L, respectively, were considered. The selected kinetic model involved the use of *T. reesei* cellulases and fermentation by *S. cerevisiae*, according to South et al. (1995). The system of five nonlinear ordinary differential equations was solved by fourth-order Runge–Kutta method using Matlab™ (MathWorks, Inc., USA) with the initial values mentioned above and for a process time of 72 h. Ethanol productivity and product yield were calculated for this type of regime. The parameter values used for solving the kinetic model can be found in South et al. (1995).

TABLE 9.3
Nomenclature of the Variables and Kinetic Parameters Involved in Equations Derived from the Model of South et al. (1995)

Symbol	Remark	Symbol	Remark
B	Input stream to the pervaporator, l/h	n	Exponent of the declining substrate reactivity, dimensionless
Bg	β-glucosidase concentration in solution, U/L	P	Ethanol concentration, g/L
c	Conversion independent component in rate function, 1/h	Po	Initial ethanol concentration, g/L
C	Cellobiose concentration, g/L	Q	Output stream for pervaporation (permeate), L/h
Ci	Concentration of the i-th component	R	Recirculation (retentate) stream from pervaporation unit to reactor, L/h
Co	Initial cellobiose concentration, g/L	r_i	Rate of formation of compound i, $g/(L \times h)$
ES	Concentration of cellulose-cellulase complex, U/L	S	Cellulose component of the biomass substrate remaining, g/L
F	Feed reactor stream, L/h	So	Initial cellulose component of the biomass substrate, g/L
G	Glucose concentration, g/L	V	Reaction volume, L
Go	Initial glucose concentration, g/L	W	Residual flow, L/h
k	Hydrolysis rate constant, g/L	x	Fractional reactor cellulose conversion, dimensionless
kc	Rate constant for hydrolysis of cellobiose to glucose, $g/(U \times h)$	X	Cell concentration, g/L
k_G	Monod constant, g/L	Xo	Initial cell concentration, g/L
$k_{C/G}$	Inhibition of cellobiose hydrolysis by glucose, g/L	$Y_{X/G}$	Cell yield per substrate consumed, dimensionless
$k_{S/C}$	Inhibition of cellulose hydrolysis by cellobiose, g/L	$Y_{P/G}$	Ethanol yield per substrate consumed, dimensionless
$k_{S/P}$	Inhibition of cellulose hydrolysis by ethanol, g/L	α	Separation factor in pervaporation
K_S	Adsorption constant for cellulosic fraction of biomass, L/U	ς_S	Specific capacity of cellulosic component for cellulose, U/g
K_m	Adsorption constant for β-glucosidase for cellobiose, g/L	μ_{max}	Maximum cell growth rate, 1/h

Subindex

i	Any of the substances involved in the fermentation
0	Initial concentration in batch processes or feed concentration in continuous processes
P	Product (ethanol)

Source: Adapted from South, C.R., D.A.L. Hogsett, and L.R. Lynd. 1995. *Enzyme and Microbial Technology* 17:797–803.

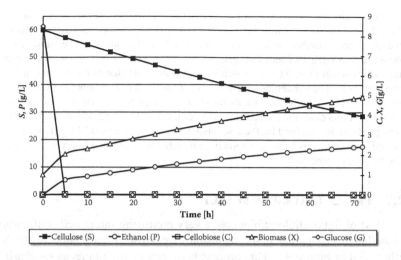

FIGURE 9.5 Behavior of batch SSF process for ethanol production from cellulose.

The results for batch SSF process can be seen in Figure 9.5. The final cellulose concentration was 28.6 g/L and the ethanol concentration reached at the end of fermentation was 17.5 g/L. The amounts of cellobiose and glucose when the cultivation was finished were near zero, which shows the efficiency of the combined process and the neutralization of the inhibitory effects of glucose on cellulases. In the case of the SHF process, the accumulating glucose in the medium during cellulose saccharification leads to reduced conversion of cellulose and hydrolyzates with lower concentrations of fermentable sugars. In contrast, during the SSF process, the accumulation of ethanol in the medium can inhibit the growth rate and, therefore, the ethanol production rate according to the kinetic expressions on which the model was based. The productivity attained by the batch SSF process was 0.292 g/(L × h) and the ethanol yield was 0.454 g/g, calculated at 48 h of cultivation.

For solving the model of an SSF process in a CSTR, the mass balance for each of the i substances involved in the process (cellulose, cellobiose, biomass, glucose, and ethanol) was considered according to following equation:

$$FC_{io} - WC_i + Vr_i = 0 \qquad (9.7)$$

Taking into consideration that Equations (9.1) through (9.5) describe the formation or consumption rate of each component, a system of five nonlinear algebraic equations with five unknowns was obtained by applying equation (9.7). For solving this system, the Newton–Raphson algorithm was used with the same initial concentrations used for the SSF process in batch regime. Equation (9.1) includes a term for cellulose conversion (x) that in the original paper of South et al. (1995) is a function of mean residence time of particulate matter of cellulose. In this case study, the conversion was set to a value of 0.70 with the use of a CSTR for carrying out both transformations (cellulose hydrolysis and ethanol fermentation) and, therefore, assuming an intensive mixing of the reaction volume.

The results obtained for continuous SSF process with a mean residence time of 72 h showed that the cellulose had a more complete conversion and that the ethanol was produced in higher amounts. The concentrations of cellulose and ethanol in the outlet stream were 10.7 and 24.9 g/L, respectively. The biomass concentration in the exiting stream was 5.6 g/L, which is comparable with that of the corresponding batch process. The concentrations of the other involved components in the effluent of the fermenter were near zero, demonstrating the good performance of the SSF process. In this case, the concentration of cellulose in the feed stream was 60 g/L. The productivity attained by the continuous SSF process was 0.345 g/(L × h) and the ethanol yield was 0.506 g/g showing favorable performance indexes related to the batch SSF process.

9.2.3 Process Integration by SSCF

In the case of lignocellulosic biomass, a very promising integrated configuration for bioethanol production is the inclusion of pentose fermentation in the SSF, as illustrated in Figure 9.6. This process is known as simultaneous saccharification and co-fermentation (SSCF). This configuration implies a higher degree of intensification through its reaction–reaction integration. In this case, the hydrolysis of cellulose, the fermentation of glucose released, and the fermentation of pentoses present in the feed stream is simultaneously accomplished as a single unit. Besides the effectiveness of employed cellulases, the key factor in SSCF is the utilization of an efficient ethanol-producing microorganism with the ability of assimilating not only hexoses (mainly glucose), but also pentoses (mainly xylose) released during the pretreatment step as a result of the hemicellulose hydrolysis. In nature, there exist several microorganisms able to assimilate both types of sugars, but their ethanol yields are low as is their growth rate. Therefore, genetically modified microorganisms have been developed and successfully proven in SSCF processes for ethanol production from lignocellulosic materials.

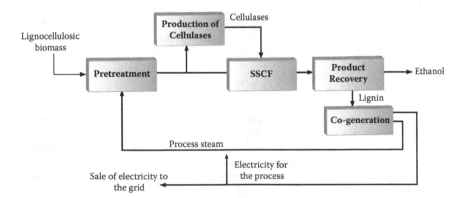

FIGURE 9.6 Simplified diagram of the integrated process for fuel ethanol production from lignocellulosic biomass by simultaneous saccharification and co-fermentation (SSCF).

In an initial stage, the co-fermentation of mixed cultures was studied (Cardona and Sánchez, 2007). For example, the co-culture of *Pichia stipitis* and *Brettanomyces clausennii* has been employed for the SSCF of aspen at 38°C and pH of 4.8 yielding 369 L EtOH per ton of aspen during 48 h batch process, as reported by Olsson and Hahn-Hägerdal (1996). In this configuration, it is necessary that both fermenting microorganisms have compatible in terms of operating pH and temperature. Chandrakant and Bisaria (1998) suggest that a combination of *C. shehatae* and *S. cerevisiae* is suitable for this kind of process.

The actual SSCF process has been demonstrated in the case of ethanol production from yellow poplar through a bench-scale integrated process that included the dilute-acid pretreatment of feedstock, conditioning of hydrolyzate for fermentation, and a batch SSCF (McMillan et al., 1999). In this case, the recombinant bacterium *Zymomonas mobilis* assimilating xylose was used. SSCF is the process on which is based the technology designed as a model process by the National Renewable Energy Laboratory (NREL) for production of fuel ethanol from aspen wood chips (Wooley et al., 1999b) and corn stover (Aden et al., 2002). In this design, the utilization of recombinant *Z. mobilis* exhibiting a glucose conversion to ethanol of 92% and a xylose conversion to ethanol of 85% is proposed. It is projected that SSCF can be carried out in a continuous regime with a residence time for the entire system of cascade fermenters of 7 d at 30°C (Cardona and Sánchez, 2007).

As in the case of SSF of biomass, the development of microbial strains able to grow at elevated temperatures can improve the technoeconomic indicators of the process (Cardona and Sánchez, 2007). Thus, ethanol-producing microorganisms capable of assimilating both types of sugars at temperatures higher than 50°C could reduce the cellulase costs by half, taking into account that a 20°C increase during saccharification can lead to double cellulose hydrolysis rate (Wooley et al., 1999a). Three examples of SSCF of lignocellulosic materials pretreated are presented in Table 9.4.

9.2.4 PROCESS INTEGRATION BY CONSOLIDATED BIOPROCESSING

The logical culmination of reaction–reaction integration for the transformation of biomass into ethanol is the consolidated bioprocessing (CBP), known also as direct microbial conversion (DMC). The key difference between CBP and the other strategies of biomass processing is that only one microbial community is employed for both the production of cellulases and fermentation, i.e., cellulase production, cellulose hydrolysis, and fermentation are carried out in a single step (see Figure 9.2). This configuration implies that no capital or operation expenditures are required for enzyme production within the process. Similarly, part of the substrate is not deviated for the production of cellulases, as shown in Figure 9.7 (compare to the more complex configurations depicted in Figure 7.6 for the SHF process with the addition of commercial cellulases, and Figure 9.4 for SSF of biomass with *in situ* production of cellulases). Moreover,

TABLE 9.4

Integration of Reaction–Reaction Processes by Simultaneous Saccharification and Co-Fermentation (SSCF) for Fuel Ethanol Production from Different Feedstocks

Technology	Bioagent	Feedstock/Medium	Remarks	References
Batch SSCF	Recombinant *Zymomonas mobilis* + *Trichoderma reesei* cellulases	Dilute-acid pretreated yellow popplar	EtOH produced 17.6–32.2 g/L; yield 0.39 g/g; productivity 0.11–0.19 g/(L.h)	McMillan et al. (1999)
Continuous SSCF	Recombinant *Z. mobilis* + *T. reesei* cellulases	Dilute-acid pretreated wood chips	Cascade of reactors; model process of the NREL; 92% glucose conversion, 85% xylose conversion	Wooley et al. (1999b)
Continuous SSCF/SHCF	Recombinant *Z. mobilis* + *T. reesei* cellulases	Dilute-acid pretreated corn stover	Cascade of reactors; previous presaccharification of biomass at 65°C; model process of the NREL; 95% glucose conversion, 85% xylose conversion	Aden et al. (2002)

Source: Modified from Cardona, C.A., and Ó.J. Sánchez. 2007. *Bioresource Technology* 98:2415–2457. Elsevier Ltd.

Note: SHCF = separate hydrolysis and co-fermentation.

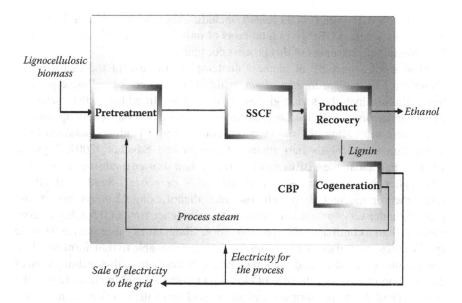

FIGURE 9.7 Simplified diagram of the integrated process for fuel ethanol production from lignocellulosic biomass by consolidated bioprocessing (CBP).

the enzymatic and fermentation systems are entirely compatible (Cardona and Sánchez, 2007).

The extended concept of CBP involves four biologically mediated transformations: (1) the production of saccharolytic enzymes (cellulases and hemicellulases); (2) the hydrolysis of carbohydrate components present in pretreated biomass to form sugars; (3) the fermentation of hexose sugars (glucose, mannose, and galactose); and (4) the fermentation of pentose sugars (xylose and arabinose). These four transformations occur in a single step. In this case, a dedicated process for production of cellulases is not required to make CBP a highly integrated configuration (Cardona and Sánchez, 2007; Lynd et al., 2005). This process is conceptually depicted in Chapter 6, Figure 6.5 for the case of ethanol production from lignocellulosic biomass.

Process integration through CBP represents a considerable improvement of technologies for conversion of lignocellulosic biomass into ethanol. The enhancement of the conversion technology contributes by far the most reduction of ethanol production costs (Cardona and Sánchez, 2007). According to projections of Lynd (1996), the reduction of production costs due to an advanced configuration involving the CBP is three times greater than the reduction related to the scale economy of the process and 10 times greater than the reduction associated with a lower cost of the feedstock. This diminish would be accomplished thanks to the reduction of more than eight times in the costs of biological conversion (Lynd et al., 1996). Lynd et al. (2005) reported the comparative simulation of SSCF and CBP processes assuming aggressive performance parameters intended to be representative of mature technology. Their results indicate that production costs of ethanol for SSCF reach US4.99 cents/L including the costs of dedicated cellulase production, whereas CBP gives total costs of only US1.11 cents/L demonstrating the potential effectiveness of this process configuration.

Most studies on CBP of biomass contemplate the use of the thermophilic bacterium *Clostridium thermocellum*, which is employed for cellulase production, cellulose hydrolysis, and glucose fermentation. In addition, the bacterium *Thermoanaerobacter thermosaccharolyticum* can be co-cultured along with *C. thermocellum* to allow the simultaneous conversion of pentoses obtained from hemicellulose hydrolysis into ethanol (Cardona and Sánchez, 2007; Wyman, 1994). In particular, the CBP using *C. thermocellum* showed a substrate conversion 31% higher than a system using *T. reesei* and *S. cerevisiae*. South et al. (1993) tested the continuous CBP of cellulose into ethanol using *C. thermocellum* and showed, under very specific conditions with a residence time of 0.5 d, higher conversions than a continuous SSF process. Some filamentous fungi such as *Monilia* sp., *Neurospora crassa,* and *Paecilomyces* sp. are also able to transform cellulose into ethanol (Szczodrak and Fiedurek, 1996). Nevertheless, this technique faces the problem of the low tolerance of clostridia to ethanol and the reduction in the ethanol yield due to the formation of acetic acid and salts of other organic acids like lactates (Baskaran et al., 1995; Cardona and Sánchez, 2007; McMillan, 1997; Wyman, 1994). This means that the final ethanol concentration is low in comparison with the traditionally used yeasts (0.8 to 60 g/L) with very large cultivation times of 3 to 12 d (Szczodrak and Fiedurek, 1996).

To date, there is no microorganism known that can exhibit the whole combination of features required for the development of a CBP, as the one shown in Figure 6.5 (Chapter 6). However, there are realistic expectations about the possibility of overcoming the limitations of current CBP organisms. In Section 6.3.2.2, the main strategies for developing engineered microorganisms that can be used in technological configurations involving CBP were disclosed. In this way, the huge possibilities of CBP are based on the development of genetically modified microorganisms allowing such a high degree of reaction–reaction integration that can make possible the direct conversion of pretreated lignocellulosic biomass into ethanol at elevated yields under industrial conditions. Some examples of CBP, not only of lignocellulosic materials but also of starch, are presented in Table 9.5.

9.3   REACTION–SEPARATION INTEGRATION FOR BIOETHANOL PRODUCTION

Reaction–reaction integration allows for the increase of process efficiency through the improvement of reaction processes. However, separation is the step where major costs are generated in the process industry. Therefore, reaction–separation integration could have the highest impact on the overall process in comparison with homogeneous integration of processes (reaction–reaction, separation–separation; Cardona and Sánchez, 2007). The reaction–separation integration is a

TABLE 9.5

Integration of Reaction–Reaction Processes by Consolidated Bioprocessing (CBP) for Fuel Ethanol Production from Different Feedstocks

Technology	Bioagent	Feedstock/Medium	Remarks	References
Batch CBP	*Clostridium thermocellum* + *C. thermosaccharolyticum*	Lignocellulosic biomass	First bacterium produces cellulases and converts formed glucose into ethanol; second bacterium converts pentoses into ethanol; by-products formation; EtOH conc. 30 g/L; low ethanol tolerance	Claassen et al. (1999) Lynd et al. (2002) McMillan (1997) Wyman (1994)
	Fusarium oxysporum	Cellulose	Anaerobic conditions; yield 0.35g/g cellulose, productivity 0.044 g/(L.h)	Panagiotou (2005a)
Continuous CBP	Recombinant *Saccharomyces cerevisiae*	Starch-containing medium	Immobilized cells in calcium alginate; yeast expresses glucoamylase and converts starch into ethanol; EtOH conc. 7.2 g/L; 200 h cultivation	Kobayashi and Nakamura (2004)

Source: Modified from Cardona, C.A., and Ó.J. Sánchez. 2007. *Bioresource Technology* 98:2415–2457. Elsevier Ltd.

particularly attractive alternative for the intensification of alcoholic fermentation processes. When ethanol is removed from the culture broth, its inhibition effect on growth rate is diminished or neutralized leading to a substantial improvement in the performance of ethanol-producing microorganisms. This improved performance can permit the increase of substrate conversion into ethanol. In particular, higher conversions make possible the utilization of concentrated culture media (with sugar content greater than 150 g/L) resulting in increased process productivities. From an energy viewpoint, this type of integration allows the increase of ethanol concentration in the culture broth. This fact has a direct effect on distillation costs since more concentrated streams feeding the columns imply lower steam demands for the reboilers and, therefore, lower energy costs.

For these reasons, most of the proposed configurations using reaction–separation integration are related to the ethanol removal by different means including the coupling of different unit operations to the fermentation or the accomplishment of simultaneous processes for favoring the *in situ* removal of ethanol from culture broth.

9.3.1   ETHANOL REMOVAL BY VACUUM

This type of reaction–separation integration is carried out by coupling the fermentation tank to a vacuum chamber that allows extracting ethanol due to its higher volatility in comparison to the rest of components of the culture broth. It has been reported that a 12-fold increase in ethanol productivity can be reached using vacuum fermentation (Cysewski and Wilke, 1977). However, for reaching this productivity, the addition of oxygen was required that negatively influenced the costs, which were already high enough due to the creation of vacuum conditions. Nevertheless, da Silva et al. (1999) point out that vacuum fermentation using a flash chamber coupled to the bioreactor can demonstrate better technical indexes than extractive fermentation or fermentation coupled to pervaporation (see Figure 9.8). Ishida and Shimizu (1996) proposed a novel regime for carrying out the repeated-batch alcoholic fermentation coupled with batch distillation obtaining ethanol concentrations of 400 g/L. Some examples of this type of reaction–separation integration are shown in Table 9.6.

9.3.2   ETHANOL REMOVAL BY GAS STRIPPING

Ethanol can be removed from the culture broth through absorption employing a stripping gas. This makes possible the increase of sugar concentration in the feed stream entering the fermenter. This process has been studied in the case of corn mashes obtained by the dry-milling process. Taylor et al. (1998) studied this integrated process in the case of the dry-milling ethanol process in a pilot plant integrating a 30 L fermenter with a 10 cm packed column for ethanol removal by the CO_2 (stripping gas) released during the fermentation. A simplified scheme of this process is presented in Figure 9.9 where two circulation loops are employed. In this scheme, concentrated solutions of the product are obtained

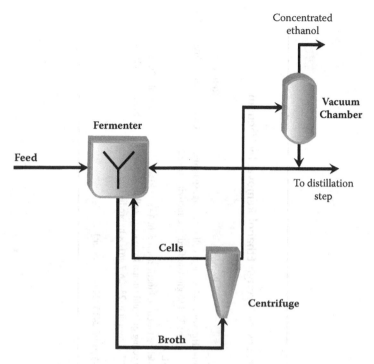

FIGURE 9.8 Simplified diagram of fermentation process with ethanol removal by vacuum and cell recirculation. (Adapted from Cardona, C.A., and Ó.J. Sánchez. 2007. *Bioresource Technology* 98:2415–2457. Elsevier Ltd.)

from the condensation of ethanol. The model proposed by these authors showed that ethanol inhibition influences especially the cell yield reaching a value of 60 g/L of ethanol in the broth above which the inhibition is very strong. The authors point out that the values of kinetic parameters depend in a high degree on the type of fermentation: batch or continuous. Later, Taylor et al. (2000) employed a saccharified corn mash containing high levels of suspended solids as a feed and compared the results obtained with a state-of-the-art dry-milling process using Aspen Plus. Savings of US0.8 cents/L of ethanol can be attained in comparison with the state-of-the-art process for which saccharification and fermentation are carried out separately (Cardona and Sánchez, 2007).

Other variants of this type of integrated configuration have been proposed as evidenced in Table 9.7. Gong et al. (1999) report the simultaneous variant of the fermentation-stripping process using an air-lift reactor with a side arm (external loop) that improves liquid circulation and mass transfer. A more complex configuration integrating the fermentation and stripping was developed by Bio-Process Innovation, Inc. (West Lafayette, IN, USA). A pilot plant was designed and built for ethanol production from lignocellulosic biomass using a 130 L multistage, continuous-stirred, reactor separator (MSCRS) for the SSF of cellulose and hemicellulose (Dale and Moelhman, 2001). The MSCRS consists of a series of six stirred

TABLE 9.6

Reaction–Separation Integration for Alcoholic Fermentation Processes through Ethanol Removal by Vacuum

Technology	Bioagent/Unit Operation	Feedstock/Medium	Remarks	References
Continuous vacuum fermentation	*Saccharomyces cerevisiae*/vacuum system	Glucose-containing medium	50 mm Hg; with and without cell recycling; sparging of oxygen; 33.4 % glucose feed; productivity 40–82 g/(L × h)	Cysewski and Wilke (1977)
Continuous fermentation coupled with vacuum flashing	*S. cerevisiae*/extractive vacuum flash chamber	Sugarcane molasses	Modeling based on kinetic approach; 4–5.33 kPa; recycling of liquid stream from flash; cell recycling; 98% conversion; 23–26.7 g/(L × h) productivity	Costa et al. (2001) da Silva et al. (1999)

Source: Modified from Cardona, C.A., and Ó.J. Sánchez. 2007. *Bioresource Technology* 98:2415–2457. Elsevier Ltd.

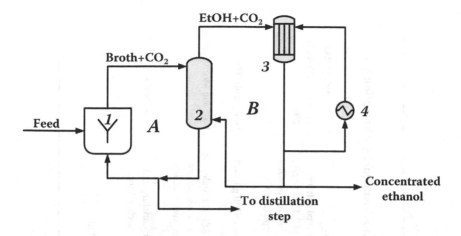

FIGURE 9.9 Simplified diagram of the fermentation process with ethanol removal by using CO_2 as a stripping gas: (A) liquid circulation loop, (B) gas circulation loop, (1) fermenter, (2) stripping column, (3) condenser, (4) refrigerator. (Adapted from Cardona, C.A., and Ó.J. Sánchez. 2007. *Bioresource Technology* 98:2415–2457. Elsevier Ltd.)

stages in which the SSF of biomass is carried out. Each stage has a stirred tank for the reaction and a gas–liquid separation contactor. In the three upper stages, SSF of cellulose is carried out at 42°C using a thermotolerant *K. marxianus*, while in three lower stages, the fermentation of xylose is achieved using the yeast *P. stipitis* at 30°C. In addition, part of the broth containing enzymes is recirculated from the last stage to the first upper stage in order to favor the reaction with the fresh pretreated biomass. Reaction–reaction integration is implemented because commercial cellulases were added for the saccharification of pretreated biomass. This defines the process temperature helping the generation of ethanol vapors. The broth overflowing from one stage into the next stage is contacted with a stripping stream of CO_2 that entraps the ethanol. A gas stream passes across the reactor and then through an absorption tower where water is used for removing ethanol vapors. CO_2 is again recirculated to the reactor. In this way, reaction–separation integration is verified through the *in situ* removal of ethanol produced in each stage. The same company has installed a pilot plant reactor using MSCRS technology (Dale, 1992) in the plant of Permeate Refining, Inc., located in Hopkinton, IA (USA), that produces 11.36 million liters per year of ethanol from starch. This unit has been in operation since September 1995 and employs starch dextrins, thought its use for lignocellulosic biomass has been proposed. Unfortunately, no reports are available describing the modeling and performance of this type of configuration (Cardona and Sánchez, 2007).

9.3.3 ETHANOL REMOVAL BY MEMBRANES

The application of membrane technology has been aimed at the design of membranes that allow the recovery of either ethanol from water (as in the case of

TABLE 9.7

Reaction–Separation Integration for Alcoholic Fermentation Processes through Ethanol Removal by Absorption Using a Stripping Gas

Technology	Bioagent/Unit Operation	Feedstock/Medium	Remarks	References
Continuous fermentation coupled with stripping	*Saccharomyces cerevisiae*/ethanol stripping with CO_2	Glucose-containing medium/saccharified corn mash	Fermenter coupled with a packed column; 60–185 d of operation; yield 0.48–0.50 g/g; EtOH conc. 55.8–64.4 g/L in fermenter and 257–364 g/L in condensate; productivity 7.5–15.8 g/(L.h)	Taylor et al (1996, 1998, 2000)
ALSA	*S. uvarum*/ethanol stripping with CO_2	Glucose	Fed-batch regime; EtOH conc. >130 g/L within 24 h	Gong et al. (1999)
MSCRS	*Kluyveromyces marxianus/ Pichia stipitis*/ethanol stripping with CO_2	Lignocellulosic biomass (oat hulls)	Six-stage reactor separator; SSF of cellulose in first three stages with *K. marxianus*; xylose fermentation in last 3 stages with *P. stipitis*; recycle of broth	Dale and Moelhman (2001)
MSCRS	*S. cerevisiae* or *Zymomonas mobilis*/ethanol stripping with CO_2	Gelatinized starch	Four-stage reactor separator; enriching and stripping sections; SSF of starch; ethanol stripping	Dale (1992)

Source:　Modified from Cardona, C.A., and Ó.J. Sánchez. 2007. *Bioresource Technology* 98:2415–2457. Elsevier Ltd.

Note:　ALSA = Airlift reactor with a side arm, MSCRS = Multistage continuous reactor–separator.

membrane modules coupled to fermenters) or water from ethanol (as in the operation of pervaporation during ethanol dehydration). Membranes make possible the removal of ethanol from a culture broth, which neutralizes the inhibitory effect of ethanol on microorganisms. Most of integrated schemes of this kind correspond to membrane modules coupled to fermenters. The use of ceramic membranes located inside the fermenter has been proposed, although most of these systems have been studied only on a laboratory scale. These laboratory configurations have shown interesting results, but their implementation at an industrial scale can be very difficult. The utilization of ceramic membranes has been proposed for the filtration of cell biomass and the removal of ethanol during the fermentation (Ohashi et al., 1998). The removed ethanol is distilled and the obtained bottoms are recycled back to the culture broth resulting in a drastic reduction of generated wastewater. This configuration uses a stirred ceramic membrane reactor (SCMR). In the same way, immobilized cells can be used in order to allow an easier separation of ethanol and the recirculation of distillation bottoms to the reactor (Kishimoto et al., 1997). Kobayashi et al. (1995) developed a mathematical model for optimization of temperature profiling during the batch operation of a fermenter coupled with a hollow-fiber module. The temperature was kept initially at 30°C descending later to 20°C and attaining higher ethanol concentration and productivity. However, it is necessary to analyze the scalability of these configurations due to their complexities (immobilization, presence of membranes, recirculation, repeated batches) and taking into account that no mathematical description has been presented (Cardona and Sánchez, 2007). The utilization of liquid membranes (porous material with an organic liquid) in schemes involving the extraction of ethanol by the organic phase and the reextraction with a liquid stripping phase used as an extractant (perstraction or membrane-aided solvent extraction) or gaseous stripping phase have been also coupled to the fermentation process showing the increased effectiveness of the latter configuration (Cardona and Sánchez, 2007; Christen et al., 1990). Some of these configurations are summarized in Table 9.8.

Pervaporation has offered new possibilities for integration, as evidenced in Table 9.8. The coupling of fermentation with the pervaporation allows the removing of produced ethanol (Figure 9.10), reducing the natural inhibition of the cell growth caused by high concentrations of ethyl alcohol (Cardona and Sánchez, 2007). Nomura et al. (2002) observed that the separation factor of silicalite zeolite membranes used for continuous pervaporation of fermentation broth was higher than the corresponding value for ethanol–water mixtures due to the presence of salts that enhance the ethanol selectivity. Ikegami et al. (2003, 2004) employed this same kind of membrane coated with two types of silicone rubber or covered with a silicone rubber sheet as a hydrophobic material for obtaining concentrated solutions of ethanol. The coupling of *T. thermohydrosulfuricum* that directly converts uncooked starch into ethanol with pervaporation has also been tested obtaining ethanol concentrations in the permeate of 27 to 32% w/w (Mori and Inaba, 1990).

O'Brien et al. (2000) employed process simulation tools (Aspen Plus) for evaluating the costs of the global process involving fermentation–pervaporation

TABLE 9.8
Reaction–Separation Integration for Alcoholic Fermentation Processes through Ethanol Removal by Using Membranes

Technology	Bioagent/Unit Operation	Feedstock/Medium	Remarks	References
Continuous fermentation coupled with filtration	Inhibitor-tolerant *Saccharomyces cerevisiae*/ cross flow microfiltration unit with stirring	Undetoxified dilute acid spruce hydrolyzate	Supplementation with complete mineral medium; 90% cell recirculation; microaerobic cond.; productivity up to 1.44 g/(L.h); 96 h of operation	Brandberg et al. (2005)
Continuous membrane–filtration bioreactor	*S. cerevisiae*/internal ceramic tubes inside the fermentor	Wood hydrolyzate	High cell retention; EtOH conc. 58.8–76.9 g/L; yield 0.43 g/g; productivity 12.9–16.9 g/(L.h); 55 h of operation	Lee et al. (2000)
Batch fermentation coupled with continuous perstraction	*S. bayanus*/Teflon sheet soaked with isotridecanol	Glucose-containing medium	Water was used as an extractant; EtOH conc. in the broth 75–61 g/L, in the extractant 38 g/L; yield 0.46; aver. productivity 1.2 g/(L.h)	Christen et al. (1990)
Continuous fermentation coupled with continuous perstraction	Immobilized *S. cerevisiae* in alginate/membrane of the type of artificial kidneys	Glucose-containing medium	Tri-*n*-butylphosphate was used as an extractant; glucose feed conc. 506 g/L; aver. EtOH conc. in broth 67 g/L, in the extractant 53 g/L; productivity 48 g/ (L.h); up to 430 h of operation	Matsumura and Märkl (1986)
Batch fermentation coupled with distillation	Immobilized *S. cerevisiae* in Ca alginate/distillation	Glucose-containing medium	Distillation was carried out periodically; recycling of distillation bottoms; 500 h of cultivation; yield 92%; EtOH conc. 10–80 g/L; reduced wastewater	Kishimoto (1997)

SCMR coupled with distillation	Free or immobilized S. cerevisiae/distillation	Glucose-containing medium	High cellular retention by ceramic membrane; recycling of distillation bottoms; 100 h of cultivation; without wastewater; productivity 13.1–14.5 g/(L.h); EtOH conc. 20–50 g/L	Ohashi et al. (1998)
Batch fermentation coupled with continuous pervaporation	S. cerevisiae/silicalite zeolite membrane	Glucose-containing medium	For 4.6 wt.% EtOH in the broth, EtOH in permeate reaches 81.7 wt.%; separation factor of membrane 88; up to 48 h of operation	Nomura et al. (2002)
Batch co-fermentation coupled with continuous pervaporation	Pichia stipitis/polytetrafluoro-ethylen membrane	Enzymatically saccharified exploded rice straw	For 10 g/L EtOH in the broth, EtOH in permeate reaches 50 g/L; yield 0.43 g/g; 100 h of operation	Nakamura (2001)
Fed-batch fermentation coupled with pervaporation	Immobilized S. cerevisiae in Ca alginate/microporous polypropylene membrane		72 h cultivation; EtOH conc. 50 g/L; yield 0.49 g/g; productivity 2.9 g/(L.h); 61.5% reduction in wastewater	Kaseno et al. (1998)
Continuous fermentation coupled with pervaporation	S. cerevisiae/commercial polydimetylsiloxane membranes	Starch from dry milling plant	Aspen Plus simulation based on fermentation–pervaporation lab experiments; EtOH conc. in permeate 420 g/L; recycling of retentate to fermenter, reduction of cost associated with fermentation by 75%	O'Brien et al. (2000)
	Immobilized S. cerevisiae on beads of PAAH gel coated with Ca alginate/ membrane of silicone composite on a polysulfone support	Glucose/molasses	For 4 wt.% EtOH conc. in the broth, EtOH in permeate reaches 12–20% wt.%; yield 0.36–0.41 g/g; productivity 20–30 g/(L.h); over 40 d of operation	Shabtai et al. (1991)

Continued

TABLE 9.8 (*Continued*)
Reaction–Separation Integration for Alcoholic Fermentation Processes through Ethanol Removal by Using Membranes

Technology	Bioagent/Unit Operation	Feedstock/Medium	Remarks	References
Continuous SSF coupled with pervaporation	*S. cerevisiae* + *Trichoderma reesei* cellulases/silicate membrane	Cellulose	Modeling based on kinetic approach; yield 0.44 g/g; EtOH conc. 248.3 g/L in permeate and 4.1 g/L in broth; reduced product inhibition effect; residence time of 72 h; 60–99% substrate conversion	Sánchez et al. (2005)
Batch fermentation coupled with membrane distillation	*S. cerevisiae*/capillary polypropylene membrane	Sucrose-containing medium	2–3 d cultivation; periodic flow of broth through membrane distillation module during 5–6 h per day or continuous coupling to bioreactor; yield 0.47–0.51 g/g; EtOH conc. 50 g/L in broth; productivity 2.5–5.5 g/(L.h)	Gryta (2002, 2001) Gryta et al. (2000)
Continuous fermentation coupled with membrane distillation	*S. cerevisiae* and *S. uvarum*/ polypropylene and poly(tetrafluoro-ethylene) membranes	Glucose and molasses solutions	430–695 h cultivation; EtOH conc. 60 g/L in broth and 200–400 g/L in cold trap; high concentrated medium (316 g/L molasses)	Calibo et al. (1989)

HFMEF	*S. cerevisiae*/hydrophobic microporous hollow fibers/ oleyl alcohol or dibutyl phtalate	Glucose	Yeast cells are immobilized on the shell side; solvent flows in fiber lumen; feed glucose conc. 300 g/L; productivity 31.6 g/(L.h)	Kang et al. (1990)
CMFS	*S. cerevisiae*/membrane bioreactor with continuous removal of ethanol by pervaporation/coupling with cell separator	Not specified	Modeling study; higher dilution rates and productivity (up to 13.5 g/(L.h); recycle ratio 0–2.0; pervaporation factor 0-2.5 h^{-1}; EtOH conc. 10–47 g/L; cell conc. increased from 1.9 to 14.6 g/L due to recycle and pervaporation.	Kargupta et al. (1998)

Source: Modified from Cardona, C.A., and Ó.J. Sánchez. 2007. *Bioresource Technology* 98:2415–2457. Elsevier Ltd.

Note: CMFS = continuous membrane fermentor–separator, HFMEF = hollow-fiber membrane extractive fermentor, PAAH = polyacrylamide hydrazide, SCMR = stirred ceramic membrane reactor.

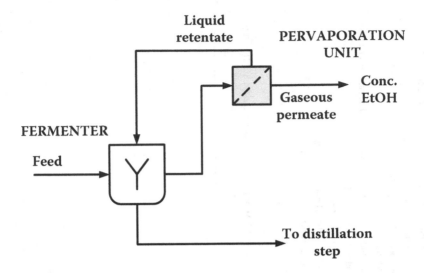

FIGURE 9.10 Simplified diagram of fermentation process with ethanol removal using a pervaporation unit coupled to the fermenter. (Adapted from Cardona, C.A., and Ó.J. Sánchez. 2007. *Bioresource Technology* 98:2415–2457. Elsevier Ltd.)

in comparison with the conventional batch process from starch. Fermentation–pervaporation was simulated based on experimental data from tests carried out during more than 200 h using commercial membranes of polydimethylsiloxane. Performed simulations revealed costs slightly higher for the coupled fermentation–pervaporation process due to the capital and membrane costs. Nevertheless, fermentation costs were reduced 75% and distillation costs decreased significantly. Sensitivity analysis indicated that few improvements in membrane flux or selectivity could make this integrated process competitive (Cardona and Sánchez, 2007). Wu et al. (2005) have investigated the mass transfer coefficients for this type of membrane in the case of pervaporation of fermentation broths showing that active yeast cells were favorable for ethanol recovery. Kargupta et al. (1998) carried out the simulation of continuous membrane fermenter separator (CMFS) removing ethanol by pervaporation in a membrane reactor, which is coupled with a cell separator in order to increase the concentration of cells inside the reactor by recycling them. The models predicted an increase in productivity because this system could be operated at high dilution rates as a consequence of *in situ* product removal and higher cell concentrations.

Besides pervaporation, membrane distillation has been studied (see Table 9.8). In this type of distillation, aqueous solution is heated for the formation of vapors, which go through a hydrophobic porous membrane favoring the pass of vapors of ethanol (which is more volatile) over the vapors of water. The process's driving force is the gradient of partial pressures mainly caused by the difference of temperatures across the membrane (Cardona and Sánchez, 2007). Gryta et al. (2000) implemented a batch fermenter coupled with a membrane distillation module leading to the ethanol removal from culture broth diminishing the inhibition effect and obtaining an increase in ethanol yield and productivities. Gryta (2001) points out that when a tubular fermenter working in a continuous regime is coupled with the membrane distillation module, higher increases in ethanol productivity can be achieved (up to 5.5 g/(L × h)). This author determined that the number of yeast cells that are deposited on the membrane is practically zero during the operation of these modules (Gryta, 2002). Calibo et al. (1989) also demonstrated the possibility of coupling the continuous fermentation with membrane distillation. They used a column fermenter, a cell settler, and a membrane module. This system operated during almost 700 h with a feed of molasses. García-Payo et al. (2000) studied the influence of different parameters for the case of air gap membrane distillation based on the model of temperature polarization. It was observed that permeate flux increases in a quadratic way when ethanol concentration increases in the membrane distillation module. Similarly, Banat and Simandl (1999) indicate that the effects of concentration and temperature polarization should be accounted for during the modeling of this process and highlight the need for optimizing it with respect to feed stream temperature. Banat et al. (1999) also analyzed different models based on Fick's law and on the solution of Maxwell–Stefan equations for this type of distillation. Likewise, the characteristics of the vacuum membrane distillation (Izquierdo-Gil and Jonsson, 2003) and direct contact membrane distillation have been studied for the concentration of aqueous solutions of ethanol (Fujii et al., 1992a, 1992b). Without a doubt, these studies are of great interest considering the simulation of these integrated configurations (Cardona and Sánchez, 2007).

In the case of the bioethanol production from sugarcane, the integration of fermentation with pervaporation or vacuum membrane distillation can allow the recovery of a valuable product: the fructose. For this, mutant strains of yeasts without the capacity of assimilating this monosaccharide should be used. Thus, continuous ethanol removal through the membranes coupled to the fermenter makes possible the accumulation of fructose in the culture medium that can be recovered in an extraction column (Cardona and Sánchez, 2007). According to Di Luccio et al. (2002), the simulation of this process based on experimental data and semiempiric models for the evaluation of the required membranes area allowed performing of a preliminary economic analysis. This analysis showed that variable costs involving membrane area influence in a higher degree the viability of the process. The process is viable only if the cost of membranes is not greater than US\$550/m^2 for a new plant or US\$800/m^2 for an adapted plant considering an internal return rate of 17%.

CASE STUDY 9.3   MODELING OF SSF COUPLED WITH PERVAPORATION FOR ETHANOL PRODUCTION

The modeling of reaction–separation processes involving membrane technology can provide important insight into the suitability of such hybrid processes during conceptual design of fuel ethanol production. In Case Study 9.2, the kinetic

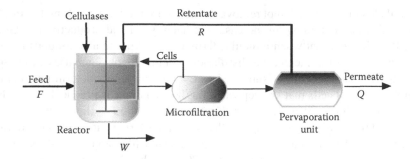

FIGURE 9.11 Scheme of an integrated reaction–separation process consisting of a CSTR for the SSF of cellulose coupled with a pervaporation unit.

modeling of processes involved during the SSF of cellulose was presented. In this case study, the SSF model presented, which was based on the mathematical description of South et al. (1995), was applied to a reaction–separation integrated process in which the SSF bioreactor was coupled with a pervaporation module.

The schema used to simulate the SSF process coupled with pervaporation appears in Figure 9.11. The CSTR has two outlet streams, one of which is directed to a microfiltration unit where biomass and other nonsoluble components are removed and recycled back to the reactor. The soluble components (mainly glucose, cellobiose, and ethanol) are sent to the pervaporation unit where the main part of the ethanol is selectively removed through a silicate membrane and the retentate is recirculated to the reactor. A cell-separating device was included in order to avoid the cell biomass and other related solids to enter the pervaporation unit. In the stream W all of the components of the culture medium are present, including the cell biomass. Unlike the membranes employed for ethanol dehydration, the silicate membranes have a higher selectivity for the ethanol, causing the concentration of this alcohol in the retentate to be reduced, which allows the recirculation of this stream to the reactor where the biological transformation takes place. It should be noted that this process implies the application of the integration principle in two ways: reaction–reaction integration through the SSF of cellulose and reaction–separation integration through the coupling of a membrane device (the pervaporation unit). The objective of this case study, presented in a previous work (Sánchez et al., 2005), was to assess the effect of ethanol removal from culture broth by using membranes. Both semicontinuous and continuous regimes were considered.

In the case of the SSF coupled with pervaporation in semicontinuous regime, the permeate stream Q is continuously removed from the system causing a decrease in the total liquid volume in the reactor. This situation can be described by the following mass balance equations for all i components:

$$\frac{d(VC_i)}{dt} = Vr_i \tag{9.8}$$

The nomenclature of this and subsequent equations can be found in Table 9.3. In the specific case of ethanol, the concentration in the reactor was affected by the continuous removal of this component so the following balance equations, including an expression for the volume change with the time, were used:

$$\frac{d(VP)}{dt} = -QP_P + Vr_P \tag{9.9}$$

$$\frac{dV}{dt} = -Q \tag{9.10}$$

To model the simultaneous saccharification and fermentation process coupled with pervaporation (SSF + PV), an equation describing a separation factor α as a result of the pervaporation was added to the described system of equations:

$$\alpha = \frac{[Ethanol \, / \, Water]_{Perm}}{[Ethanol \, / \, Water]_{Feed}} \tag{9.11}$$

where *Ethanol* is the concentration of ethanol and *Water* the concentration of water (in g/L) and the subindices *Perm* and *Feed* correspond to the permeate and feed, respectively. The formulation of α was taken from Nomura et al. (2002) for a silicate membrane. These authors reported also different values of this separation factor for water–ethanol mixtures ($\alpha = 41$), solutions containing yeast ($\alpha = 38$), salts ($\alpha = 62$), and fermentation broth ($\alpha = 88$). The last value was used in this case study.

The simulation of the batch SSF process coupled with pervaporation for the same initial concentrations and a residence time identical to those of the batch process presented in Case Study 9.2 gave a final ethanol concentration in the fermentation broth of 0.04 g/L as a consequence of the product removal in the pervaporation module. The concentration of ethanol in the permeate varied with the time due to the changes in the broth volume and reached relatively high values of 92 g/L (Figure 9.12). The cellulose consumption was 45.3% higher (considering the volume change) than in the case of batch SSF without pervaporation indicating the favorable effect of product removal on the diminishing of inhibition effect of ethanol on cellulose uptake rate, growth rate, and product biosynthesis rate. The cell biomass had a similar profile in the batch SSF process with and without pervaporation, but had a better growth in the former case because of the minor ethanol concentration in the broth.

In the case of continuous SSF of cellulose coupled with a pervaporation unit and taking into account the scheme shown in Figure 9.11, the mass balance equations were formulated assuming that the stream of permeate Q only contains ethanol and water. Therefore, the global material balances are the same as those used for the SSF process in a CSTR (see Case Study 9.2), except for ethanol balance that now includes a term considering the separation achieved in the pervaporation unit:

$$FP_o - WP - QP_q + Vr_P = 0 \tag{9.12}$$

Ethanol and biomass productivities and product yield were evaluated considering the variation in the mean residence time (or dilution rate) and in the fraction of

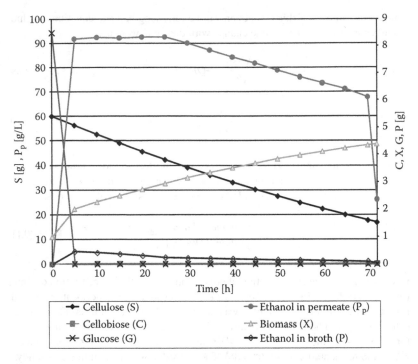

FIGURE 9.12 Behavior of semicontinuous SSF of cellulose coupled with pervaporation.

outlet stream that is sent to the pervaporation unit. For the base case, the value of the feed stream was 100 L/h and the mean residence time was 72 h.

When the analysis of the continuous SSF process coupled with pervaporation was performed, the concentration of ethanol in the permeate increased significantly reaching 248.3 g/L, whereas the ethanol concentration in the fermentation broth was 4.1 g/L. This low concentration has no inhibition effects on the bioprocess for this fermentation regime. In addition, the cellulose concentration in the broth was 3.9 g/L revealing an important increase in the substrate conversion. The biomass concentration was higher than in all the other cases (7.2 g/L). The concentration of cellobiose and glucose in the broth were negligible. In this case, a flow rate of permeate of 12 L/h and a separation factor of 88 were considered and the concentration of cellulose in the feed stream was 60 g/L. The results were obtained for a mean residence time of 72 h and a feed flow rate of 120 L/h. Comparing the obtained data with those of the previous configuration, it can be observed a better production of ethanol, a greater consumption of cellulose, an improved production of cell biomass, and similar results for cellobiose and glucose.

If the permeate flowrate is changed, an increase of the productivity can be observed due to the higher removed amounts of ethanol and to the corresponding reduced end product inhibition. On the other hand, productivity decreases with the increase of residence time and the subsequent rise in the reactor volumes. The cellulose conversion was highly affected by the residence time as expected, but a

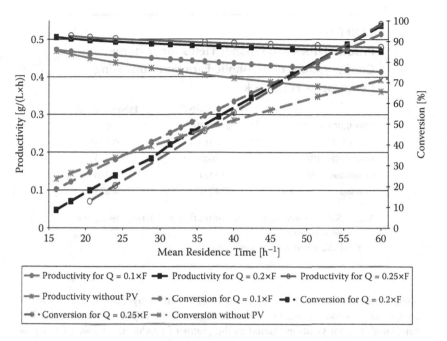

FIGURE 9.13 Productivity and cellulose conversion for SSF process with and without pervaporation (PV) for various values of permeate flow rates (Q) expressed as a fraction of feed flow rate (F).

reduce conversion was observed for the case when pervaporation was not included in the configuration. These results are shown in Figure 9.13.

Four models for comparing the consumption of cellulosic biomass and the production of ethanol were solved (considering the two models of Case Study 9.2) and analyzed considering configurations with and without pervaporation and for batch and continuous regimes. From the results, it can be observed that better conversion to ethanol of cellulose is obtained when the pervaporation is included in the SSF process. This can be explained by a lower ethanol concentration in the fermentation broth. This reduced ethanol concentration implies that the value of the inhibition terms in the rates equations be lesser leading to the increase of the cellulose and cellobiose hydrolysis, the growth of biomass, the consumption of glucose, and consequently, the production of ethanol. For the two configurations involving pervaporation, the selective separation effect of the membrane module allowed the effective removal of the product from the reaction–fermentation zone.

Although the pervaporation model employed is a simple representation, it is a good estimation for analyzing the behavior of a simultaneous saccharification and fermentation process along with this separation method. The proposed models allowed observing how the inhibition effects by ethanol influence its production, and how its continuous removal causes a considerable increase of cellulose conversion and, for this reason, an extra production of ethanol. The productivities for the evaluated configurations are presented in Table 9.9. The ethanol produced per

TABLE 9.9

Productivities and Ethanol Yields for Different Configurations of Alcoholic Fermentation Using Cellulose as Feedstock

Configuration	Productivity/ g/(L × h)	Ethanol Yield/ g/g
Batch SSF	0.2916	0.454[a]
Batch SSF+ PV	0.3645	0.493
Continuous SSF	0.3451	0.506
Continuous SSF + PV	0.4791	0.525

Note: SSF = simultaneous saccharification and fermentation, PV = pervaporation.

[a] Yield calculated at 48 h of cultivation.

hour, considering, in batch cases, a break of 12 h to discharge, clean, and charge the equipment was considered. The better configuration is SSF + PV in a CSTR. The calculated ethanol yields evaluated as the grams of produced ethanol from 1 g of consumed cellulose showed better results again for the simultaneous saccharification and fermentation process coupled with pervaporation in a CSTR. In contrast, the minimum yield corresponded to the batch SSF.

The ethanol in the permeate must be passed through separation equipment (i.e., distillation) to get rid of the excess water to obtain an appropriate ethanol for fuel use. At this point, it is important to remark that the increased concentration of ethanol allows the reduction of energy costs during the distillation of permeate in comparison with those corresponding to the distillation of fermentation broth where the ethanol concentration is very low (approximately 25 g/L for a continuous SSF process, according to the performed simulations). This case study allows comparing these four systems and obtaining results that can be used for future studies involving more meticulous models of pervaporation developed from rigorous mass, momentum, and energy balances and disregards many common assumptions, such as constant pressure, constant temperature, binary mixture, steady-state conditions, and constant physical properties. In addition, the proposed approach for modeling of this process can be used for the design of laboratory and pilot plant experiments. The proposed configuration of simultaneous saccharification and fermentation coupled with pervaporation in a CSTR has advantages over other batch and continuous systems in terms of productivities and product yield. Therefore, this process becomes attractive considering that the use of biomass to produce ethanol is not economically feasible on a large commercial scale yet, and should be taken into consideration due to the significant qualitative and quantitative improvements demonstrated through modeling of this complex bioprocess. However, the limitations inherent in membrane technologies in real practice should be thoroughly assessed.

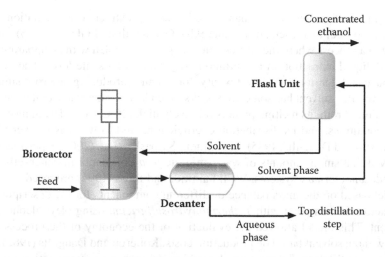

FIGURE 9.14 Simplified diagram of fermentation process with ethanol removal by liquid extraction (extractive fermentation). (Adapted from Cardona, C.A., and Ó.J. Sánchez. 2007. *Bioresource Technology* 98:2415–2457. Elsevier Ltd.)

9.3.4 ETHANOL REMOVAL BY LIQUID EXTRACTION

A reasonable approach for increasing the productivity of alcoholic fermentation is the removal of the product that causes the inhibition through an extractive biocompatible agent (solvent) that favors the migration of ethanol to solvent phase, a process known as extractive fermentation. Some fatty alcohols have the ability of being used as extractive agents thanks to their ethanol selectivity and low toxicity for cells of ethanol-producing microorganisms. In an early work, Minier and Goma (1982) showed that primary aliphatic alcohols with a chain length having fewer than 12 carbons inhibit the growth of yeast cells. They chose the fatty alcohol *n*-dodecanol as a solvent for *in situ* extraction of ethanol in a special continuous pulse-packed column with immobilized cells of *S. cerevisiae*. This configuration allowed the utilization of very concentrated glucose feed due to the reduction of ethanol in the culture broth. In addition, immobilization seems to protect the cells against solvent toxicity (Aires Barros et al., 1987; Cardona and Sánchez, 2007; Tanaka et al., 1987).

The simultaneous scheme of the integrated extractive fermentation process is shown in Figure 9.14. In this configuration, the solvent is directly added to the culture broth where it comes in contact with ethanol. Ethanol migrates to the organic (solvent) phase, whereas the other components of the culture broth remain in the aqueous phase (cells, substrates). Liquid medium from this bioreactor is continuously removed in order to carry out the decantation of both phases. An aqueous phase with a reduced content of ethanol is sent through the distillation step and the ethanol-rich solvent phase is flashed for the regeneration of the solvent, which is recycled back to the bioreactor, and the production of a concentrated solution of ethanol.

Some configurations for ethanol production by extractive fermentation have been proposed, as presented in Table 9.10. Gyamerah and Glover (1996) implemented a process where the fermentation stage was coupled with an apparatus for liquid–liquid extraction in a continuous regime at pilot-scale level. They chose n-dodecanol for its very low toxicity for ethanol-producing microorganisms. However, this solvent has some drawbacks: it tends to form a stable emulsion with the culture broth, its melting point is relatively high (26°C) considering fermentation conditions, and its distribution coefficient related to water is not very high (Kollerup and Daugulis, 1985). In addition, Kirbaşlar et al. (2001) experimentally showed that small amounts of water migrate to n-dodecanol in water–ethanol–n-dodecanol ternary systems. Weilnhammer and Blass (1994) proposed a simple model based on the mass balance of different components for the description of extractive fermentation with T. thermohydrosulfuricum using oleyl alcohol as a solvent. This model allowed the evaluation of the economy of the process with and without solvent based on production costs. Kollerup and Daugulis (1985) proposed a mathematical model for describing an extractive fermentation process for continuous production of ethanol from a glucose-containing medium. In this model, a simple relationship between ethanol concentration in aqueous phase and ethanol content in solvent phase was considered. Additionally, kinetic description of microbial growth did not take into account the inhibition effect due to high concentrations of substrate. Fournier (1986) developed a rigorous description considering the use of UNIversal Functional Activity Coefficient (UNIFAC) equations for continuous extractive fermentation (Cardona and Sánchez, 2007).

Further integration could include a process where simultaneous saccharification and extractive fermentation may be carried out, as reported by Moritz and Duff (1996). These authors used oleyl alcohol and demonstrated that this solvent produced no effect either on cellulases or cell biomass, i.e., it was biocompatible for both biological agents involved in the process. Oliveira et al. (1998, 2001) proposed an extractive biocatalytic process in which ethanol produced by yeasts is extracted by oleic acid and used as substrate for lipase-catalyzed esterification reaction with this same acid. In this way, the combination of the enzymatic reaction and the extractive fermentation in a single vessel improves the product extraction. Acceptable results for broths with a high concentration of glucose (300 g/L) were obtained, although experiments without lipase (only extractive fermentation) indicated that oleic acid is not a good extracting agent for ethanol because its concentration in the aqueous phase was higher than in the solvent phase at the end of fermentation. Kang et al. (1990) employed a hollow-fiber membrane reactor for carrying out extractive fermentation with yeast using oleyl alcohol and dibutyl phthalate, obtaining productivities of 31.6 g/(L × h). Fournier (1988) proposed a mathematical model for this type of hollow-fiber membrane extractive fermenters that predicts significant improvements in productivity related to conventional CSTR with solvent extraction. The utilization of a liquid-lift external loop bioreactor where the solvent is sparged into the base of the column containing a second liquid (the broth) of higher density has been proposed for the extraction of ethanol as well (Modaressi et al., 1997; Stang et al., 2001).

TABLE 9.10

Reaction–Separation Integration for Alcoholic Fermentation Processes through Ethanol Removal by Liquid–Liquid Extraction

Technology	Bioagent/Unit Operation	Feedstock/Medium	Remarks	References
Continuous fermentation coupled with liquid–liquid extraction	Immobilized yeast/n-dodecanol	Glucose-containing medium	18 d of operation; use of very concentrated feedstocks (10–48% w/w); 78% reduction of aqueous effluents	Gyamerah and Glover (1996)
Batch extractive co-fermentation	Zymomonas mobilis/n-dodecanol	Glucose and xylose	Modeling based on kinetic approach and liquid–liquid equilibrium; solvent is regenerated by flashing; productivity 2.2–3.0 g/(L.h); solvent volume/aqueous volume ratio 1.33–3.0	Gutiérrez et al. (2005)
Continuous extractive fermentation	Immobilized Saccharomyces cerevisiae/n-dodecanol	Glucose	Pneumatically pulsed packed reactor; flowrates: solvent 1–2.55 L/h, medium 0.057–0.073 L/h; feed glucose conc. 261–409 g/L; EtOH conc. in solvent 3.37–10 g/L, in broth 9.4–33 g/L; yield 0.51; productivity 1.03 g/(L.h)	Minier and Goma (1982)
	Trichoderma thermohydrosulfuricum/ oleyl alcohol	Glucose	Flowrates: broth 0.15–0.55 L/h, solvent 0–18 L/h; feed glucose conc. 12.5–100 g/L; EtOH conc. in the broth < 4.47 g/L, in the reextraction water 3–14 g/L; 65°C; productivity < 0.128 g/(L.h)	Weilnhammer and Blass (1994)

Continued

TABLE 9.10 (Continued)

Reaction–Separation Integration for Alcoholic Fermentation Processes through Ethanol Removal by Liquid–Liquid Extraction

Technology	Bioagent/Unit Operation	Feedstock/Medium	Remarks	References
Fed-batch SSEF	S. cerevisiae/commercial cellulases/oleyl alcohol	Primary clarifier sludge from chemical pulping process/cellulose	Reactor with up to 2.5% aqueous phase; 50% substrate conversion; 48–275 h cultivation; 65% increase in productivity in comparison with conventional fed-batch process	Moritz and Duff (1996)
HFMEF	S. cerevisiae/hydrophobic microporous hollow fibers/oleyl alcohol or dibutyl phtalate	Glucose	Yeast cells are immobilized on the shell side; solvent flows in fiber lumen; feed glucose conc. 300 g/L; productivity 31.6 g/(L.h)	Kang et al. (1990)

Source: Modified from Cardona, C.A., and Ó.J. Sánchez. 2007. *Bioresource Technology* 98:2415–2457. Elsevier Ltd.

Note: HFMEF = hollow-fiber membrane extractive fermenter, SSEF = simultaneous saccharification and extractive fermentation.

The selection of the solvent is a crucial factor for extractive fermentation technology. Bruce and Daugulis (1991) developed a solvent screening program for evaluating possible extracting agents to be used in extractive fermentation configurations. This program considered the biocompatibility of the solvent and utilized the UNIFAC activity model for predicting liquid–liquid equilibrium data. Using this program, these authors (Bruce and Daugulis, 1992) identified the solvent mixture of oleyl alcohol with 5% (v/v) 4-heptanone as a promising extractant due to its reduced inhibitory effect and increased distribution coefficient. A large amount of alcohols have been tested in order to examine their ethanol extracting properties in water–ethanol–solvent systems (Offeman et al., 2005a, 2005b). The collected data have allowed insight into the relationship between the structure of the solvent and its extracting characteristics. For this type of work, molecular simulation can provide a deeper understanding on solvent conformation and associations among water, ethanol, and solvent. Although these works are intended to the selection of proper solvent for ethanol dehydration, the obtained results are of great value for extractive fermentation studies, provided the needed biocompatibility tests are done (Cardona and Sánchez, 2007).

Another approach for extractive fermentation that has been used is the application of aqueous two-phase fermentation where two phases are formed in the bioreactor as a result of adding two or more incompatible polymers (Banik et al., 2003). In this way, ethanol can be partitioned between both phases accumulating in the upper layer, whereas cell biomass is accumulated in the lower phase. This allows the separation of the ethanol-rich phase and the distillation of this alcohol reducing its inhibition effect. Nevertheless, the complexity of the process in the case of continuous regime and the high cost of polymers have delayed further development in this ethanol production technology (Cardona and Sánchez 2007).

CASE STUDY 9.4   RIGOROUS MODELING OF EXTRACTIVE CO-FERMENTATION

Modeling of extractive fermentation processes for fuel ethanol production plays a crucial role when different process alternatives are being analyzed in the framework of conceptual process design, especially when process synthesis procedures are applied. Considering that technologies for ethanol production from lignocellulosic biomass are not currently mature, the analysis of different options intended to reduce lignocellulosic ethanol production costs is a task of great significance. In a previous work (Sánchez et al., 2006), the co-fermentation process integrated by means of reaction–separation approach was assessed. The objective of that work was to model the extractive fermentation process for fuel ethanol production from lignocellulosic biomass analyzing cultivation kinetics coupled with liquid extraction.

To describe the continuous process of extractive fermentation for fuel ethanol production, n-dodecanol was selected as an extracting agent (solvent). A feed aqueous stream containing sugars and nutritive components is added to a CSTR where a solvent stream is continuously fed as well. Fed sugars are generated during the pretreatment of lignocellulosic biomass in which major polysaccharides are broken down into elementary sugars like hexoses (glucose) and pentoses (mainly xylose). Formed sugars are converted into ethanol in the reactor. Ethanol is distributed between aqueous and organic (solvent) phases, diminishing its concentration in the culture aqueous broth and allowing reduction of the product inhibition effect on the microorganisms. The ethanol-enriched solvent phase is continuously removed from the reactor through a decanting unit. This stream is sent to a flash unit in order to recover the obtained ethanol and to regenerate the solvent, which can be recycled to the CSTR.

With the aim of developing a rigorous model that describes both the fermentation and liquid–liquid extraction processes, the kinetics cultivation is coupled with an extraction model. The liquid–liquid equilibrium was described through an algorithm based on the mass balance equations developed for the isothermal flash in the case of two liquid immiscible phases. Activities of components in each phase were calculated by means of the UNIFAC model, since this model has demonstrated to be the most appropriate for description of equilibrium when two or more liquid phases are present for this case. This algorithm was integrated into the ModELL-R software, which was especially designed by the research group to which the authors of this book belong. The software couples two convergence algorithms (Newton–Raphson and False Position Method) in order to calculate the liquid fraction of each phase. ModELL-R was developed in Delphi package v7.0 (Borland Software Corp., Austin, TX, USA).

The kinetic model of alcoholic fermentation was taken from Leksawasdi et al. (2001; see Chapter 7, Case Study 7.1). This model describes the simultaneous consumption by a recombinant strain of Z. *mobilis* of two main substrates contained in the lignocellulosic hydrolyzates: glucose and xylose. The following assumptions were considered for the development of the overall model of extractive fermentation:

- The substrate uptake, biomass formation, and product biosynthesis are carried out only in the aqueous phase; hence, no reactions occur in the organic (solvent) phase.
- Ethanol is the main component migrating to the solvent phase; small amounts of water can migrate to the organic phase depending on the solvent.
- No migration of substrates and biomass to the solvent phase takes place.
- Solvent is biocompatible with the microorganisms and does not have effect on the fermentation process.
- Stirring of bioreactor ensures total mixing between liquid phases and does not produce damage to the growing cells.

The configuration corresponding to continuous extractive fermentation involves the continuous feeding of culture medium and solvent to the reactor and the continuous removal of the liquid aqueous phase and solvent phase from the reactor in a separate way with the help of a decanter (see Figure 9.14). In this case, the flowrate (in L/h) of influent aqueous stream (F_A) is greater than the flowrate of effluent aqueous stream (Q_A) because of the migration of ethanol to the solvent phase. Similarly, the flowrate of influent solvent stream (F_E) is less than the flowrate of effluent solvent stream (Q_E). Mass balance equations representing this process are as follows:

$$-Q_A X + V_A r_X = 0 \tag{9.13}$$

$$F_A S_{10} - Q_A S_1 - V_A r_{S1} = 0 \tag{9.14}$$

$$F_A S_{20} - Q_A S_2 - V_A r_{S2} = 0 \tag{9.15}$$

$$F_E P_0^* - Q_A P - Q_E P^* + V_A r_P = 0 \tag{9.16}$$

$$\rho_A F_A + \rho_E F_E - \rho_A Q_A - \rho_E Q_E = 0 \tag{9.17}$$

$$P^* = k_{EtOH} P \tag{9.18}$$

where r_X is the cell growth rate (in g/(L × h)), r_{S1} and r_{S2} are the glucose and xylose consumption rates, respectively (in g/(L × h)), and r_P is the ethanol formation rate (in g/(L × h)); X, S_1, S_2, and P are the concentrations of cell biomass, glucose, xylose, and ethanol in the aqueous effluent from the bioreactor (in g/L), and X_0, S_{10}, S_{20}, and P_0 are the corresponding concentrations in the aqueous feed stream (in g/L); P^* and P_0^* are the ethanol concentrations in the solvent effluent and in the solvent feed streams, respectively (in g/L). The balance can be applied for the case when the solvent contains small amounts of ethanol as a result of noncomplete regeneration of the extracting agent. This system of equations is nonlinear due to the equations describing the process kinetics and is solved through multivariate Newton–Raphson algorithm. For this, a constant solvent volume/aqueous volume ratio is assumed. The relationship between both effluent flow rates is fixed. The ethanol concentration in the solvent phase (P^*) that is in equilibrium with ethanol concentration in the aqueous phase is determined using the distribution coefficient k_{EtOH} as shown by equation (9.18), which is calculated by the algorithm for liquid–liquid equilibrium. The determination of all variables involved in the model is performed using the algorithm shown in Figure 9.15 incorporated into the software ModELL-R. For specified inlet aqueous dilution rate ($D_{Ai} = F_A/V_A$) and solvent feed flow rate/aqueous feed flow rate ratio ($R = F_E/F_A$), the program requires concentrations of cell biomass, substrates, and ethanol in feed streams.

The simulation of alcoholic extractive fermentation from biomass was performed with a glucose concentration of 100 g/L and an xylose concentration of 50 g/L in the feed aqueous stream. These concentrations correspond to those of lignocellulosic hydrolyzates. The behavior of this process for $R = 2$ in dependence of inlet dilution rate (D_{Ai}) is shown in Figure 9.16. The cell washout occurs at dilution rates near 0.33 h^{-1}. These results were obtained for a solvent volume/aqueous volume ratio (V_E/V_A) of 2. Higher ethanol productivities are found in the range 0.25 to 0.30 h^{-1}. In order to elucidate the best operating value of D_{Ai}, GAMS software (General Algebraic Modeling System, GAMS Development Corp., Washington, DC, USA) was used for maximizing total ethanol productivity. For this, a simple linear relationship for Equation (9.18) was considered. The optimal D_{Ai} was 0.265 h^{-1}.

The coupled algorithm was used for process simulation varying the solvent feed flow rate/aqueous feed flow rate ratio (R) for an inlet dilution rate of 0.265 h^{-1}. Best results were obtained for values greater than 4 that correspond to an increased

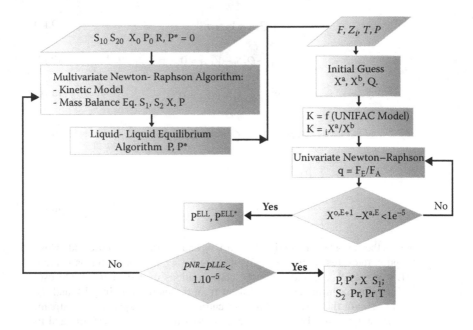

FIGURE 9.15   Algorithm for calculation of extractive (co-)fermentation process.

amount of consumed substrates. Both total productivity and productivity for ethanol recovered from solvent phase are approaching constant values. For R higher than 8, the model predicts the formation of homogeneous mixture without extraction. The simulation of this process modifying both R and D_{Ai} shows that the zone of manipulating variables with higher productivities (inlet concentrations of glucose and xylose in the aqueous stream of 100 g/L and 50 g/L, respectively) corresponded to dilution rates near washout conditions, and to higher values of R within the range 1.29 to 7.9. If the concentration of both substrates in the feed stream is varied, the problem becomes more complex. Physically, the increase in substrate can be achieved as a result of evaporation of initial hydrolyzate obtained from biomass pretreatment. For this reason, the proportion of glucose and xylose in the inlet aqueous stream should be constant and equal to 2:1. Best values of total productivity and ethanol productivity recovered from solvent phase correspond to an inlet concentration of total sugars of about 600 g/L. The simulation was carried out until the concentration of sugars was less than or equal to 600 g/L, which corresponds to maximum solubility of these sugars in water.

In this way, a model describing extractive co-fermentation was developed. This model allows for doing the analysis of this reaction–separation integration process in order to consider it in subsequent process synthesis methodologies. This is especially valuable for such process synthesis approaches as the hierarchical decomposition (see Chapter 2) that requires the development of proper models for simulation of alternative configurations during each hierarchical level of analysis.

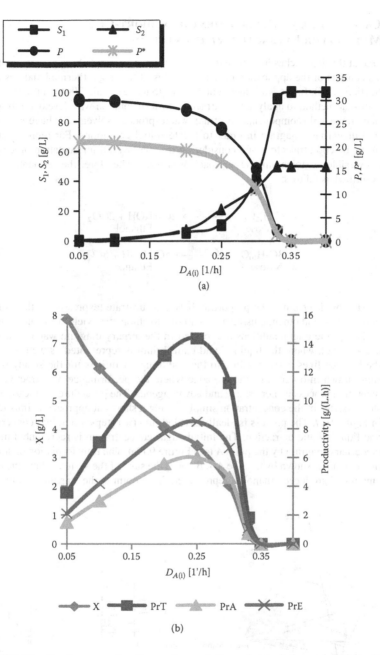

FIGURE 9.16 Continuous extractive co-fermentation using *n*-dodecanol. Effect of inlet aqueous dilution rate (D_{Ai}) on (a) effluent concentrations of glucose (S_1), xylose (S_2), ethanol in aqueous phase (P), and ethanol in solvent phase (P^*); (b) total ethanol productivity (*PrT*), productivity for ethanol recovered from aqueous phase (*PrA*), productivity for ethanol recovered from solvent phase (*PrE*), and effluent concentration of cells (*X*). Concentration of sugars in feed aqueous stream: glucose, 100 g/L; xylose, 50 g/L.

CASE STUDY 9.5 DEVELOPMENT OF A SHORT-CUT METHOD FOR EXTRACTIVE FERMENTATION

One of the approaches highlighted in Chapter 2 for accomplishing process synthesis consists of the application of the principles of topologic thermodynamics (see Section 2.2.6). In the work mentioned above (Sánchez et al., 2006), a preliminary short-cut method to analyze the extractive co-fermentation was developed. In particular, the main components (substrate–water–product–solvent) can be represented in a quaternary diagram in order to locate initial conditions. For representation of the reaction trajectory and considering that the overall fermentation process is irreversible, a stoichiometric approach was used. Therefore, the fermentation is described as follows:

$$C_6H_{12}O_6 \xrightarrow{\text{Z. mobilis}} 2C_2H_5OH + 2CO_2$$
$$\text{Glucose} \qquad\qquad\qquad \text{Ethanol}$$

$$3C_5H_{10}O_5 \xrightarrow{\text{Z. mobilis}} 5C_2H_5OH + 5CO_2$$
$$\text{Xylose} \qquad\qquad\qquad \text{Ethanol}$$

Having determined the proportion between substrate (expressed as the sum of both sugars) and ethanol, using a maximum stoichiometric yield from substrate of 0.511 g/g, the inlet conditions are located in the ternary diagram water–ethanol–n-dodecanol, where the liquid–liquid equilibrium is represented as well. Finally, the balance lines corresponding to the operating conditions for the steady-states were drawn in order to define the zone where the performance of extractive fermentation process is more stable and advantageous. The procedure for locating the steady-states in the concentration simplex using a short-cut approach is illustrated in Figure 9.17. The process is ideally divided into two steps: microbial conversion and liquid–liquid extraction. The initial sugar concentration is represented in the quaternary diagram by the point A (see Figure 9.17a). The transformation of sugars into ethanol is shown by the line AB, B being the state of the system where the total amount of produced ethanol is represented. This point is the starting mixture for

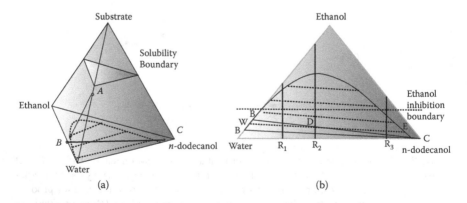

FIGURE 9.17 Representation of extractive fermentation: (a) quaternary diagram, (b) ternary diagram.

the liquid–liquid equilibrium. The line BC represents the addition of n-dodecanol to the aqueous medium containing ethanol (Figure 9.17b). This line lies in the ternary diagram water–ethanol–n-dodecanol, where the zone of heterogeneous mixtures is also drawn. Vertical lines represent the geometric place of points that represent the operating conditions related to the solvent feed stream/aqueous feed stream (R). The intersection of these vertical lines with the line BC (point D) represents the theoretical conditions corresponding to the mixtures before the separation in phases (the equivalent to the feed mixture in a liquid extractor). Through tie lines, the compositions of the extract (E) and the raffinate (W) are obtained. These correspond to the compositions of solvent phase effluent and aqueous phase effluent, respectively. For identical inlet concentrations of sugars in the aqueous stream, the position of the starting point B changes when the inlet dilution rate varies. For example, if D_{Ai} increases, the new line $B'C$ will lie below the original line BC, which it is explained by a major dilution of ethanol and, therefore, the line approaches the bottom edge of the ternary diagram.

Using this short-cut approach, the zone of feasible operating points can be easily determined. Let us analyze the extreme case when the feed aqueous stream has the maximum allowable concentration of sugars. From the fermentation stoichiometry, this condition corresponds to an inlet concentration in the feed aqueous stream of about 600 g/L of total sugars. Assuming a 95% yield, the total amount of ethanol that could be produced is 0.486 g/g, which implies a theoretical starting ethanol concentration of 291.6 g/L (approximately an ethanol mass fraction of 0.42). This value determines the position of the substrate solubility boundary (point H in Figure 9.18). Because the concentration of ethanol in the aqueous phase (raffinate) should not be above the ethanol inhibition boundary (approximately 10% w/w), the operation conditions represented by the line R_3 should be such that the ethanol content in the raffinate corresponding to point D'' be equal to the ethanol content of the point I to avoid product inhibition. In this way, the area delimited by the points $R_3D''EK$ is the zone of feasible steady-states for given conditions of the process

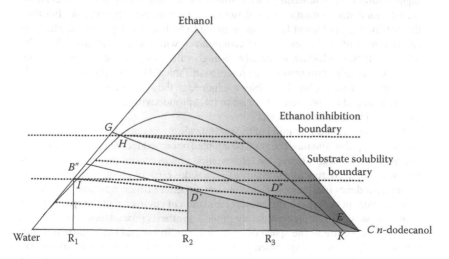

FIGURE 9.18 Representation of extractive fermentation process for different concentrations of sugars in the inlet aqueous streams.

TABLE 9.11

Preliminary Optimum Results for Manipulating Variables Calculated by GAMS and Corresponding Values Calculated by ModELL-R Software

Variable		D_{Ai}	S_{10}	S_{20}	PrT	P	P*	S_1	S_2
Units	R	1/h	g/L	g/L	g/(L×h)	g/L	g/L	g/L	g/L
ModELL-R	3.038	0.185	400	200	54.79	40.31	73.88	5.33	16.77
GAMS	3.038	0.185	400	200	51.46	40.52	73.55	5.39	19.95

with the maximum concentrations of sugars in the culture broth. For a given inlet dilution rate of 0.1 h^{-1}, an R ratio of 3.6 (line R_2), and a working volume of 1 L, the location of the point D' and the corresponding composition of the extract, can be found. In this case, the ethanol mass content of the extract and raffinate is of 7.5% and 9.0%, respectively, resulting in a total ethanol productivity of 26.4 g/(L × h). The productivity calculated by the rigorous model using ModELL-R is 28.86 g/(L × h). Hence, the developed short-cut method allows one to determine the feasibility of operating parameters and an estimation of the productivities.

The delimited zones can be taken into account for the development of a preliminary strategy of optimization. Since the region of feasible steady-states was determined in the ternary diagram (see Figure 9.18), the values range of such manipulating variables as inlet dilution rate, the R ratio, and the concentration of the sugars in the inlet aqueous stream is known and can be bounded for solving an optimization algorithm. The GAMS system was employed to find the optimal value of above-mentioned variables that maximize the total ethanol productivity using the nonlinear programming (NLP) solver CONOPT3. With this aim, liquid–liquid equilibrium relationships were simplified for generating a way to evaluate the ethanol concentration in both phases during extractive fermentation. For this, the distribution coefficient k_{EtOH} was assumed to be linearly dependent of the total concentration of substrates. A good concordance with the data obtained from the ModELL-R was achieved, especially for high values of substrates concentration. The results of this optimization are presented in Table 9.11. From this table, it is evident that calculated optimal variables are indeed in the zone predicted by the short-cut method, and the predicted increase in total productivity is effectively attained, as is shown by the rigorous model.

For a more accurate solution of the optimization problem, the rigorous description of the equilibrium model should be coupled or embedded into the GAMS code. Further analysis of this extractive fermentation process could include the formulation of an objective function that considers, besides ethanol productivity, other performance indexes like the conversion of sugars (better utilization of the feedstock), or the amount of generated wastewater (evaluation of environmental impact). These issues are very significant considering process synthesis procedures. After obtaining a global picture of the space of operating conditions and their optimal values for the studied process, experimental runs should be performed in order to confirm the validity of the given theoretical approach. In this manner, the acquired insight of the process will make possible the reduction of expensive experimental work in the search of the optimal operation.

9.4 SEPARATION–SEPARATION INTEGRATION FOR BIOETHANOL PRODUCTION

The development of technologies for separation–separation integration has been linked to the development of the different unit operations involved during downstream processes and to new approaches for process intensification. The examples of separation–separation integration in the case of ethanol production mostly correspond to integration of the conjugated type, i.e., when integrated processes are carried out in different equipments closing the flowsheet by fluxes or refluxes.

Integration possibilities are particularly important for ethanol dehydration. In Chapter 8, the features and advantages of the integrated process of extractive distillation were emphasized. In the specific case of fuel ethanol production, the utilization of salts as extractive agents (saline extractive distillation) has demonstrated certain energetic advantages compared to other dehydration schemes according to some reports (Barba, et al. 1985; Llano-Restrepo and Aguilar-Arias, 2003). Cited results indicate that energy costs of saline distillation were lower than is the case of azeotropic distillation (using benzene, pentane, or diethyl ether), extractive distillation (using ethylene glycol or gasoline), or solvent extraction, being almost the same as the costs of pervaporation. Pinto et al. (2000) employed Aspen Plus for the simulation and optimization of the saline extractive distillation for several substances (NaCl, KCl, KI, and $CaCl_2$). This configuration was compared to the simulated scheme of conventional extractive distillation with ethylene glycol and with data for azeotropic distillation. Obtained results showed considerably lower energy consumption for the process with salts. However, for this latter case, the recovery of salts was not simulated. Thus, if evaporation and recrystallization of salts is contemplated, energy requirements could significantly increase, taking into account the energetic expenditures. In this way, the utilization of commercial simulators shows the viability for predicting the behavior of a given process configuration provided the appropriate thermodynamic models of studied systems has been completed.

Gros et al. (1998) describe the process synthesis for ethanol dehydration using near critical propane. To this end, these authors combined thermodynamic models for the description of ethanol recovery under supercritical conditions based on Group Contribution Associating Equation of State (GCA-EOS) with robust methods of simulation and optimization (integrating the SQP with MINLP). Considering the energy consumption as the objective function, the developed software analyzed the main units required by the configuration: high-pressure multistage extractors, distillation columns, and multiphase flash separators. Obtained results showed that configurations involving vapor recompression and feed preconcentration are competitive alternatives in comparison to azeotropic distillation (Cardona and Sánchez, 2007).

The utilization of pervaporation for the production of absolute (anhydrous) ethanol through its coupling with the previous distillation step has been reported (Cardona and Sánchez, 2007). The modeling and optimization of the process using MINLP tools showed 12% savings in the production costs with a 32% increase in membrane area and a reduction in both reflux ratio and ethanol concentration in the distillate of the column (Lelkes et al., 2000; Szitkai et al., 2002). Through pilot plant studies, the integration of distillation process with the pervaporation has been attained resulting in good indexes in terms of energy savings. These savings are due to the low operation costs of pervaporation and to the high yield of dehydrated ethanol, typical of pervaporation processes (Tsuyomoto et al., 1997). The comparison between azeotropic distillation using benzene and the pervaporation system using multiple membrane modules showed that, at same ethanol production rate and quality (99.8 %), operation costs, including the membrane replacement every two to four years, for the pervaporation system are approximately one-third to one-quarter those of azeotropic distillation.

REFERENCES

Aden, A., M. Ruth, K. Ibsen, J. Jechura, K. Neeves, J. Sheehan, B. Wallace, L. Montague, A. Slayton, and J. Lukas. 2002. Lignocellulosic biomass to ethanol process design and economics utilizing co-current dilute acid prehydrolysis and enzymatic hydrolysis for corn stover. Technical Report NREL/TP-510-32438, National Renewable Energy Laboratory, Washington, D.C.

Aires Barros, M.R., J.M.S. Cabral, and J.M. Novais. 1987. Production of ethanol by immobilized *Saccharomyces bayanus* in an extractive fermentation system. *Biotechnology and Bioengineering* 29:1097–1104.

Alkasrawi, M., T. Eriksson, J. Börjesson, A. Wingren, M. Galbe, F. Tjerneld, and G. Zacchi. 2003. The effect of Tween-20 on simultaneous saccharification and fermentation of softwood to ethanol. *Enzyme and Microbial Technology* 33:71–78.

Ballesteros, I., J.M. Oliva, M.J. Negro, P. Manzanares, and M. Ballesteros. 2002. Ethanol production from olive oil extraction residue pretreated with hot water. *Applied Biochemistry and Biotechnology* 98-100:717–732.

Ballesteros, I., J.M. Oliva, F. Sáez, and M. Ballesteros. 2001. Ethanol production from lignocellulosic byproducts of olive oil extraction. *Applied Biochemistry and Biotechnology* 91–93:237–252.

Ballesteros, M., J.M. Oliva, M.J. Negro, P. Manzanares, and I. Ballesteros. 2004. Ethanol from lignocellulosic materials by a simultaneous saccharification and fermentation process (SSF) with *Kluyveromyces marxianus* CECT 10875. *Process Biochemistry* 39:1843–1848.

Banat, F.A., F.A. Al-Rub, and M. Shannag. 1999. Modeling of dilute ethanol–water mixture separation by membrane distillation. *Separation and Purification Technology* 16:119–131.

Banat, F.A., and J. Simandl. 1999. Membrane distillation for dilute ethanol separation from aqueous streams. *Journal of Membrane Science* 163:333–348.

Banik, R.M., A. Santhiagu, B. Kanari, C. Sabarinath, and S.N. Upadhyay. 2003. Technological aspects of extractive fermentation using aqueous two-phase systems. *World Journal of Microbiology & Biotechnology* 19:337–348.

Barba, D., V. Brandani, and G. di Giacomo. 1985. Hyperazeotropic ethanol salted-out by extractive distillation. Theoretical evaluation and experimental check. *Chemical Engineering Science* 40:2287–2292.

Baskaran, S., H.-J. Ahn, and L.R. Lynd. 1995. Investigation of the ethanol tolerance of *Clostridium thermosaccharolyticum* in continuous culture. *Biotechnology Progress* 11:276–281.

Bothast, R.J., and M.A. Schlicher. 2005. Biotechnological processes for conversion of corn into ethanol. *Applied Microbiology and Biotechnology* 67 (19–25).

Brandberg, T., N. Sanandaji, L. Gustafsson, and C.J. Franzén. 2005. Continuous fermentation of undetoxified dilute acid lignocellulose hydrolysate by *Saccharomyces cerevisiae* ATCC 96581 using cell recirculation. *Biotechnology Progress* 21:1093–1101.

Bruce, L.J., and A.J. Daugulis. 1991. Solvent selection strategies for extractive biocatalysis. *Biotechnology Progress* 7:116–124.

Bruce, L.J., and A.J. Daugulis. 1992. Extractive fermentation by *Zymomonas mobilis* and the use of solvent mixtures. *Biotechnology Letters* 14:71–76.

Calibo, R.L., M. Matsumura, and H. Kataoka. 1989. Continuous ethanol fermentation of concentrated sugar solutions coupled with membrane distillation using a PTFE module. *Journal of Fermentation and Bioengineering* 67 (1):40–45.

Cardona, C.A., L.F. Gutiérrez, and Ó.J. Sánchez. 2008. Process integration: Base for energy saving. In *Energy efficiency research advances*, ed. D.M. Bergmann. Hauppauge, NY: Nova Science Publishers.

Cardona, C.A., and Ó.J. Sánchez. 2007. Fuel ethanol production: Process design trends and integration opportunities. *Bioresource Technology* 98:2415–2457.

Cardona, C.A., Ó.J. Sánchez, M.I. Montoya, and J.A. Quintero. 2005a. Analysis of fuel ethanol production processes using lignocellulosic biomass and starch as feedstocks. Paper presented at the 7th World Congress of Chemical Engineering, July 10–14 2005, Glasgow, Scotland, U.K.

Cardona, C.A., Ó.J. Sánchez, M.I. Montoya, and J.A. Quintero. 2005b. Simulación de los procesos de obtención de etanol a partir de caña de azúcar y maíz (Simulation of processes for ethanol production from sugarcane and corn, in Spanish). *Scientia et Technica* 28:187–192.

Chandrakant, P., and V.S. Bisaria. 1998. Simultaneous bioconversion of cellulose and hemicellulose to ethanol. *Critical Reviews in Biotechnology* 18 (4):295–337.

Christen, P., M. Minier, and H. Renon. 1990. Ethanol extraction by supported liquid membrane during fermentation. *Biotechnology and Bioengineering* 36:116–123.

Claassen, P.A.M., J.B. van Lier, A.M. López Contreras, E.W.J. van Niel, L. Sijtsma, A.J.M. Stams, S.S. de Vries, and R.A. Weusthuis. 1999. Utilisation of biomass for the supply of energy carriers. *Applied Microbiology and Biotechnology* 52:741–755.

Costa, A.C., D.I.P. Atala, F. Maugeri, and R. Maciel. 2001. Factorial design and simulation for the optimization and determination of control structures for an extractive alcoholic fermentation. *Process Biochemistry* 37:125–137.

Cysewski, G.R., and C.R. Wilke. 1977. Rapid ethanol fermentations using vacuum and cell recycle. *Biotechnology and Bioengineering* 19:1125–1143.

da Silva, F.L.H., M.A. Rodrigues, and F. Maugeri. 1999. Dynamic modelling, simulation and optimization of an extractive continuous alcoholic fermentation process. *Journal of Chemical Technology and Biotechnology* 74:176–182.

Dale, M.C. 1992. Method of use of a multi-stage reactor-separator with simultaneous product separation. U.S. Patent US5141861.

Dale, M.C., and M. Moelhman. 2001. Enzymatic simultaneous saccharification and fermentation (SSF) of biomass to ethanol in a pilot 130-liter multistage continuous reactor separator. Paper presented at Bioenergy 2000, Moving Technology into the Marketplace, Buffalo, NY.

De Bari, I., E. Viola, D. Barisano, M. Cardinale, F. Nanna, F. Zimbardi, G. Cardinale, and G. Braccio. 2002. Ethanol production at flask and pilot scale from concentrated slurries of steam-exploded aspen. *Industrial & Engineering Chemistry Research* 41:1745–1753.

Di Luccio, M., C. Borges, and T. Alves. 2002. Economic analysis of ethanol and fructose production by selective fermentation coupled to pervaporation effect of membrane costs on process economics. *Desalination* 147:161–166.

Fan, Z., S. Sout, K. Lyford, J. Munsie, P. van Walsum, and L.R. Lynd. 2003. Conversion of paper sludge to ethanol in a semicontinuous solids-fed reactor. *Bioprocess and Biosystems Engineering* 26:93–101.

Fournier, R.L. 1986. Mathematical model of extractive fermentation: Application to the production of ethanol. *Biotechnology and Bioengineering* 28:1206–1212.

Fournier, R.L. 1988. Mathematical model of microporous hollow-fiber membrane extractive fermentor. *Biotechnology and Bioengineering* 31:235–239.

Fujii, Y., S. Kigoshi, H. Iwatani, and M. Aoyama. 1992a. Selectivity and characteristics of direct contact membrane distillation type experiment. I. Permeability and selectivity through dried hydrophobic fine porous membranes. *Journal of Membrane Science* 72:53–72.

Fujii, Y., S. Kigoshi, H. Iwatani, M. Aoyama, and Y. Fusaoka. 1992b. Selectivity and characteristics of direct contact membrane distillation type experiment. II. Membrane treatment and selectivity increase. *Journal of Membrane Science* 72:73–89.

García-Payo, M.C., M.A. Izquierdo-Gil, and C. Fernández-Pineda. 2000. Air gap membrane distillation of aqueous alcohol solutions. *Journal of Membrane Science* 169:61–80.

Gauss, W.F., S. Suzuki, and M. Takagi. 1976. Manufacture of alcohol from cellulosic materials using plural ferments. U.S. Patent US3990944.

Gong, C.S., N.J. Cao, J. Du, and G.T. Tsao. 1999. Ethanol production from renewable resources. *Advances in Biochemical Engineering/Biotechnology* 65 (207–241).

Gros, H.P., S. Díaz, and E.A. Brignole. 1998. Near-critical separation of aqueous azeotropic mixtures: Process synthesis and optimization. *Journal of Supercritical Fluids* 12:69–84.

Grossmann, I.E., J.A. Caballero, and H. Yeomans. 2000. Advances in mathematical programming for the synthesis of process systems. *Latin American Applied Research* 30:263–284.

Gryta, M. 2001. The fermentation process integrated with membrane distillation. *Separation and Purification Technology* 24:283–296.

Gryta, M. 2002. The assessment of microorganism growth in the membrane distillation system. *Desalination* 142:79–88.

Gryta, M., A.W. Morawski, and M. Tomaszewska. 2000. Ethanol production in membrane distillation bioreactor. *Catalysis Today* 56:159–165.

Gutiérrez, L.F., Ó.J. Sánchez, and C.A. Cardona. 2005. Modeling of batch extractive fermentation for the fuel ethanol production. Paper presented at the 8th Conference on Process Integration, Modelling and Optimisation for Energy Saving and Pollution Reduction, PRES 2005, in Giardini Naxos, Italy.

Gyamerah, M., and J. Glover. 1996. Production of ethanol by continuous fermentation and liquid–liquid extraction. *Journal of Chemical Technology and Biotechnology* 66:145–152.

Hari Krishna, S., G.V. Chowdary, D. Srirami Reddy, and C. Ayyanna. 1999. Simultaneous saccharification and fermentation of pretreated *Antigonum leptopus* (Linn) leaves to ethanol. *Journal of Chemical Technology and Biotechnology* 74:1055–1060.

Hari Krishna, S., K. Prasanthi, G. Chowdary, and C. Ayyanna. 1998. Simultaneous saccharification and fermentation of pretreated sugar cane leaves to ethanol. *Process Biochemistry* 33 (8):825–830.

Hari Krishna, S., T.J. Reddy, and G.V. Chowdary. 2001. Simultaneous saccharification and fermentation of lignocellulosic wastes to ethanol using a thermotolerant yeast. *Bioresource Technology* 77:193–196.

Ikegami, T., D. Kitamoto, H. Negishi, K. Haraya, H. Matsuda, Y. Nitanai, N. Koura, T. Sano, and H. Yanagishita. 2003. Production of high-concentration bioethanol by pervaporation. *Journal of Chemical Technology and Biotechnology* 78 (9):1006–1010.

Ikegami, T., D. Kitamoto, H. Negishi, K. Iwakabe, T. Imura, T. Sano, K. Haraya, and H. Yanagishita. 2004. Reliable production of highly concentrated bioethanol by a conjunction of pervaporation using a silicone rubber sheet-covered silicalite membrane with adsorption process. *Journal of Chemical Technology and Biotechnology* 79 (8):896–901.

Ingram, L.O., and J.B. Doran. 1995. Conversion of cellulosic materials to ethanol. *FEMS Microbiology Reviews* 16:235–241.

Ishida, K., and K. Shimizu. 1996. Novel repeated batch operation for flash fermentation system: Experimental data and mathematical modelling. *Journal of Chemical Technology and Biotechnology* 66:340–346.

Izquierdo-Gil, M.A., and G. Jonsson. 2003. Factors affecting flux and ethanol separation performance in vacuum membrane distillation (VMD). *Journal of Membrane Science* 214:113–130.

Kádár, Zs., Zs. Szengyel, and K. Réczey. 2004. Simultaneous saccharification and fermentation (SSF) of industrial wastes for the production of ethanol. *Industrial Crops and Products* 20:103–110.

Kang, W., R. Shukla, and K.K. Sirkar. 1990. Ethanol production in a microporous hollow-fiber-based extractive fermentor with immobilized yeast. *Biotechnology and Bioengineering* 36:826–833.

Kargupta, K., S. Datta, and S.K. Sanyal. 1998. Analysis of the performance of a continuous membrane bioreactor with cell recycling during ethanol fermentation. *Biochemical Engineering Journal* 1:31–37.

Kaseno, M.I. Miyazawa, and T. Kokugan. 1998. Effect of product removal by a pervaporation on ethanol fermentation. *Journal of Fermentation and Bioengineering* 86 (5):488–493.

Kirbaşlar, Ş.İ., S. Çehreli, D. Üstün, and E. Keskinocak. 2001. Equilibrium data on water–ethanol–1-dodecanol ternary system. *Turkish Journal of Engineering and Environmental Sciences* 25:111–115.

Kishimoto, M., Y. Nitta, Y. Kamoshita, T. Suzuki, and K-I. Suga. 1997. Ethanol production in an immobilized cell reactor coupled with the recycling of effluent from the bottom of a distillation column. *Journal of Fermentation and Bioengineering* 84 (5):449–454.

Kobayashi, F., and Y. Nakamura. 2004. Mathematical model of direct ethanol production from starch in immobilized recombinant yeast culture. *Biochemical Engineering Journal* 21:93–101.

Kobayashi, M., K. Ishida, and K. Shimizu. 1995. Efficient production of ethanol by a fermentation system employing temperature profiling and recycle. *Journal of Chemical Technology and Biotechnology* 63:141–146.

Kollerup, F., and A.J. Daugulis. 1985. A mathematical model for ethanol production by extractive fermentation in a continuous stirred tank fermentor. *Biotechnology and Bioengineering* 27:1335–1346.

Krishnan, M.S., N.P. Nghiem, and B.H. Davison. 1999. Ethanol production from corn starch in a fluidized-bed bioreactor. *Applied Biochemistry and Biotechnology* 77–79:359–372.

Latif, F., and M.I. Rajoka. 2001. Production of ethanol and xylitol from corn cobs by yeasts. *Bioresource Technology* 77:57–63.

Lee, W.G., B.G. Park, Y.K. Chang, H.N. Chang, J.S. Lee, and S.C. Park. 2000. Continuous ethanol production from concentrated wood hydrolysates in an internal membrane-filtration bioreactor. *Biotechnology Progress* 16:302–304.

Leksawasdi, N., E.L. Joachimsthal, and P.L. Rogers. 2001. Mathematical modeling of ethanol production from glucose/xylose mixtures by recombinant *Zymomonas mobilis*. *Biotechnology Letters* 23:1087–1093.

Lelkes, Z., Z. Szitkai, E. Rev, and Z. Fonyo. 2000. Rigorous MINLP model for ethanol dehydration system. *Computers and Chemical Engineering* 24:1331–1336.

Li, X., and A. Kraslawski. 2004. Conceptual process synthesis: Past and current trends. *Chemical Engineering and Processing* 43:589–600.

Llano-Restrepo, M., and J. Aguilar-Arias. 2003. Modeling and simulation of saline extractive distillation columns for the production of absolute ethanol. *Computers and Chemical Engineering* 27 (4):527–549.

Lynd, L.R. 1996. Overview and evaluation of fuel ethanol from cellulosic biomass: Technology, economics, the environment, and policy. *Annual Review of Energy and the Environment* 21:403–465.

Lynd, L.R., R.T. Elander, and C.E. Wyman. 1996. Likely features and costs of mature biomass ethanol technology. *Applied Biochemistry and Biotechnology* 57/58:741–761.

Lynd, L.R., K. Lyford, C.R. South, G.P. van Walsum, and K. Levenson. 2001. Evaluation of paper sludges for amenability to enzymatic hydrolysis and conversion to ethanol. *TAPPI Journal* 84:50–69.

Lynd, L.R., W.H. van Zyl, J.E. McBride, and M. Laser. 2005. Consolidated bioprocessing of cellulosic biomass: An update. *Current Opinion in Biotechnology* 16:577–583.

Lynd, L.R., P.J. Weimer, W.H. van Zyl, and I.S. Pretorious. 2002. Microbial cellulose utilization: Fundamentals and biotechnology. *Microbiology and Molecular Biology Reviews* 66 (3):506–577.

Madson, P.W., and D.A. Monceaux. 1995. Fuel ethanol production. In *The alcohol textbook*, ed. T.P. Lyons, D.R. Kelsall, and J. E. Murtagh. Nottingham, U.K.: University Press.

Mamma, D., P. Christakopoulos, D. Koullas, D. Kekos, B.J. Macris, and E. Koukios. 1995. An alternative approach to the bioconversion of sweet sorghum carbohydrates to ethanol. *Biomass and Bioenergy* 8 (2):99–103.

Mamma, D., D. Koullas, G. Fountoukidis, D. Kekos, B.J. Macris, and E. Koukios. 1996. Bioethanol from sweet sorghum: Simultaneous saccharification and fermentation of carbohydrates by a mixed microbial culture. *Process Biochemistry* 31 (4):377–381.

Matsumura, M., and H. Märkl. 1986. Elimination of ethanol inhibition by perstraction. *Biotechnology and Bioengineering* 28:534–541.

McMillan, J.D. 1997. Bioethanol production: Status and prospects. *Renewable Energy* 10 (2/3):295–302.

McMillan, J.D., M.M. Newman, D.W. Templeton, and A. Mohagheghi. 1999. Simultaneous saccharification and co-fermentation of dilute-acid pretreated yellow poplar hardwood to ethanol using xylose-fermenting *Zymomonas mobilis*. *Applied Biochemistry and Biotechnology* 77–79:649–665.

Minier, M., and G. Goma. 1982. Ethanol production by extractive fermentation. *Biotechnology and Bioengineering* 24:1565–1579.

Modaressi, H., B. Milne, and G.A. Hill. 1997. Hydrodynamic behavior of a liquid-lift, external-loop bioreactor using a spinning sparger. *Industrial & Engineering Chemistry Research* 36:4681–4687.

Montesinos, T., and J-M. Navarro. 2000. Production of alcohol from raw wheat flour by amyloglucosidase and *Saccharomyces cerevisiae*. *Enzyme and Microbial Technology* 27:362–370.

Mori, Y., and T. Inaba. 1990. Ethanol production from starch in a pervaporation membrane bioreactor using *Clostridium thermohydrosulfuricum*. *Biotechnology and Bioengineering* 36:849–853.

Moritz, J.W., and S.J.B. Duff. 1996. Simultaneous saccharification and extractive fermentation of cellulosic substrates. *Biotechnology and Bioengineering* 49 (5):504–511.

Nakamura, Y., T. Sawada, and E. Inoue. 2001. Enhanced ethanol production from enzymatically treated steam-exploded rice straw using extractive fermentation. *Journal of Chemical Technology and Biotechnology* 76:879–884.

Nomura, M., T. Bin, and S. Nakao. 2002. Selective ethanol extraction from fermentation broth using a silicalite membrane. *Separation and Purification Technology* 27:59–66.

Novozymes & BBI International. 2005. *Fuel ethanol. A technological evolution.* Novozymes & BBI International. http://www.bbibiofuels.com/ethanolevolution/FuelEthanol-lr-05.pdf (accessed February 2007).

O'Brien, D., L. Roth, and A.J. McAloon. 2000. Ethanol production by continuous fermentation-pervaporation: A preliminary economic analysis. *Journal of Membrane Science* 166:105–111.

Offeman, R.D., S.K. Stephenson, G.H. Robertson, and W.J. Orts. 2005a. Solvent extraction of ethanol from aqueous solutions. I. Screening methodology for solvents. *Industrial & Engineering Chemistry Research* 44:6789–6796.

Offeman, R.D., S.K. Stephenson, G.H. Robertson, and W.J. Orts. 2005b. Solvent extraction of ethanol from aqueous solutions. II. Linear, branched, and ring-containing alcohol solvents. *Industrial & Engineering Chemistry Research* 44:6797–6803.

Ohashi, R., Y. Kamoshita, M. Kishimoto, and T. Suzuki. 1998. Continuous production and separation of ethanol without effluence of wastewater using a distiller integrated SCM-reactor system. *Journal of Fermentation and Bioengineering* 86 (2):220–225.

Oliveira, A.C., M.F. Rosa, J.M.S. Cabral, and M.R. Aires-Barros. 1998. Improvement of alcoholic fermentations by simultaneous extraction and enzymatic esterification of ethanol. *Journal of Molecular Catalysis B: Enzymatic* 5:29–33.

Oliveira, A.C., M.F. Rosa, J.M.S. Cabral, and M.R. Aires-Barros. 2001. Effect of extraction and enzymatic esterification of ethanol on glucose consumption by two *Saccharomyces cerevisaie* strains: A comparative study. *Journal of Chemical Technology and Biotechnology* 76:285–290.

Olsson, L., and B. Hahn-Hägerdal. 1996. Fermentation of lignocellulosic hydrolysates for ethanol production. *Enzyme and Microbial Technology* 18:312–331.

Otto, E., and J. Escovar-Kousen. 2004. Ethanol production by simultaneous saccharification and fermentation (SSF). World Intellectual Property Organization Patent WO2004046333.

Panagiotou, G., P. Christakopoulos, and L. Olsson. 2005a. Simultaneous saccharification and fermentation of cellulose by *Fusarium oxysporum* F3 —Growth characteristics and metabolite profiling. *Enzyme and Microbial Technology* 36 (5-6):693–699.

Panagiotou, G., P. Christakopoulos, S.G. Villas-Bôas, and L. Olsson. 2005b. Fermentation performance and intracellular metabolite profiling of *Fusarium oxysporum* cultivated on a glucose-xylose mixture. *Enzyme and Microbial Technology* 36 (1):100–106.

Panagiotou, G., S.G. Villas-Bôas, P. Christakopoulos, J. Nielsen, and L. Olsson. 2005c. Intracellular metabolite profiling of *Fusarium oxysporum* converting glucose to ethanol. *Journal of Biotechnology* 115 (4):425–434.

Pinto, R.T.P., M.R. Wolf-Maciel, and L. Lintomen. 2000. Saline extractive distillation process for ethanol purification. *Computers and Chemical Engineering* 24:1689–1694.

Quintero, J.A., M.I. Montoya, Ó.J. Sánchez, O.H. Giraldo, and C.A. Cardona. 2008. Fuel ethanol production from sugarcane and corn: Comparative analysis for a Colombian case. *Energy* 33 (3):385–399.

Rivera, M., and C.A. Cardona. 2004. Análisis de procesos simultáneos reacción-extracción a nivel productivo. Generalidades del proceso, equilibrios fásico y químico simultáneos (Analysis of simultaneous reaction-extraction processes at productive level. Process generalities, simultaneous phase and chemical equilibria, in Spanish). *Ingeniería y Competitividad* 6 (1):17–25.

Ryabova, O.B., O.M. Chmil, and A.A. Sibirny. 2003. Xylose and cellobiose fermentation to ethanol by the thermotolerant methylotrophic yeast *Hansenula polymorpha*. *FEMS Yeast Research* 4:157–164.

Saha, B.C., L.B. Iten, M.A. Cotta, and Y.V. Wu. 2005. Dilute acid pretreatment, enzymatic saccharification, and fermentation of rice hulls to ethanol. *Biotechnology Progress* 21:816–822.

Sánchez, Ó.J., and C.A. Cardona. 2008. Trends in biotechnological production of fuel ethanol from different feedstocks. *Bioresource Technology* 99:5270–5295.

Sánchez, Ó.J., C.A. Cardona, and D.C. Cubides. 2005. Modeling of simultaneous saccharification and fermentation process coupled with pervaporation for fuel ethanol production. Paper presented at the 2nd Mercosur Congress on Chemical Engineering and 4th Mercosur Congress on Process Systems Engineering, Rio de Janeiro, Brazil.

Sánchez, Ó.J., L.F. Gutiérrez, C.A. Cardona, and E.S. Fraga. 2006. Analysis of extractive fermentation process for ethanol production using a rigorous model and a short-cut method. In *Computer aided methods in optimal design and operations*, ed. I.D.L. Bogle and J. Žilinskas. Singapore: World Scientific Publishing Co.

Shabtai, Y., S. Chaimovitz, A. Freeman, E. Katchalski-Katzir, C. Linder, M. Nemas, M. Perry, and Kedem. O. 1991. Continuous ethanol production by immobilized yeast reactor coupled with membrane pervaporation unit. *Biotechnology and Bioengineering* 38:869–876.

South, C.R., D.A. Hogsett, and L.R. Lynd. 1993. Continuous fermentation of cellulosic biomass to ethanol. *Applied Biochemistry and Biotechnology* 39/40:587–600.

South, C.R., D.A.L. Hogsett, and L.R. Lynd. 1995. Modeling simultaneous saccharification and fermentation of lignocellulose to ethanol in batch and continuous reactors. *Enzyme and Microbial Technology* 17:797–803.

Stang, G.D., D.G. Macdonald, and G.A. Hill. 2001. Mass transfer and bioethanol production in an external-loop liquid-lift bioreactor. *Industrial & Engineering Chemistry Research* 40:5074–5080.

Stankiewicz, A., and J.A. Moulijn. 2002. Process intensification. *Industrial & Engineering Chemistry Research* 41:1920–1924.

Stenberg, K. , M. Bollók, K. Réczey, M. Galbe, and G. Zacchi. 2000a. Effect of substrate and cellulase concentration on simultaneous saccharification and fermentation of steam-pretreated softwood for ethanol production. *Biotechnology and Bioengineering* 68 (2):204–210.

Stenberg, K., M. Galbe, and G. Zacchi. 2000b. The influence of lactic acid formation on the simultaneous saccharification and fermentation (SSF) of softwood to ethanol. *Enzyme and Microbial Technology* 26:71–79.

Szczodrak, J., and J. Fiedurek. 1996. Technology for conversion of lignocellulosic biomass to ethanol. *Biomass and Bioenergy* 10 (5/6):367–375.

Szitkai, Z., Z. Lelkes, E. Rev, and Z. Fonyo. 2002. Optimization of hybrid ethanol dehydration systems. *Chemical Engineering and Processing* 41:631–646.

Takagi, M., S. Abe, S. Suzuki, G.H. Emert, and N. Yata. 1977. A method for production of ethanol directly from cellulose using cellulase and yeast. Paper presented at the Proceedings of Bioconversion Symposium, Delhi, India.

Tanaka, H., S. Harada, H. Kurosawa, and M. Yajima. 1987. A new immobilized cell system with protection against toxic solvents. *Biotechnology and Bioengineering* 30:22–30.

Taylor, F., M.J. Kurantz, N. Goldberg, and J.C. Craig, Jr. 1996. Control of packed column fouling in the continuous fermentation and stripping of ethanol. *Biotechnology and Bioengineering* 51:33–39.

Taylor, F., M.J. Kurantz, N. Goldberg, and J.C. Craig, Jr. 1998. Kinetics of continuous fermentation and stripping of ethanol. *Biotechnology Letters* 20:67–72.

Taylor, F., M.J. Kurantz, N. Goldberg, A.J. McAloon, and J.C. Craig, Jr. 2000. Dry-grind process for fuel ethanol by continuous fermentation and stripping. *Biotechnology Progress* 16:541–547.

Tsuyomoto, M., A. Teramoto, and P. Meares. 1997. Dehydration of ethanol on a pilot plant scale, using a new type of hollow-fiber membrane. *Journal of Membrane Science* 133:83–94.

Varga, E., H.B. Klinkle, K. Réczey, and A.B. Thomsen. 2004. High solid simultaneous saccharification and fermentation of wet oxidized corn stover to ethanol. *Biotechnology and Bioengineering* 88 (5):567–574.

Weilnhammer, C., and E. Blass. 1994. Continuous fermentation with product recovery by *in-situ* extraction. *Chemical Engineering and Technology* 17:365–373.

Wooley, R., M. Ruth, D. Glassner, and J. Sheejan. 1999a. Process design and costing of bioethanol technology: A tool for determining the status and direction of research and development. *Biotechnology Progress* 15:794–803.

Wooley, R., M. Ruth, J. Sheehan, K. Ibsen, H. Majdeski, and A. Galvez. 1999b. Lignocellulosic biomass to ethanol process design and economics utilizing co-current dilute acid prehydrolysis and enzymatic hydrolysis. Current and futuristic scenarios. Technical Report NREL/TP-580-26157, National Renewable Energy Laboratory, Washington, D.C.

Wu, Y., Z. Xiao, W. Huang, and Y. Zhong. 2005. Mass transfer in pervaporation of active fermentation broth with a composite PDMS membrane. *Separation and Purification Technology* 42:47–53.

Wyman, C.E. 1994. Ethanol from lignocellulosic biomass: Technology, economics, and opportunities. *Bioresource Technology* 50:3–16.

Wyman, C.E., D.D. Spindler, and K. Grohmann. 1992. Simultaneous saccharification and fermentation of several lignocellulosic feedstocks to fuel ethanol. *Biomass and Bioenergy* 3 (5):301–307.

Zhang, Y.-H.P., and L.R. Lynd. 2004. Toward an aggregated understanding of enzymatic hydrolysis of cellulose: Noncomplexed cellulose systems. *Biotechnology and Bioengineering* 8 (7):797–824.

10 Environmental Aspects of Fuel Ethanol Production

This chapter deals with the wastes generated during the different processing steps for fuel ethanol production. The influence of the feedstock is recognized by the type of wastes produced. Main methods for the treatment of stillage (the major polluting effluent of all ethanol production processes) are discussed. In addition, the environmental issues related to the overall fuel ethanol production process are highlighted emphasizing some methodologies for assessment of its environmental performance.

10.1 EFFLUENT TREATMENT DURING FUEL ETHANOL PRODUCTION

10.1.1 RESIDUES GENERATED IN THE PROCESS OF BIOETHANOL PRODUCTION

Fuel ethanol production generates solid wastes, atmospheric emissions, and liquid effluents. The atmospheric emissions correspond mostly to the gas outlet stream from fermenters, which are washed with water in the scrubbers in order to recover the volatilized ethanol. The gases exiting from the scrubbers contain mainly carbon dioxide that is released into the atmosphere. The CO_2 can be used for production of dry ice and beverages.

However, if these gases are not utilized, they should be considered in the calculation of the environmental impact of ethanol-producing facilities. In this regard, it should be emphasized that the bioethanol presents net emissions of nearly zero CO_2 because the plant biomass already fixed the CO_2 during its growth and only this carbon dioxide is released during the combustion of fuel ethanol in the engines. In contrast, the burning of fossil fuels releases into the atmosphere additional amounts of carbon dioxide that was fixed by the plant biomass millions years ago.

The solid wastes formed during production of fuel ethanol are strongly linked to the raw material from which it is produced. When sugarcane is used, huge amounts of sugarcane bagasse are produced. Fortunately, this solid material has multiple uses and applications. Its most important utilization is as solid biofuel due to its high energy content. In fact, bagasse combustion allows the generation of the thermal energy (steam) required not only during the conversion of sugarcane into ethanol, but also during cane sugar production. The bagasse also can cover the electricity needs if co-generation units are used (see Chapter 11). In the corn-to-ethanol process by the dry-milling technology, most solid residues are concentrated in the so-called distiller's dried grains with solubles (DDGS), so the generation of solid wastes is limited. If wet-milling technology is employed, the solids generated produce part of the different co-products in the framework of a corn biorefinery, as corn gluten meal and feed. When cassava roots are used, the solids produced correspond to the root peels that can be utilized as feedstock for production of mushrooms as well as the fibrous residue contained in the stillage stream. This fibrous residue is obtained after centrifugation of the whole stillage to obtain the thin stillage and this solid material. In addition, this residue can be used for animal feed or as a substrate in solid-state fermentations.

The production of ethanol from lignocellulosic biomass, in turn, generates lignin as the most important solid residue. This polymer can be isolated during the pretreatment step if some pretreatment methods like the pretreatment with solvents (organosolv process) or oxidative delignification are employed (see Chapter 4, Section 4.3.3). Nevertheless, most pretreatment methods allow the lignin to remain in the solid fraction resulting from this processing step along with the cellulose. After enzymatic hydrolysis using cellulases, the lignin remains in the liquid suspension until the end of the process where it can be recovered from the stillage stream. The lignin has a high energy value and, therefore, is used as a solid biofuel for feeding boilers or co-generation units, as in the case of sugarcane bagasse.

The stillage (or vinasses) is the major effluent of all flowsheets for ethanol production involving the submerged fermentation of streams containing sugars or carbohydrate polymers. This is valid for feedstocks such as sugarcane juice, molasses, starchy materials, and pretreated lignocellulosic biomass. The stillage represents the residual liquid material obtained after distillation of ethanol from the fermented wort (wine) and contains both solid and soluble matter. The elevated organic load of the stillage is most responsible for the high polluting properties of this burden, thus this stream should undergo treatment to reduce this load and minimize the environmental impact during its discharge into the water streams.

10.1.2 METHODS FOR TREATMENT AND UTILIZATION OF STILLAGE

The stillage properties depend on the type of feedstock used for ethanol production as well as the conversion technology employed. The stillage presents high values of biological oxygen demand (BOD_5 = 30,000–60,000 mg/L) that is, along with the chemical oxygen demand (COD), a measure of the content of organic matter in an effluent sample. Wilkie et al. (2000) have reviewed the characteristics of several types of stillage for various ethanol production processes, including the stillage yields. The stillage generated in the process employing sugarcane molasses has high polluting loads in terms of both BOD (45,000 mg/L) and COD (113,000 mg/L) as well as in its yield (2.52 liters of stillage per kilogram of molasses) and content of potassium, phosphorous, and sulfates. The stillage derived from the process using sugarcane juice exhibits lower levels of BOD, COD, and yield (12,000 mg/L, 25,000 mg/L, and 1.33 L/kg, respectively). The stillage from corn process shows intermediate indicators for BOD and COD (37,000 mg/L and 56,000 mg/L, respectively) but a higher stillage yield (6.20 L/kg). Finally, in the case of lignocellulosic ethanol produced from hardwood, the values of BOD, COD, and yield are 13,200 mg/L, 25,500 mg/L, and 20.4 L/kg, respectively (Wilkie et al., 2000).

If stillage is directly discharged into natural water streams, the microorganisms present in them will degrade the organic matter discharged, which uses the oxygen dissolved in the water. This leads to a critical diminution of dissolved oxygen provoking the death of most aquatic organisms. In addition, the utilization of nutrients contained in molasses can lead to the massive propagation of algae on the surface of ponds and lakes causing the blocking of solar light leading to the death of fish and other living organisms. This phenomenon is known as eutrophication. Thus, the stillage discharge can seriously alter the ecological equilibrium of the aquatic ecosystems affected. Residual sugars from fermentation cause stillage COD to reach high values. For every 1% of residual sugar, a stillage COD increment of 16,000 mg/L can be expected (Wilkie et al., 2000). For this reason, full completion of fermentation is desirable in order to reduce stillage volume (greater ethanol concentrations) and minimize the concentration of remaining sugars. For each 1% ethanol left in the stillage (in the case of nonefficient distillation), the stillage COD is incremented by more than 20,000 mg/L.

Due to its elevated organic matter content of stillage, methods for its treatment and economic utilization should be implemented in the industry. Among the most used methods for stillage treatment, irrigation, recycling, evaporation, incineration, and composting should be highlighted. Some of the new treatment methods produce value-added substances. The most important stillage treatment methods are presented in the next sections.

10.1.2.1 Stillage Recycling
Usually, in ethanol producing plants, the stillage obtained from the distillation step (whole stillage) undergoes centrifugation in order to recover organic solids, especially yeast debris. The liquid fraction is known as *thin stillage*. In a typical distillery, more than 20 liters of thin stillage can be generated per each liter of ethanol produced. To minimize the effluent treatment costs, a portion of thin stillage (from 25 to 75%) is recycled in different process steps as fermentation or saccharification (in the case of conversion of starchy materials). In a similar way, the whole stillage can be recycled back to the fermentation step when cane molasses is used. Although this procedure decreases the volume of fresh water employed, it decreases, in turn, the volume of stillage generated; the total amount of organic matter in the stillage (measured as COD) does not change because its concentration increases with the amount of recycled stillage (Wilkie et al., 2000).

The North American ethanol industry that uses corn as feedstock makes use of a fraction of thin stillage to replace a percentage of water required during the mashing process (saccharification step). The thin stillage stream recycle is called *backset*. The drawback to this practice is that there is an accumulation of undesirable substances, such as lactic acid, produced by contaminant bacteria as well as minerals and nonutilized substrates. In this way, the continuous backset usage results in the buildup of compounds inhibiting the yeast growth and ethanol production. On a laboratory scale, it has been demonstrated that the use of a backset in recycling 50% of thin stillage during the wheat mashing for five successive fermentation batches will produce a 60% loss of yeast viability, though the fermenting ability of these microorganisms and the glucoamylase activity is not affected (Chin and Ingledew, 1994). The backset utilization has been evaluated in the case of very high gravity (VHG) fermentations of wheat starch. As noted in Chapter 7, Section 7.1.3.2, VHG fermentation requires an available nitrogen source, but the recycled thin stillage contains no significant amounts of free amino nitrogen. In this sense, the use of fresh yeast autolyzates are obtained by centrifugation of the fermented wort before its distillation. The lysis of yeasts is induced through their resuspension in water and by increasing the temperature up to 48°C for 48 h, followed by pasteurization at 75°C. In this way, the costs associated with the treatment of cell biomass can be reduced in ethanol-producing plants as well (Jones and Ingledew, 1994).

The possibility of implementing a system with zero-discharge of stillage for the process employing starch or starch residues from wet-milling process has been proposed. The stillage is decanted and the resulting thin stillage undergoes ultrafiltration. The solids produced are recycled back to the decanter while the permeate is recirculated to the cooking step of starch. After eight-fold recycling of the permeate, the yield is similar to that of the conventional process, though this system increases the fermentation time from 60 to 90 h (Nguyen, 2003).

For the process based on sugarcane molasses, a bioconcentration system of vinasses has been proposed. This system consists of the recycling of 60% stillage to the fermentation step by replacing part of the water used during the preparation of the culture medium (Navarro et al., 2000). This stillage recycling percentage can increase the ethanol production without provoking inhibitory effects resulting from the accumulation of by-products released by the yeasts. In the process, 46.2% reduction of fresh water, 66% decrease of nutrients, and 50% reduction of sulfuric acid are attained. The energy balance of the process can be improved if subsequent stillage incineration is implemented since the liquid effluent contains about 24% solids. Moreover, the obtained stillage contains significant amounts of glycerol. In this way, it is possible to generate only five liters of stillage per liter of ethanol produced. These systems require the use of osmotolerant yeasts, such as *Schizosaccharomyces pombe,* able to grow under high solids concentration in culture medium (Goyes and Bolaños, 2005). The distilleries of CSR (Australia) utilize the biostill process (see Chapter 7, Section 7.1.2.3) implementing such types of recycling and using *S. pombe.* With this technology, one can obtain stillage with 28 to 30% solids content making it suitable for its direct utilization as a fertilizer (Bullock, 2002). In the distilleries co-located in cane sugar mills, the vinasses can be alternatively used as part of the water for washing the sugarcane or recycled to the molasses dilution step (Sheehan and Greenfield, 1980).

10.1.2.2   Stillage Evaporation

The evaporation represents an intermediate step during the stillage treatment. The evaporated stillage is the starting point for solids recovery, fertilization, and incineration of stillage. Thus, during ethanol production from corn by the dry-milling technology, the thin stillage obtained can be evaporated in order to produce syrup (also called *distiller's solubles*). This syrup is combined with the solids from corn to produce a co-product used for animal feed (DDGS). The condensed water generated in the multiple-effect evaporators contains small amounts of volatile organic compounds and can be employed during cooking and liquefaction steps of the corn suspension, but the accumulation of inhibitors impedes the 100% recycling of these condensates. The condensates from evaporators can also undergo aerobic or anaerobic treatment, thus providing all the nutrients required (Wilkie et al., 2000). Palmqvist et al. (1996) proposed, from bench-scale data, to fraction the stillage by evaporation and recirculate only those fractions that have demonstrated having no inhibitory effects on fermentation using *Saccharomyces cerevisiae* in the case of willow wood hydrolyzates pretreated by steam explosion. The nonvolatile fraction has high inhibitory effects presumably caused by the presence of lignin degradation products. Other fractions have low levels of BOD and COD so they can be directly discharged without any treatment. Similar studies were carried out for pine and spruce (softwood) performing the simulation of evaporation step using a six-effect evaporation train to optimize the energy consumption (Larsson et al., 1997).

10.1.2.3   Solids Recovery

The stillage obtained from the process employing sugarcane molasses can be centrifuged to collect the yeast debris, which has a significant nutritive value, remaining the thin stillage. This stillage is later evaporated leaving up to 50 to 65% solids to increase its stability against the microbial action. In this way, it can be commercialized for animal feed. However, its high salts content, especially potassium, has limited its use to only up to 10% of the ruminant diet (less than 2% in swine diet) to avoid laxative effects (Nguyen, 2003). The stillage derived from other types of sucrose-containing feedstocks, such as cane juice or beet molasses, can be used in the same way.

10.1.2.4   Stillage Incineration

The stillage obtained in processes employing sugarcane, acid hydrolyzates of wood, or spent sulfite liquors as feedstocks can be incinerated, which offers a positive energy return as well as the recovery of minerals. Before its incineration, the stillage should be concentrated up to 50 to 60% solids in 4- or 5-effect evaporators. Just the combustion of stillage offers the energy needed for this concentration process. The incineration ash has about 30 to 40% K_2O and 2 to 3% P_2O_5, which converts it into a fertilizer after its dilution in water and neutralization with sulfuric acid (Nguyen, 2003; Olguín et al., 1995; Sheehan and Greenfield, 1980). The burners of stillage require special designs in order to support its high ash content.

10.1.2.5   Fertilization

The stillage from sugarcane molasses retains potassium and the ash contained in the original raw materials. Where economically viable, this stillage is transported by trucks to irrigate the surrounding sugarcane plantations. One option to directly apply the stillage is to combine this effluent with 10% total solids, as done in Colombia (Gnecco, 2003). To reduce the transport costs during the fertilization of cane fields, evaporated stillage (up to 60° Brix) is employed allowing the reincorporation of potassium into the soil, though this process means high energy consumption (Goyes and Bolaños, 2005). Thus, the evaporation of stillage up to 55% solids and its mixing with a nitrogen source allows the production of a liquid fertilizer that is quite appropriate for cane plantations (Gnecco, 2003). In particular, the fertilization experiences using this evaporate stillage in Colombia have shown signs of productivity increases of both sugarcane and cane sugar of nearly 6%, and reductions in fertilization of 2%. In addition, this evaporated stillage can undergo drying in order to produce a solid stillage that is also commercialized as a fertilizer (Gnecco, 2006).

The stillage aids in the formation of an initial buffer in the soil with calcium and magnesium elevating its pH. In addition, it improves the physical properties of the soil and increases the retention of water and salts, as well as the soil microflora, increasing the land yield. Among the drawbacks of using stillage as a fertilizer are the strong odors, insect invasion, increase of the soil acidity, salts lixiviation, putrescibility, manganese deficiency, and inhibition of seed germination. In addition, the accumulation of sulfates reduces them to hydrogen sulfide, which is, in turn, transformed by soil bacteria into sulfuric acid (Navarro et al., 2000; Nguyen, 2003; Olguín et al., 1995). Furthermore, inadequate handling of irrigation techniques can increase the pollution of groundwater. The use of denitrifying bacteria for detoxification of the distillery effluents through the removal of potassium, chlorides, and sulfates has been referenced (Olguín et al., 1995).

In Australia, the wheat-milling residue is washed with water to obtain an effluent containing starch (starch waste) that is employed as a feedstock for fuel ethanol production. The stillage generated in this process is centrifuged to recover yeast fragments and then neutralized with lime and used for irrigating plantations. In a preliminary analysis of different treatment alternatives, Nguyen (2003) shows that the irrigation is a valid option, but with the augment of the capacities of ethanol production facilities, the soil can reach its maximum nutrients load, thus other options have to be considered. In principle, the stillage from the process directly employing starch and the stillage derived from starch waste have the same advantages and drawbacks.

10.1.2.6   Anaerobic Digestion

The stillage that cannot be used as animal feed and that has high BOD and COD values can be treated by anaerobic digestion. During this process, the transformation of the organic matter by a mixed bacterial culture is carried out in such a way that the BOD is reduced and the resulting sludge can be more easily disposed of, i.e., it is stabilized. Wilkie et al. (2000) have reviewed the conditions for accomplishing the anaerobic digestion of different stillage types under both mesophilic and thermophilic conditions and have reported the main configurations of the equipment required by this process. COD reductions of 97% for corn stillage in an up-flow anaerobic sludge blanket (UASB) reactor, as well as 94% reduction of the soluble COD of sugarcane stillage, have been achieved. For stillage derived from wheat starch, some studies at pilot plant scale have shown 89 to 92% reductions of COD (Nguyen, 2003).

The production of sludge by anaerobic digestion is about 10% lower than the sludge produced by aerobic digestion. This anaerobic sludge can be used as a feedstock for producing components for balanced animal feed (Wilkie et al., 2000). The effluents from anaerobic digestion contain plant macronutrients (nitrogen, phosphorous, and potassium), as well as micronutrients (iron, zinc, manganese, copper, and magnesium), so they could be applied to plantations ensuring an appropriate dosage. These nutrients should be removed when the liquid effluents are to be directly discharged into the natural water streams. If they are not removed, eutrophication problems can arise. These nutrients can be removed through the cultivation of algae in special ponds containing these effluents. In general, the stillage treated by anaerobic digestion should undergo color removal through different methods, including flocculation and coagulation, photovoltaic removal, or microbial removal (Wilkie et al., 2000). One alternative for degradation of the organic matter contained in the stillage is the use of aerobic processes like high-rate oxidation ponds for those cases when low cost lands are available (Olguín et al., 1995).

10.1.2.7 Composting

The stillage can be employed in composting processes as well. In particular, a mixture containing the organic fraction of the municipal solid waste (MSW) and stillage with a 3:1 ratio has been proposed to produce compost. In this case, the stillage supplies the nutrients needed for the process. The aerobic fermentation process inherent to the composting can be carried out during 30 to 50 days. Good results have been reported for this type of stillage treatment (Vaccari et al., 2005). In a similar way, the by-product from sugarcane cropping can be mixed with the stillage for their composting with a residence time of 35 days (Goyes and Bolaños, 2005). In Colombia, most of the sugar mills with co-located distilleries use the stillage, evaporated up to 30% solids, in mixtures with press mud for producing a solid fertilizer through composting for cane plantations (Gnecco, 2003). The use of concentrated stillage with 55% solids is also possible for its direct utilization as a fertilizer or for composting processes. The compost produced from the press mud, ash from boilers using cane bagasse, and concentrated stillage with 30% solids can supply a third of the nitrogen and much more phosphorous and potassium required by the sugarcane. In addition, the composting reduces by 50% the volume of the used materials since the oxidation reactions occurring during the aerobic solid-state fermentation are exothermal and, therefore, contribute to the vaporization of moisture contained in the stillage (Gnecco, 2003).

Nandy et al. (2002) describe the implementation of an effluent treatment process in an ethanol-producing plant integrated to a sugar mill in India. The plant produces 46,000 L/d of industrial rectified ethanol from the molasses generated in the same sugar mill, which processes 4,000 ton/d of sugarcane. The stillage undergoes anaerobic digestion in two fixed-film anaerobic reactors. The methane produced allows the reduction of the fossil fuel consumption for the generation of the steam required during the distillation step. Then, a 4-effect evaporation train is employed to concentrate the effluent from anaerobic digestion toward 50% solids. The evaporators employed are the down-flow film type. Finally, the effluent of the evaporation step is used for composting the press mud generated in the sugar mill (140 ton/d). In this process of aerobic conversion through thermophilic microorganisms whose reactor has a residence time of 45 days, a biomanure that can be sold is obtained at a rate of 64 ton/d. In this way, the burdens are practically eliminated. This plant represents an example of the integrated utilization of the bioresources generated and available in the joint production of cane sugar and ethanol.

10.1.2.8   Stillage as a Culture Medium

Considering the great amount of different organic compounds contained in the stillage, its use as a feedstock in other fermentation processes such as the production of single-cell protein (SCP) has been proposed (de la Cruz et al., 2004). SCP is employed as a component of animal feed due to its high content of proteins and vitamins. However, ethanol producers consider that both the anaerobic digestion and SCP production require high investments, have very prolonged residence times, and have difficult operation (Goyes and Bolaños, 2005). For Cuban situations, de la Cruz et al. (2004) propose the combination of stillage irrigation and recycling to the fermentation step in ethanol production plants using sugarcane with a capacity of 55,000 L/d. For plants with a capacity of 70,000 L/d ethanol, the combination of stillage recycling and SCP production is suggested. For plants producing 90,000 L/d or more of ethanol, the installation of a plant producing SCP with a capacity of 15 ton/d other than the irrigation with stillage is proposed. These authors also point out the use of the entire stillage as fluidizing agents during the production of cement.

10.1.2.9   Stillage Oxidation

The oxidation of stillage from sugarcane in supercritical water (temperatures higher than 550°C and pressures above 25 MPa) is chemically equivalent to incineration. When the organic compounds make contact with the supercritical water in the presence of excess oxygen, the total oxidation reaction prevails allowing the degradation of the organic matter, which is converted into CO_2, water, and nitrogen. When the oxygen is not in excess, the pyrolysis and hydrolysis reactions prevail, which can generate some value-added products. The main drawbacks of this technology are related to the corrosion and presence of insoluble inorganic salts that generate incrustations in the equipment. Goyes and Bolaños (2005) have carried out tests for total oxidation of cane stillage using hydrogen peroxide in a batch reactor achieving a 97% conversion of the organic matter with residence times lower than 3.5 min. Moreover, a carbonaceous material with an elevated surface area under partial oxidation conditions of stillage was obtained.

It is estimated that in the fuel ethanol production process from lignocellulosic biomass, 15 liters of wastewater from each liter of ethanol produced are generated. Besides the conventional options for treating this liquid effluent by concentration–incineration and anaerobic digestion, catalytic oxidation has been proposed. The feasibility of oxidizing the generated stillage through heterogeneous catalysts based on mixed metallic oxides (MnO_2/CeO_2) within a process for producing ethanol from timothy grass pretreated by steam explosion has been demonstrated (Belkacemi et al., 2000).

10.1.2.10   Wastewater Treatment of Biomass-to-Ethanol Process

In general, it is considered that the characteristics of the stillage produced from lignocellulosic materials are comparable to those of the stillage obtained when conventional feedstocks are employed, thus the treatment and utilization methods of this latter type of stillage can be applied to the lignocellulosic stillage (Wilkie et al., 2000). The flowsheets proposed for the treatment of the stillage from biomass contemplate its evaporation and centrifugation in order to obtain the thin stillage and solid materials (Wooley et al., 1999). The solids recovered have high lignin content so they are sent to the burner of the boiler for generation of the steam required by the overall ethanol production process. The thin stillage is partially recirculated, along with the bottoms of the rectification column, as water for washing the pretreated lignocellulosic biomass (see Chapter 11). The remaining thin stillage is directed to the evaporation train in order to concentrate it into a syrup containing about 60% water. Due to this high moisture content, this syrup can undergo anaerobic digestion though its incineration.

Besides the stillage, other liquid waste streams are generated in the process for fuel ethanol production from lignocellulosic biomass. For the treatment of this wastewater, different alternatives have been evaluated that include its anaerobic digestion to remove 90% of its organic matter content, as well as the utilization of the biogas produced for steam generation. The effluent of anaerobic digestion can be sent to an aerobic digestion pond where another 90% of the organic matter content is removed. The aerobic sludge formed in this process is withdrawn by clarification and filtration. The filtered sludge can also be sent to the burner for steam generation. The water from the clarifier can be recirculated in the process (Merrick & Company, 1998; Wooley et al., 1999).

CASE STUDY 10.1   PRELIMINARY COMPARISON OF
EFFLUENT TREATMENT ALTERNATIVES

Process synthesis procedures should provide insight on the more suitable technologies for effluent treatment during fuel ethanol production. In a previous work (Cardona et al., 2006), three effluent treatment configurations were analyzed. The effluent studied was the thin stillage generated during the production of fuel ethanol from lignocellulosic biomass as well as other liquid effluents (wastewater from

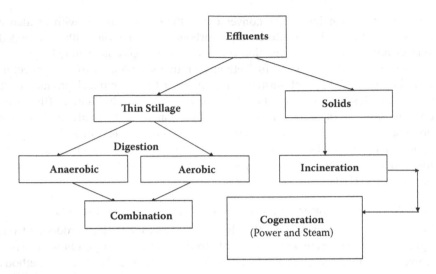

FIGURE 10.1 Option for effluent treatment analyzed in Case Study 10.1.

pretreatment and detoxification). In addition, the solid residues generated during the centrifugation of the whole stillage were also studied. These residues are mainly represented by the lignin nontransformed during the process, residual cellulose and hemicellulose, and the cell biomass (in this case *Zymomonas mobilis* fragments). The configurations were simulated using Aspen Plus® v11.1 applying the general guideline for simulations described in Chapter 8, Case Study 8.1. To analyze the biological transformations taking place during the degradation of the organic matter contained in the stillage, the stoichiometric approach was employed in the framework of the process simulator. The percentage of organic matter removal (degradation) was evaluated based on the mass balance results calculated. In addition, preliminary data on generation of electricity was also taken into account. The aim of this case study was to investigate different flowsheet configurations and their combinations for the treatment of the liquid and solid effluents obtained during fuel ethanol production from lignocellulosic biomass.

The treatment procedures analyzed consider the anaerobic digestion of the thin stillage, the aerobic digestion of this same stream, the combination of these two treatments, and the incineration of solid residues in order to obtain electricity. These options are depicted in Figure 10.1. The anaerobic digestion was analyzed through the following chemical reactions mediated by a consortium (mixed culture) of anaerobic bacteria in the corresponding reactor:

$$(C_6H_{10}O_5)_n + H_2O \rightarrow C_6H_{12}O_6$$

Cellulose   Water   Glucose

$$(C_5H_8O_4)_n + H_2O \rightarrow C_5H_{10}O_5$$

Hemicellulose (xylan)   Water   Xylose

$$CH_{1.8}O_{0.5}N_{0.2} + 0.5\ H_2O \rightarrow 0.167\ C_6H_{12}O_6 + 0.2\ NH_3 + 0.1\ H_2$$
Z. mobilis   Water   Glucose   Ammonia   Hydrogen

$$C_6H_{12}O_6 \rightarrow 2\ C_2H_5OH + 2\ CO_2$$
Glucose   Ethanol   Carbon dioxide

$$3\ C_5H_{10}O_5 \rightarrow 5\ C_2H_5OH + 5\ CO_2$$
Xylose   Ethanol   Carbon dioxide

$$2\ C_2H_5OH\ +\ CO_2\ \rightarrow\ CH_4\ +\ 2\ CH_3COOH$$
Ethanol   Carbon dioxide   Methane   Acetic acid

$$CH_3COOH\ \rightarrow\ CH_4\ +\ CO_2$$
Acetic acid   Methane   Carbon dioxide

$$4H_2\ +\ CO_2\ \rightarrow\ CH_4\ +\ 2\ H_2O$$
Hydrogen   Carbon dioxide   Methane   Water

These reactions represent some of the main stages of anaerobic digestion of organic matter. The first three reactions correspond to the hydrolytic processes, the following three reactions are carried out by fermentative bacteria, the sixth reaction represents the acetogenesis stage, and the last two reactions globally describe the action of methanogenic bacteria. The formation of anaerobic bacteria is not shown in this reaction scheme although it has been taken into account.

The aerobic digestion was simulated considering the oxidation reaction for each one of the organic compounds present in the liquid effluent. As an example, the complete oxidation of glucose is presented as follows:

$$C_6H_{12}O_6\ +\ O_2\ \rightarrow\ 6\ CO_2\ +\ 6\ H_2O$$
Glucose   Oxygen   Carbon dioxide   Water

Under aerobic conditions, a large portion of the organic matter contained in wastewater may be oxidized biologically by microorganisms to carbon dioxide and water, thus the formation of activated sludge (aerobic bacteria) was also considered. Approximately 50% reduction in solids content can be achieved through this treatment.

Finally, the incineration of the solid residues was simulated through a stoichiometric approach where all the organic compounds are completely oxidized into carbon dioxide and water without formation of any cell biomass. The products of this process are the combustion gases and the remaining ash. The main combustion reaction is the burning of lignin whose energy content achieves 25.4 MJ/kg.

The first configuration contemplates the anaerobic digestion of the liquid effluent (mostly thin stillage) in an anaerobic reactor producing biogas and a suspension of anaerobic sludge that is settled in a decanter. Then, the sludge is mixed with the predried solids residue in order to be burnt in the co-generation system. This

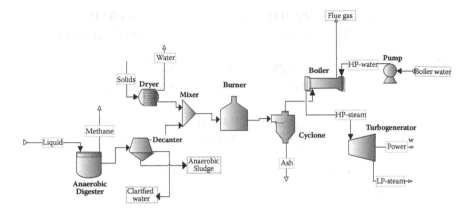

FIGURE 10.2 Effluent treatment for liquid and solid wastes from biomass-to-ethanol process by anaerobic digestion and co-generation.

system consists of a burner where the combustion reaction is carried out, a cyclone for removing the ash, a heat exchanger representing the boiler, and a turbogenerator where the electricity is produced employing the high-pressure (HP) steam from the boiler. The low-pressure (LP) steam can be used to cover the thermal energy required in the overall ethanol production process. This configuration is schematically depicted in Figure 10.2.

The second configuration is based on the aerobic digestion of the wastewater. As in the previous case, the aerobic sludge generated and the solids residue are sent to the cogeneration system (Figure 10.3). Finally, the third configuration comprises the combination of anaerobic and aerobic digestion as well as the delivery of the generated sludge and solids residue to the cogeneration unit, as can be observed in Figure 10.4.

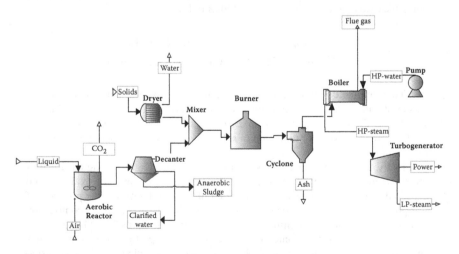

FIGURE 10.3 Effluent treatment for liquid and solid wastes from biomass-to-ethanol process by aerobic digestion and co-generation.

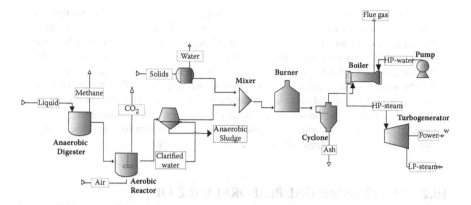

FIGURE 10.4 Effluent treatment for liquid and solid wastes from biomass-to-ethanol process by co-generation and anaerobic digestion followed by aerobic digestion.

The preliminary results obtained are presented in Table 10.1. The scheme involving the anaerobic digestion of the liquid effluents followed by the aerobic digestion of the liquid stream exiting the anaerobic digester represents higher removal of the organic matter contained in these effluents, as suggested in the work of Merrick & Company (1998). If only one type of digestion is considered, the anaerobic digestion gives better results. In the Table 10.1, the sole incineration of the two types of effluent streams (solid and liquid) is also considered as a fourth configuration. In this case, the power regenerated was considered as the comparison standard

TABLE 10.1
Reduction of Organic Load for Different Effluent Treatment Configurations

Compound	Configuration I[a]	Configuration II[b]	Configuration III[c]	Configuration IV[d]
Glucose/%	40.55	51.89	100.00	N/A
Cellulose/%	100.00	60.00	100.00	N/A
Hemicellulose/%	100.00	70.00	100.00	N/A
Xylose/%	100.00	55.00	100.00	N/A
Ammonia/%	100.00	100.00	100.00	N/A
Acetic acid/%	91.13	100.00	100.00	N/A
Ethanol/%	100.00	50.00	100.00	N/A
Furfural/%	100.00	49.81	100.00	N/A
Power generated/%	94.30	92.02	90.68	100.00

[a] Anaerobic digestion + co-generation.
[b] Aerobic digestion + co-generation.
[c] Anaerobic digestion followed by aerobic digestion + co-generation.
[d] Incineration of all the effluent streams considered.

(100%). The configurations involving the separate treatment of the wastewater show a reduced generation of electricity due to the lower availability of organic compounds entering the cogeneration unit. However, the incineration implies that the liquid streams are evaporated before their mixing with the solid residues. For this process, energy is required and the need for thermal energy should be supplied by the low-pressure steam extracted from the turbogenerator. This means that the net energy (thermal and electric) produced can be lower than the net energy of the schemes involving the separate treatment of wastewater. Undoubtedly, process simulation tools are invaluable to assess all these configurations and the energy surplus that may be obtained from these treatment schemes.

10.2   ENVIRONMENTAL PERFORMANCE OF FUEL ETHANOL PRODUCTION

The assessment of fuel ethanol production processes from an environmental point of view implies not only the quantification and analysis of the polluting burdens generated by the conversion technologies, but also the evaluation of the environmental performance of such processes considering their influence on the planet in terms of the depletion of natural resources, contamination of the ecosystems, and global environmental impacts generated.

The starting point to satisfy the standard of clean production is the environmental diagnosis in order to determine the opportunities for preventing or reducing the contamination sources and the viable alternatives to carry out such reduction. In the last four decades, the design of processes and the production of chemicals have experienced a great evolution (Montoya et al., 2006). Initially, reaction and separation systems were designed and optimized according to only one economic objective. In the 1970s and 1980s due to the global energy crisis, the utility system (thermal energy, power, cooling water) was included in the process design and optimization procedures. Nowadays, under the clean production scheme, the environmental objective should be taken into account along with the economic objective. In this sense, the environmental objective should be to consider not only the specific environmental impact of the process, but also its impacts in other steps of its life cycle.

The waste minimization as a means to attain the clean production and contribute to the sustainable development has been extensively studied in the process industry and in academic circles. The waste minimization incorporates both the waste reduction in the source and the use of recycling to reduce the amount and risk of the residues. Nevertheless, the difference between hazardous and nonhazardous wastes is not considered. In this connection, the minimization of the environmental impact is a much stricter norm and, although its scope is similar to the waste minimization, emphasizes in a higher degree the different impacts of the chemical species on the environment (Yang and Shi, 2000). The measurement of the environmental performance of a process can be viewed as a decision problem involving two levels: screening indices on the process level and environmental performance indicators on the chemical species level. The latter indicators are the base of the screening indexes and offer enough flexibility to consider the fate of all the components involved and their subsequent impacts. Such impacts are in the framework of a group of categories, i.e., environmental performance indicators show the impact with which the process contributes to a given category. Among these categories, the following should be highlighted, as reported by Yang and Shi (2000): energy consumption, resource consumption, greenhouse effect, ozone depletion, acidification, eutrophication, photochemical smog, human toxicity, ecotoxicity, area used and species diversity, odor, and noise. To consider the environmental impact of a process, several methodologies have been developed, such as the life cycle assessment (LCA) and the waste reduction (WAR) algorithm, among others.

10.2.1 WAR ALGORITHM

In Chapter 2, Table 2.1, the different steps involved in the development of an industrial process during its life cycle were highlighted. During the conceptual process design step, the information available is quite limited and uncertain. In this case, the screening indexes and environmental performance indicators should be based on simple mass and energy balances. Thus, the environmental indicators developed by Heinzie et al. (1998) that take into account the different decision levels during the conceptual design are based on mass losses. In a similar way, the ecotoxicological models developed by Elliot et al. (1996) are also based in mass units instead of concentration units (Yang and Shi, 2000).

The waste reduction algorithm originally proposed by Hilaly and Sikdar (1994) was based on the concept of pollution balance. Currently, the WAR algorithm is one of the most practical methodologies for assessing the environmental impact of a process, especially at the conceptual design step. It allows assessing and comparing the environmental friendliness of many different industrial processes. The methodology of the WAR algorithm developed by the National Risk Management Research Laboratory of the U.S. Environmental Protection Agency (EPA) uses the concept of potential environmental impact (PEI) and proposes to add a conservation relationship over the PEI based on the input and output impact flows of the process. In this context, the PEI of a given quantity of mass and energy is understood as the effect that this mass and energy would have on average on the environment if they were to be discharged into the environment from this process (Cardona et al., 2004; Young and Cabezas, 1999). This definition implies that the impact is a quantity still not realized and the PEI has a probabilistic nature. Thus, the PEI of a chemical process is usually caused by the mass and energy that this process acquires or emits into the environment. In this way, the WAR algorithm is a tool to perform the evaluation of the PEI of a process based on the product streams, outlet wastes, and the feed to the process.

The overall PEI of a chemical k is determined by summing the specific PEI of the chemical over all the possible impact categories:

$$\psi_k = \sum_l \alpha_l \psi_{kl} \qquad (10.1)$$

where ψ_{kl} represents every impact category and α_l is the weighting factor, which is used to emphasize the particular areas of concern. These categories fall into two general areas of concern— global atmospheric and local toxicological—with four categories in each area. The four global atmospheric impact categories include

1. Global warming potential (GWP)
2. Ozone depletion potential (ODP)
3. Acidification or acid-rain potential (AP)
4. Photochemical oxidation or smog formation potential (PCOP)

The four local toxicological impact categories include

1. Human toxicity potential by ingestion (HTPI)
2. Human toxicity potential by either inhalation or dermal exposure (HTPE)
3. Aquatic toxicity potential (ATP)
4. Terrestrial toxicity potential (TTP; Cardona et al., 2004; Young and Cabezas, 1999)

This algorithm was modified in a previous work (Cardona et al., 2004) and the changes incorporated into the new versions of WARGUI (WAste Reduction algorithm—Graphical User Interface) software released by the EPA.

The WAR algorithm handles two classes of indexes to assess the environmental impact of a chemical process. The first class measures the PEI emitted by the process while the other one measures the PEI generated within the process. Each class has two main indexes: total output rate of PEI (expressed as PEI per time unit) and PEI leaving the system per mass of product streams. The first class characterizes the PEI emitted by the system and is used to answer questions about the external environmental efficiency of the process, i.e., the ability of the plant to produce the desired products at a minimum discharge PEI. The second class of indexes characterizes the PEI generated by the system and its importance lies in the determination of the internal environmental efficiency of the process, i.e., how much PEI is being generated or consumed inside the process. The lower the values of these indexes, the more environmentally efficient the process is. Considering the variation of the plant capacities, the index per mass of products should be employed if the goal is to assess the environmental impact of a process independent of its production rate, especially if different alternative processes are to be evaluated (process synthesis).

CASE STUDY 10.2   EVALUATION OF ENVIRONMENTAL
PERFORMANCE OF FUEL ETHANOL PRODUCTION FROM
STARCHY AND LIGNOCELLULOSIC MATERIALS

The assessment of the environmental impacts generated during the fuel ethanol
production process is a key criterion to select different conversion technologies
during early stages of the design of such process. If different feedstocks are evaluated, the analysis of the environmental performance for the proposed technological
configurations is even more important. In a previous work (Cardona et al., 2005a),
the environmental performance of several process alternatives for fuel ethanol
production from two types of feedstock was determined. The types of feedstock
corresponded to lignocellulosic materials (herbaceous biomass, wood chips, sugarcane bagasse, and waste paper) and starchy materials (corn, wheat, and cassava).
Using the commercial process simulator Aspen Plus, different configurations were
analyzed considering two options for ethanol dehydration step: adsorption using
molecular sieves and azeotropic distillation.

The results obtained for the environmental performance of these processes (in
terms of PEI per mass of product) are shown in Figure 10.5 for the three most representative process alternatives: (1) ethanol production from corn by dry-milling
technology using molecular sieves for the dehydration step, (2) ethanol production from lignocellulosic biomass (wood chips) using azeotropic distillation for the
dehydration step, and (3) ethanol production from lignocellulosic biomass (wood
chips) using molecular sieves for the dehydration step. It is evident that the production of ethanol from starch has a lower impact on the environment than the biomass ethanol process. Because processes using biomass involve a pretreatment step
where inorganic acids are used, which tends to increase the PEI. In comparing two

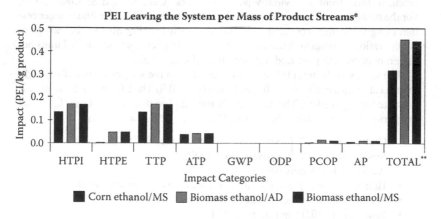

FIGURE 10.5   Potential environmental impact (PEI) per mass of product streams for
different ethanol production flowsheets according to the eight impact categories addressed
by the WAR algorithm: HTPI = human toxicity potential by ingestion, HTPE = human
toxicity potential by either inhalation or dermal exposure, TTP = terrestrial toxicity potential, ATP = aquatic toxicity potential, GWP = global warming potential, ODP = ozone
depletion potential, PCOP = photochemical oxidation or smog formation potential, AP =
acidification or acid-rain potential. Dehydration technologies: AD = azeotropic distillation, MS = molecular sieves.

kinds of separation technologies for the same feedstock (wood biomass), the utilization of molecular sieves for recovery of the product has slightly lower PEI than the process involving azeotropic distillation due to the release of relatively small amounts of the entrainer (in this case, the toxic benzene) into the output streams. In this sense, the adsorption with molecular sieves is a cleaner separation technology and it is currently being used in the bioethanol industry.

Case Study 10.3   Comparison of Environmental Performance of Fuel Ethanol Production using Sugarcane and Lignocellulosic Biomass as Feedstocks

In a previous work (Sánchez et al., 2007), the environmental impacts of fuel ethanol production processes from sugarcane or lignocellulosic biomass were estimated. The evaluation of the environmental performance was performed for three technological configurations employing sugarcane juice as feedstock (stand-alone facilities). The first one comprised the utilization of sugarcane bagasse for co-generation of process steam and power and included the production of concentrated stillage for fertilization purposes. The second option did not consider the co-generation though it did consider the production of the concentrated stillage as a fertilizer, thus the bagasse was accumulated as a solid residue. The third variant involved neither cogeneration nor production of concentrated stillage. The simulation data for these alternative flowsheets were taken from a preceding work (Cardona et al., 2005b). In addition, the production of fuel ethanol from lignocellulosic biomass through a process including the dilute-acid pretreatment of biomass, simultaneous saccharification and co-fermentation (SSCF) and ethanol dehydration using molecular sieves was also analyzed. The simulation data for production of biomass ethanol were taken, in turn, from a previously published work (Cardona and Sánchez, 2006). For the biomass process, two variants were considered: with and without cogeneration using the lignin recovered from the whole stillage. For all five technological configurations, the normalized values of the total impact expressed as PEI units per kilogram of ethanol produced were obtained (Figure 10.6).

The values of the total PEI were calculated from the weighted sum of the eight impact categories evaluated by the software WARGUI. The PEI for each category is presented in Figure 10.7. The weighted factors employed for six of the eight impact categories were taken from the work of Chen et al. (2002) as follows:

- Global warming 2.5 (equivalent to GWP in the WAR algorithm)
- Smog formation 2.5 (equivalent to PCOP)
- Acid rain 10.0 (equivalent to AP)
- Human noncarcinogenic inhalation toxicity 5.0 (equivalent to HTPE)
- Human noncarcinogenic ingestion toxicity 5.0 (HTPI)
- Fish toxicity 10.0 (equivalent to ATP)

The two remaining impact categories in the WAR algorithm methodology were assigned the following weights: TTP 2.5 and ODP 2.5.

The results obtained indicate that the use of sugarcane presents a higher environmental friendliness than the use of lignocellulosic biomass. This is explained by the complexity of the conversion process from biomass to ethanol. For such conversion, a pretreatment step involving the use of inorganic acids and high pressure is required. In addition, the hydrolysis of cellulose and fermentation of formed

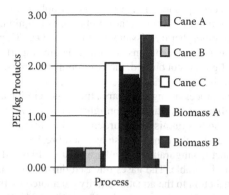

FIGURE 10.6 Total output rate of potential environmental impact (PEI) per mass of products for five process configurations for fuel ethanol production: Cane A = production of ethanol and fertilizer (concentrated stillage) from sugarcane employing bagasse for co-generation, Cane B = production of ethanol and fertilizer (concentrated stillage) from sugarcane without co-generation, Cane C = production of ethanol from sugarcane without co-generation, Biomass A = production of ethanol from lignocellulosic biomass without co-generation, Biomass B = production of ethanol from lignocellulosic biomass employing the recovered lignin for co-generation.

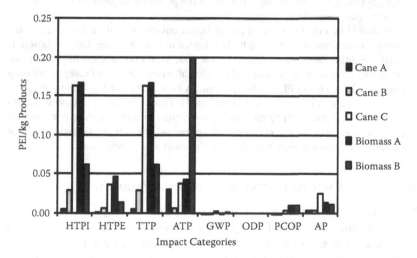

FIGURE 10.7 Potential environmental impact (PEI) per mass of product streams for different ethanol production configurations according to the eight impact categories considered by the WAR algorithm (the denominations of the categories are presented in Figure 10.5): Cane A = production of ethanol and fertilizer (concentrated stillage) from sugarcane employing bagasse for co-generation, Cane B = production of ethanol and fertilizer (concentrated stillage) from sugarcane without co-generation, Cane C = production of ethanol from sugarcane without co-generation, Biomass A = production of ethanol from lignocellulosic biomass without co-generation, Biomass B = production of ethanol from llgnocellulosic biomass employing the recovered lignin for cogeneration.

glucose should be accomplished. These processes imply higher energy expenditures. To supply this amount of energy, the combustion of lignin is considered leading to the release of atmospheric emissions containing CO_2, CO, particulate matter, and polycyclic aromatic hydrocarbons, which generate important environmental impacts. In fact, if lignin is not burned, the environmental impacts are appreciably reduced (see Figure 10.6).

In the case of the process using sugarcane, the co-generation and the production of concentrated stillage as a fertilizer (configuration Cane A in Figures 10.6 and 10.7) is a good option to diminish the burdens to the environment. However, the configuration considering the cane bagasse as a solid residue and the concentrated stillage as co-product (configuration Cane B in Figures 10.6 and 10.7) shows indicators slightly more favorable. The gases released during the bagasse combustion have a higher contribution to the aquatic toxicity calculated by the WARGUI software than the components of the bagasse itself (see Figure 10.7). This difference is mostly responsible for the better environmental performance of this configuration compared to the scheme involving the co-generation using the bagasse as a solid fuel. Nevertheless, the economic considerations do indicate the evident benefit of burning the bagasse because no money is spent for acquiring the fossil fuels needed to supply the thermal energy for the overall process. If the stillage is not treated and considered as a liquid effluent, the environmental impact potential remarkably increases, as shown in Figures 10.6 and 10.7. The total PEI per kilogram of products is increased by 430% related to the Cane A case. This fact is explained by the very high organic load of the stillage that raises the potential impacts corresponding to the four toxicological impact categories.

It should be noted that the apparent better environmental performance of the configuration Cane B is due in a higher degree to the weighting factors chosen. If equal weighting factors are selected for the four local toxicological impact categories (using a value of 10) related to the global atmospheric impact categories (using a value of 2.5), a lower PEI/kg for the Cane A case is obtained: 0.43 for configuration Cane A and 0.74 for configuration Cane B. This selection of weighting factors favors the local effects on human health, flora, and fauna during the environmental analysis more than the effects on the biosphere, which is logical considering that the former effects are more significant in the short and middle terms.

10.2.2 LIFE CYCLE ASSESSMENT OF BIOETHANOL PRODUCTION

The methodology of life cycle assessment (LCA) is a systematic analysis tool considering the environmental impacts of products, processes, or services, and provides a reference structure for the development and application of screening indices and environmental performance indicators, especially during the extension of the system boundaries to the other steps of a life cycle of a product. Therefore, LCA is a technique for assessing the environmental performance of a product, process, or activity from "cradle to grave," i.e., from extraction of raw materials to final disposal (Azapagic, 1999).

Life cycle assessment is defined as a process to evaluate the environmental burdens associated with a product, process, or activity by identifying and quantifying energy and materials used and wastes released to the environment; to assess the impact of those energy and material uses and releases to the environment;

and to identify and evaluate opportunities to effect environmental improvements (Azapagic, 1999). A fundamental feature of LCA compared to other methodologies for environmental evaluation is the analysis of a system (for instance, a process for production of a particular chemical) during its whole life cycle from the extraction and processing of the feedstocks to the disposal of the target product, co-products, and by-products considering their effect over the whole environment (over the global warming, ozone layer depletion, etc.) and including manufacturing, transport, use, reuse, maintenance, and recycling of the different materials involved. In contrast, most methods for environmental assessment are focused only on the immediate effects of the system on the surroundings, such as the effects of the emissions and burdens from the processing plant. In this sense, there exists the possibility that certain measures adopted to reduce these emissions and burdens in a given process lead to the increase of other emissions or burdens in other steps of the life cycle of this process, e.g., during the feedstock extraction. Despite these conceptual advantages, the application of LCA during the conceptual design step is quite difficult due to the limitation in the available information as mentioned above.

Applying the LCA methodology, the environmental benefits of using the excess of cane bagasse, which remains after employing the bagasse as an energy source in sugar mills, as a feedstock for fuel ethanol production instead of burning it in open fields were demonstrated (Kadam, 2002). Thus, some achievements can be attained such as the emissions reduction, less fossil fuels consumption, diminution in the rate of natural resources depletion, reduction of human toxicity, and less contribution to greenhouse gas effect. Hu et al. (2004b) performed LCA for cars fueled by blends using ethanol from cassava obtaining better results than for conventional cars in China. Kim and Dale (2002) proposed an allocation procedure based on the system expansion approach for the net energy analysis of corn ethanol. The allocation procedure is a key factor in LCA when the multi-input/output process is analyzed as in the case of ethanol production employing wet and dry milling. These same authors also determined the nonrenewable energy consumption and greenhouse gas emissions for corn ethanol production in selected counties in the United States, showing positive net energy values and the possibility for reducing greenhouse gas emissions (Kim and Dale, 2005a). Bullock (2002) cites the LCA studies based on data from Australian ethanol plants for gasoline blends with 10% ethanol content. These studies indicate that there is no greenhouse abatement when molasses, wheat starch waste, or wheat are used as feedstocks. In contrast, the hypothetical process from wood waste can lead to CO_2 abatement. Gasoline blends with 85% ethanol content present significant environmental benefits represented mainly by the greenhouse gas abatement.

Several recent studies applying the LCA methodology have been published demonstrating some advantages of fuel ethanol production even in the polemic case of corn ethanol. For example, using ethanol derived from corn dry milled as liquid fuel (E10 fuel) would reduce nonrenewable energy and greenhouse gas emissions, but would increase acidification, eutrophication, and photochemical smog, compared to using gasoline as liquid fuel (Kim and Dale, 2005a). Other studies using the cereal straw or corn stover as feedstocks for biomass ethanol have also indicated the environmental benefits, but show problems related to soil acidification and eutrophication (Gabrielle and Gagnaire, 2008; Kim and Dale, 2005b).

In a similar way, the LCA performed in France by comparing the ethyl tert-butyl ether (ETBE) obtained from beet ethanol (a partial renewable product) to the methyl tert-butyl ether (MTBE) from fossil origin showed that the energy yield of ethanol (1.18) and ETBE (0.93) are higher than the yields of gasoline (0.74 to 0.80) and MTBE (0.73). Moreover, the ETBE has a lower contribution to the greenhouse gas effect due to its renewable character, and its use as a gasoline oxygenate provokes fewer emissions of nonburnt hydrocarbons than in the case of MTBE. Nevertheless, the ethanol cost for ETBE production in France has been higher than the cost of gasoline and methanol (a feedstock for MTBE production; Poitrat, 1999).

10.2.3 OTHER METHODOLOGIES FOR THE ENVIRONMENTAL ANALYSIS OF BIOETHANOL PRODUCTION

An integral study was carried out in Brazil regarding the ecological, economic, and social aspects of fuel ethanol production including the agricultural and industrial steps. Three ethanol production plants were analyzed and the results were quantified through the environmental efficiency of each process. According to the analysis performed, none of the three plants achieved the highest level established for environmental efficiency (Borrero et al., 2003). Prakash et al. (1998), in turn, proposed an indicator called *figure of merit*, which is expressed as the ratio between the net energy yield of a fuel (the difference between the gross energy produced during its combustion and the energy needed to produce it) and the CO_2 emissions produced by that fuel. Using this figure, it was concluded that for anhydrous ethanol production from cane molasses in India the net energy yield is about 2% the potential of ethanol as a gasoline substitute in road transport has been estimated to be as high as 28%. A similar indicator based on the figure of merit has been proposed by Hu et al. (2004a) as well. Lynd and Wang (2004) developed a methodology for evaluating fossil fuel displacement for biological processing of biomass in the absence of product-specific information other than the product yield and whether fermentation is aerobic or anaerobic. With the help of this methodology, the authors answer affirmatively the question: Are there biomass-based processes and products with fossil fuel displacement sufficiently large that they could play a substantial role in a society supported by sustainable resources?

Another methodology that has been used to measure the environmental feasibility and sustainability of bioethanol production processes over a long term is the so-called emergy analysis (see Chapter 2, Section 2.2.5). Emergy is the abbreviation of embedded energy. This methodology takes into account the different system inputs, such as the renewable and nonrenewable energy sources, goods, labor, and all the materials involved in a process. Each input is assessed under the same physical condition, the equivalent solar energy (emergy) concentrated to supply the given input. In this regard, the emergy is a measure of the global convergence of energy, time, and space needed to make available a given resource. Bastanioni and Marchettini (1996) analyzed four systems for bioethanol production and concluded that mankind is far from a sustainable production of biofuels because it appears that the biomass energy cannot be the basic energy source in countries having a high level of energy consumption, with arable land being the major constraint. Other methodologies for environmental assessment of fuel ethanol production from lignocellulosic biomass have been reported as the application of the sustainable process index (SPI), a highly aggregated indicator of the environmental pressure of a process, to the bioenergy assessment model (BEAM), which have allowed showing that the bioenergy integrated systems are superior to the fossil fuel systems in terms of their environmental compatibility (Krotscheck et al., 2000).

Berthiaume et al. (2001) propose a method to quantify the renewability of a biofuel taking as an example the ethanol production from corn. This method considers the exergy (useful or available energy) as the quantitative measure of the maximum amount of work that can be obtained from an imbalance between the physical system and the environment surrounding it. The exergy was accounted for evaluating the departure from ideal behavior of a system caused by a consumption of nonrenewable resources through the so-called restoration work. These authors point out that ethanol production is not renewable, though they emphasize that this evaluation was performed employing many simplifications. Thus, further research is needed to improve the accurateness of the results and the validity of the conclusions.

Undoubtedly, the unification of the environmental assessment criteria is required in such a way that the main aspects of fuel production and utilization, including both fossil fuels and biofuels, are taken into account. The large number of consideration, supposition, assumption, approximation, and methodological approaches limit the use and comparison of the environmental indicators that have been applied to fuel ethanol production. An example is the contradictory results about the environmental and energy performance of corn ethanol that have been published (Patzek et al., 2005; Pimentel, 2003; Shapouri et al., 2003; Wang et al., 1999). This difficulty imposes some constraints when different process alternatives for bioethanol production are being evaluated, especially if the optimization of an objective function considering not only the technoeconomic indicators but also the environmental performance indexes of the different technological configurations proposed is performed.

REFERENCES

Azapagic, A. 1999. Life cycle assessment and its application to process selection, design and optimisation. *Chemical Engineering Journal* 73:1–21.
Bastanioni, S., and N. Marchettini. 1996. Ethanol production from biomass: Analysis of process efficiency and sustainability. *Biomass and Bioenergy* 11 (5):411–418.

Belkacemi, K., F. Larachi, S. Hamoudi, and A. Sayari. 2000. Catalytic wet oxidation of high-strength alcohol-distillery liquors. *Applied Catalysis A: General*
199:199–209.

Berthiaume, R., C. Bouchard, and M.A. Rosen. 2001. Exergetic evaluation of the renewability of a biofuel. *Exergy: An International Journal* 1 (4):256–268.

Borrero, M.A.V., J.T.V. Pereira, and E.E. Miranda. 2003. An environmental management
method for sugar cane alcohol production in Brazil. *Biomass and Bioenergy* 25
(3):287–299.

Bullock, G.E. 2002. *Ethanol from sugarcane*, Sugar Research Institute, Queensland, Australia.

Cardona, C.A., V.F. Marulanda, and D. Young. 2004. Analysis of the environmental impact
of butylacetate process through the WAR algorithm. *Chemical Engineering Science*
59 (24):5839–5845.

Cardona, C.A., and Ó.J. Sánchez. 2006. Energy consumption analysis of integrated
flowsheets for production of fuel ethanol from lignocellulosic biomass. *Energy*
31:2447–2459.

Cardona, C.A., Ó.J. Sánchez, M.I. Montoya, and J.A. Quintero. 2005a. Analysis of fuel
ethanol production processes using lignocellulosic biomass and starch as feedstocks.
Paper presented at the 7th World Congress of Chemical Engineering, July 10–14
2005, Glasgow, Scotland, U.K..

Cardona, C.A., Ó.J. Sánchez, M.I. Montoya, and J.A. Quintero. 2005b. Simulación de los
procesos de obtención de etanol a partir de caña de azúcar y maíz (Simulation of
processes for ethanol production from sugarcane and corn, in Spanish). *Scientia et
Technica* 28:187–192.

Cardona, C.A., Ó.J. Sánchez, and J.I. Rossero. 2006. Analysis of integrated schemas
for effluent treatment during fuel ethanol production. Paper presented at the 17th
International Congress of Chemical and Process Engineering (CHISA 2006), Prague,
Czech Republic.

Chen, H., Y. Wen, M.D. Waters, and D.R. Shonnard. 2002. Design guidance for chemical
processes using environmental and economic assessments. *Industrial & Engineering
Chemistry Research* 41:4503–4513.

Chin, P.M., and W.M. Ingledew. 1994. Effect of lactic acid bacteria on wheat mash fermentations prepared with laboratory backset. *Enzyme and Microbial Technology*
16:311–317.

de la Cruz, R., E. González, and L. López. 2004. Evaluación económica de alternativas
para la solución de los residuales de una planta de alcohol integrada a una fábrica de
azúcar de caña (Economic evaluation of alternatives for the solution of the residuals
from an ethanol plant integrated to a cane sugar mill, in Spanish). *Revista Universidad
EAFIT* 40 (135):60–72.

Elliott, A.D., B. Sowerby, and B.D. Crittenden. 1996. Quantitative environmental impact
analysis for clean design. *Computers and Chemical Engineering* 20:1377–1382.

Gabrielle, B., and N. Gagnaire. 2008. Life-cycle assessment of straw use in bio-ethanol
production: A case study based on biophysical modelling. *Biomass and Bioenergy*
32 (5):431–441.

Gnecco, J. 2006. *Situación de la producción de etanol en Colombia (Situation of ethanol
production in Colombia*, in Spanish). Sucromiles. http://www.ciat.cgiar.org/train-
ing/pdf/060315_escenarios_de_produccion_de_etanol_en_colombia.pdf (accessed
February 2007).

Gnecco, J.G. 2003. Alternativas de disposición de vinaza (Alternatives for disposal of stillage, in Spanish). Paper presented at the III Jornada Técnica de Ingeniería Química,
Manizales, Colombia.

Goyes, A., and G. Bolaños. 2005. Un estudio preliminar sobre el tratamiento de vinazas en agua supercrítica (A preliminary study on stillage treatment in supercritical water, in Spanish). Paper presented at the XXIII Congreso Colombiano de Ingeniería Química, Manizales, Colombia.

Heinzie, E., D. Weirich, F. Brogli, V.H. Hoffmann, G. Koller, M.A. Verduyn, and K. Hungerbuhler. 1998. Ecological and economical objective functions for screening in integrated development of fine chemical processes. *Industrial & Engineering Chemistry Research* 37:3395–3407.

Hilaly, A. H., and S. K. Sikdar. 1994. Pollution balance: A new methodology for minimizing waste production in manufacturing processes. *Journal of the Air & Waste Management Association* 44:1303-1308.

Hu, Z., F. Fang, D.F. Ben, G. Pu, and C. Wang. 2004a. Net energy, CO_2 emission, and life-cycle cost assessment of cassava-based ethanol as an alternative automotive fuel in China. *Applied Energy* 78:247–256.

Hu, Z., G. Pu, F. Fang, and C. Wang. 2004b. Economics, environment, and energy life cycle assessment of automobiles fueled by bio-ethanol blends in China. *Renewable Energy* 29:2183–2192.

Jones, A.M., and W.M. Ingledew. 1994. Fermentation of very high gravity wheat mash prepared using fresh yeast autolysate. *Bioresource Technology* 50:97–101.

Kadam, K.L. 2002. Environmental benefits on a life cycle basis of using bagasse-derived ethanol as a gasoline oxygenate in India. *Energy Policy* 30:371–384.

Kim, S., and B.E. Dale. 2002. Allocation procedure in ethanol production system from corn grain: I. System expansion. *International Journal of Life Cycle Assessment* 7 (4):237–243.

Kim, S., and B.E. Dale. 2005a. Environmental aspects of ethanol derived from no-tilled corn grain: Nonrenewable energy consumption and greenhouse gas emissions. *Biomass and Bioenergy* 28:475–489.

Kim, S., and B.E. Dale. 2005b. Life cycle assessment of various cropping systems utilized for producing biofuels: Bioethanol and biodiesel. *Biomass and Bioenergy* 29:426–439.

Krotscheck, C., F. König, and I. Obernberger. 2000. Ecological assessment of integrated bioenergy systems using the Sustainable Process Index. *Biomass and Bioenergy* 18:341–368.

Larsson, M., M. Galbe, and G. Zacchi. 1997. Recirculation of process water in the production of ethanol from softwood. *Bioresource Technology* 60:143–151.

Lynd, L.R., and M.Q. Wang. 2004. A product-nonspecific framework for evaluating the potential of biomass-based products to displace fossil fuels. *Journal of Industrial Ecology* 7 (3–4):17–32.

Merrick & Company. 1998. *Wastewater treatment options for the biomass-to-ethanol process.* Merrick Project No. 19013104, Merrick & Company, Aurora, CO, USA.

Montoya, M.I., J.A. Quintero, Ó.J. Sánchez, and C.A. Cardona. 2006. Evaluación del impacto ambiental del proceso de obtención de alcohol carburante utilizando el algoritmo de reducción de residuos (Environmental impact assessment of fuel ethanol production process using the waste reduction algorithm, in Spanish). *Revista Facultad de Ingeniería* 36:85–95.

Nandy, T., S. Shastry, and S.N. Kaul. 2002. Wastewater management in a cane molasses distillery involving bioresource recovery. *Journal of Environmental Management* 65:25–38.

Navarro, A.R., M. del C. Sepúlveda, and M.C. Rubio. 2000. Bio-concentration of vinasse from the alcoholic fermentation of sugar cane molasses. *Waste Management* 20:581–585.

Nguyen, M.H. 2003. Alternatives to spray irrigation of starch waste based distillery effluent. *Journal of Food Engineering* 60:367–374.

Olguín, E.J., H.W. Doelle, and G. Mercado. 1995. Resource recovery through recycling of sugar processing by-products and residuals. *Resources, Conservation and Recycling* 15:85–94.

Palmqvist, E., B. Hahn-Hägerdal, M. Galbe, M. Larsson, K. Stenberg, Z. Szengyel, C. Tenborg, and G. Zacchi. 1996. Design and operation of a bench-scale process development unit for the production of ethanol from lignocellulosics. *Bioresource Technology* 58:171–179.

Patzek, T.W., S.-M. Anti, R. Campos, K.W. Ha, J. Lee, B. Li, J. Padnick, and S.-A. Yee. 2005. Ethanol from corn: Clean renewable fuel for the future, or drain on our resources and pockets? *Environment, Development and Sustainability* 7:319–336.

Pimentel, D. 2003. Ethanol fuels: Energy balance, economics, and environmental impacts are negative. *Natural Resources Research* 12 (2):127–134.

Poitrat, E. 1999. The potential of liquid biofuels in France. *Renewable Energy* 16:1084–1089.

Prakash, R., A. Henham, and I.K. Bhat. 1998. Net energy and gross pollution from bioethanol production in India. *Fuel* 77 (14):1629–1633.

Sánchez, Ó.J., C.A. Cardona, and D.L. Sánchez. 2007. Análisis de ciclo de vida y su aplicación a la producción de bioetanol: Una aproximación cualitativa (Life cycle assessment and its application to bioethanol production: A qualitative approximation, in Spanish). *Revista Universidad EAFIT* 43 (146):59–79.

Shapouri, H., J.A. Duffield, and M. Wang. 2003. The energy balance of corn ethanol revisited. *Transactions of the ASAE* 46 (4):959–968.

Sheehan, G.J., and P.F. Greenfield. 1980. Utilisation, treatment and disposal of distillery wastewater. *Water Research* 14:257–277.

Vaccari, G., E. Tamburini, G. Sgualdino, K. Urbaniec, and J. Klemeš. 2005. Overview of the environmental problems in beet sugar processing: Possible solutions. *Journal of Cleaner Production* 13:499–507.

Wang, M., C. Saricks, and D. Santini. 1999. *Effects of fuel ethanol use on fuel-cycle energy and greenhouse gas emissions*. Report no. ANL/ESD-38, Argonne National Laboratory, Center for Transportation Research, Argonne, IL, USA. www.transportation.anl.gov/pdfs/TA/58.pdf.

Wilkie, A.C., K.J. Riedesel, and J.M. Owens. 2000. Stillage characterization and anaerobic treatment of ethanol stillage from conventional and cellulosic feedstocks. *Biomass and Bioenergy* 19:63–102.

Wooley, R., M. Ruth, J. Sheehan, K. Ibsen, H. Majdeski, and A. Galvez. 1999. *Lignocellulosic biomass to ethanol process design and economics utilizing co-current dilute acid prehydrolysis and enzymatic hydrolysis. Current and futuristic scenarios*. Technical Report NREL/TP-580-26157, National Renewable Energy Laboratory, Washington, D.C.

Yang, Y., and L. Shi. 2000. Integrating environmental impact minimization into conceptual chemical process design: A process systems engineering review. *Computers and Chemical Engineering*. 24:1409–1419.

Young, D.G., and H. Cabezas. 1999. Designing sustainable process with simulation: The waste reduction (WAR) algorithm. *Computers and Chemical Engineering* 23:1477–1491.

11 Technological Configurations for Fuel Ethanol Production in the Industry

In this chapter, the different configurations for fuel ethanol production employing the three most important types of raw materials (sucrose-containing, starchy, and lignocellulosic materials) are presented. In particular, such configurations involving integrated processes are discussed. The role played by process systems engineering during the definition and development of the diverse process flowsheets is emphasized. Finally, examples of process synthesis procedures applied to ethanol production are presented.

11.1 ETHANOL PRODUCTION FROM SUCROSE-CONTAINING MATERIALS

Average ethanol yields from sucrose-containing feedstocks based on sugarcane can reach 70 L/ton cane and 9 L/ton of C-grade molasses (in addition to about 100 kg of sugar; Moreira, 2000). The most used fermenting microorganism is *Saccharomyces cerevisiae* due to its ability to hydrolyze cane sucrose for conversion to glucose and fructose, two easily assimilable hexoses. Fermentation pH is 4 to 5 and temperature is 30° to 35°C. Ethanol in Brazil is obtained from sugarcane, and the country is the world's leading producer, followed by India. About 80% of ethanol in Brazil is produced from fresh sugarcane juice and the remaining percentage from molasses (Wilkie et al., 2000). Sugar cane molasses is the main feedstock for ethanol production in India (Cardona and Sánchez, 2007). In Colombia, different sucrose-containing streams are used to produce fuel ethanol, especially the B-grade molasses. Berg (2001) indicates that the output/input ratio of energy for ethanol production from cane is the highest among the main types of feedstocks, reaching a value of 8. This indicator expresses the ratio between the energy released during the combustion of ethanol and the energy required for its production using the whole life cycle of the product from extraction of raw materials until the transformation process producing ethanol (Sánchez and Cardona, 2005).

In general, the process for ethanol production for sugarcane includes the extraction and conditioning of cane juice to make it more assimilable by yeasts during fermentation. From the resulting culture broth, cell biomass is separated and then the concentration of ethanol and its dehydration are carried out employing different unit operations. The product is the anhydrous ethanol that is the trade form in which fuel ethanol is utilized as a gasoline oxygenate. This process can utilize not only the crushed cane but also cane molasses as a feedstock as well as other streams with high content of fermentable sugars derived from the process for cane sugar production in sugar mills, as mentioned in Chapter 3, Section 3.1.5 (for example, a fraction of the clarified syrup). In the latter case, ethanol production facilities are located next to sugar mills, as in the case of Colombian distilleries. In the former case, distilleries can operate in an independent way as in Brazil, where an important number of autonomous (stand-alone) distilleries employing cane juice are currently in operation.

The overall process for fuel ethanol production from sugarcane in autonomous distilleries is shown in Figure 11.1. Production process in a distillery co-located at a sugar mill differs from the autonomous distilleries in the first steps (cleaning, milling, clarification). After these steps, the process is almost the same, although it is necessary to condition the molasses before fermentation. Fuel ethanol production process from sugar beet is similar to the process from sugarcane. Culture broth (fermented wort) is extracted from fermenters and sent to centrifuges for cell biomass separation. The removed microorganisms can be reutilized in the fermentation step. The obtained wine is directed to the first distillation column (concentration column) where the ethanol concentration of the wine is increased. Exhaust gas from fermenters having a fraction of volatilized ethanol is collected and sent to the scrubber. In this unit, ethanol is dissolved in a water stream, where a dilute alcoholic solution is obtained that is also sent to the first distillation column. The resulting ethanol-enriched stream is fed to the second distillation column (rectification column) whose distillate has a high ethanol concentration (near

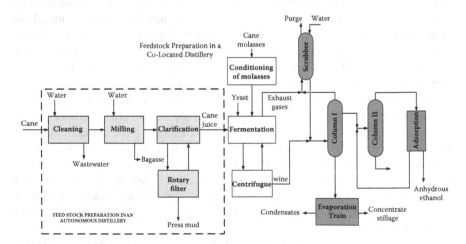

FIGURE 11.1 Simplified diagram of fuel ethanol production from sugarcane. The dotted box represents process steps carried out in an autonomous distillery.

96% by weight). This stream is sent through the dehydration step that can be carried out using different technologies. In the flowsheet depicted in Figure 11.1, it is indicated that the ethanol dehydration is performed by adsorption with molecular sieves. The bottoms of the concentration column or stillage (vinasses) are directed to the effluent treatment step where they are generally evaporated for their concentration.

The fermentation step is central to the overall process for fuel ethanol production because it represents the transformation of sugar-containing raw materials into ethyl alcohol employing yeasts or other ethanol-producing microorganisms. Ethanolic fermentation technologies of sucrose-based media, mainly cane juice and cane or beet molasses, can be considered relatively mature, especially if they are operated batchwise. However, many research efforts are being made worldwide in order to improve the efficiency of the process. In particular, these efforts are aimed at increasing conversion of the feedstock and ethanol productivity, and at reducing production costs, especially energy costs. The features of ethanolic fermentation of sucrose-containing materials are discussed in Chapter 7, Section 7.1.2.

Process simulation plays a crucial role during the analysis of the technical, economic, and environmental performance of fuel ethanol production from sucrose-containing materials. In addition, simulation tools are very significant when process synthesis procedures are being applied, particularly when the knowledge-based process synthesis approach is employed (see Chapter 2). Thus, if the hierarchical decomposition procedure is applied, process simulation can provide the necessary data on the process behavior in order to select or discard the different alternatives proposed during each hierarchical level of analysis (see, for example, the work of Sánchez, 2008).

CASE STUDY 11.1   SIMULATION OF FUEL ETHANOL PRODUCTION FROM SUGARCANE

With the aim of obtaining valuable information on the technoeconomic and environmental performance of ethanol production from sugarcane in a stand-alone facility under Colombian conditions, a characteristic technological configuration for bioethanol production was simulated in previous works (Cardona et al., 2005b; Quintero et al., 2008). For this, the process was analyzed considering five main processing steps: raw material conditioning, fermentation, separation and dehydration, effluent treatment, and co-generation. In the simulated process that is depicted in Figure 11.2, the feedstock is washed, crushed, and milled to extract the sugarcane juice and produce bagasse. The cane juice is sent to a clarification process, where pH is adjusted, some impurities are removed, and the press mud is generated. This material is the filter cake obtained during the removal of suspended solids in the rotary drum filter employed for juice clarification. The press mud is commercialized as a component of animal feed or for composting. The cane juice is sterilized and directed to the fermentation stage. Using the yeast *S. cerevisiae*, which is continuously separated by centrifugation and recycled back to the fermenter, performs the fermentation. Fermentation gases, mostly CO_2, are washed in an absorption column to recover more than 98% of the volatilized ethanol from the fermenter,

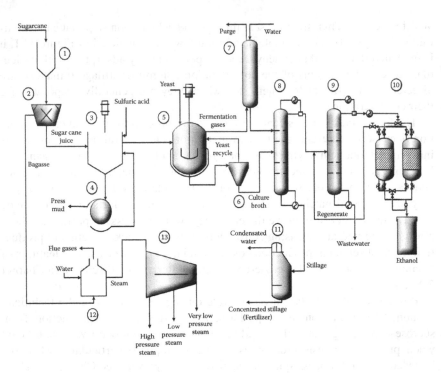

FIGURE 11.2 Simplified flowsheet of fuel ethanol production from sugarcane: (1) washing tank, (2) mill, (3) clarifier, (4) rotary drum, (5) fermenter, (6) centrifuge, (7) ethanol absorber, (8) concentration column, (9) rectification column, (10) molecular sieves. (11) evaporator train, (12) combustor, (13) turbogenerator. (From Quintero, J.A., M.I. Montoya, Ó.J. Sánchez, O.H. Giraldo, and C.A. Cardona. 2008. *Energy* 33 (3):385–399. Elsevier Ltd. With permission.)

and sent to the first distillation column. The culture broth containing 8 to 11% (by weight) ethanol is recovered in a separation step consisting of two distillation columns. In the first (concentration) column, aqueous solutions of ethanol are concentrated up to 63%. In the second (rectification) column, the concentration of the ethanolic stream reaches a composition near the azeotrope (95.6%). The dehydration of this ethanol is achieved through adsorption in vapor phase with molecular sieves by the PSA technology (see Chapter 8, Section 8.2.5). The stream obtained during the regeneration of molecular sieves containing 70% ethanol is recycled to the rectification column.

The stillage treatment consists of an evaporation step allowing the generation of a marketable by-product employed as a fertilizer of cane plantations. If the stillage is not concentrated or evaporated at a low degree, it can be used for both irrigation and fertilization of sugarcane plantations surrounding the ethanol production facility. Hence, the environmental impact of the whole process is reduced since the most important liquid effluent is converted into a value-added product. Condensed water from evaporators and bottoms from the rectification column are collected and sent to the wastewater treatment step. Part of this water can be used as feed water for the co-generation system. Currently, the bagasse obtained is employed in sugar mills and cane-based distilleries for combined generation of the steam and power required by the process. For this, co-generation units have to be installed. These units basically comprise a burner (combustor) for combustion of solid bagasse, a boiler where the feed water is converted into steam, and a turbogenerator (steam turbine), where exhausted steam for the process is obtained along with power. The electricity surplus not consumed by the plant can be sold to the energy network.

The simulation of this process was carried out employing Aspen Plus®. Main input data employed for process simulation are shown in Table 11.1. The simulation considered a production capacity of about 17,830 kg/h anhydrous ethanol. The simulation approach described in Chapter 8, Case Study 8.1 and others, was also applied for this case study. The economic analysis was performed using the Aspen Icarus Process Evaluator® (Aspen Technology, Inc., Burlington, MA, USA) package. This analysis was estimated in US dollars for a 10-year period at an annual interest rate of 16.02% (typical for the Colombian economy), using the straight-line depreciation method and a 33% income tax. The above mentioned software estimates the capital costs of process units as well as the operating costs, among other valuable data, employing the mass and energy balance information provided by Aspen Plus. In addition, specific information regarding the local conditions was used for the economic analysis in the framework of the package utilized. In this way, the net present value (NPV) of the process was determined.

Some simulation results of main streams for the process studied are shown in Table 11.2. The compositions of the streams calculated by simulation, agree very well with those reported for commercial processes. The moisture and fiber contents of bagasse and press mud are close to the contents of moisture (bagasse: 50%, press mud: 75%) and fiber (bagasse: 46%, press mud: 13%) previously reported for these co-products (ETPI, 2003; Moreira, 2000). The value of generated cane stillage per liter of ethanol obtained from simulation (11.01 L/L EtOH) is within the range reported by Wilkie et al. (2000) from experimental data (10 to 20 L/L EtOH). The stillage composition calculated by simulation is close to the stillage composition of Brazilian distilleries, as cited by Sheehan and Greenfield (1980). For instance, the content of organic matter in nonconcentrated stillage is calculated at 26 g/L, while the corresponding average values in Brazilian distilleries using cane juice and cane molasses are 19.5 g/L and 63.4 g/L, respectively. In general, streams data determined through simulation for this processes were compared to available data of existing production facilities taken from literature and personal communications. Hence, the simulation results were satisfactorily validated.

The results obtained for ethanol yield in the process analyzed, along with total operating and capital costs are shown in Table 11.3. For sugarcane in the case of the most productive zone in Colombia (the Cauca River valley), this value is 123 ton/ha for a harvesting time of 13 months (CENICAÑA, 2003). The average yield for all the country, including nontechnified cane crops, reaches 92.7 ton/ha which can be compared to the average yield of sugarcane in Brazil (73.91 ton/ha) and India (59.05 ton/ha; FAO, 2007), the major sugar producers in the world. The calculated ethanol production cost (Table 11.4) is higher than the average production cost of Brazilian ethanol (US$0.198/L in 2007; Xavier, 2007). The price of Brazilian hydrous ethanol could be even lower—about US$0.150/L (Macedo and Nogueira, 2005). This could be explained by the lower cost of the sugarcane in Brazil (about US$0.010/kg in some producing states). As with Brazil, the high productivity of sugarcane, the advantageous output/input energy ratio of the cane-to-ethanol process compared to

TABLE 11.1

Main Process Data for Simulation of Fuel Ethanol Production from Sugarcane

Feature	Value	Feature	Value
Feedstock	Sugarcane	Product	Fuel ethanol
Composition	Sugars 14%[a], fiber 13.5%, protein 0.4%, ash 1.5%, acids and fats 0.6%, moisture 70%	Composition Flow rate	Ethanol 99.5%, water 0.5% 17,822 kg/h
Feed flow rate	292,619 kg/h		
Co-product 1	*Cachaza*	Co-product 2	Concentrated stillage
Pretreatment		Ethanol dehydration	
Milling		Technology	PSA with molecular sieves
Number of mills	2	Number of units	2
Water flow rate	75,640 kg/h	Temperature	116°C
pH conditioning and sucrose hydrolysis		Pressure	1.7 atm (adsorption) 0.14 atm (desorption)
Agent	Dilute H_2SO_4	Cycle time	10 min
Temperature	65°C	Co-generation system	
Residence time	5 min	Solid fuel	Cane bagasse
Number of units	1	Solid fuel flow rate	77,623 kg/h
Sucrose conversion	90%	Flue gases temperature	176°C
Fermentation		Temperature of steam from boiler	510°C
Bioagent	*Saccharomyces cerevisiae*	Pressure of the exhausted steam from turbines	
Temperature	31°C	High	13 atm
Residence time	48 h	Low	4.42 atm
Number of units	16	Very low	1.68 atm
Ethanol content	6%	Stillage concentration	
Conventional distillation		Number of evaporators	5
Number of columns	2	Average area of each evaporation unit	2458 m²
Pressure of columns	1 atm	Involved components	21
Ethanol content at distillate (1st column)	58%	Blocks	34

TABLE 11.1 (*Continued*)
Main Process Data for Simulation of Fuel Ethanol Production from Sugarcane

Feature	Value	Feature	Value
Ethanol content at distillate (2nd column)	90%	Streams	137
		Substreams in streams	3

Source: Quintero, J.A., M.I. Montoya, Ó.J. Sánchez, O.H. Giraldo, and C.A. Cardona. 2008. *Energy* 33 (3):385–399. Elsevier Ltd. With permission.

[a] All the percentages are expressed by weight.

TABLE 11.2
Flow Rates and Composition of Some Streams for Sugarcane-Based Ethanol Process

	Streams					
Compounds	Sugarcane wt.%	Bagasse wt.%	Press mud wt.%	Purge wt.%	Ethanol wt.%	Concentrated stillage wt.%
Ethanol	—	—	—	0.02	99.62	—
Sugars	14.00	1.02	—	—	—	28.84
Fiber	13.50	47.33	16.28	—	—	—
CO_2	—	—	—	98.25	—	—
Protein	0.40	0.12	1.94	—	—	—
Water	70.00	49.90	67.80	1.67	0.38	44.93
Ash	1.50	1.53	7.30	—	—	10.38
Others	0.60	0.10	6.68	0.06	—	15.85
Total flow rate (kg/h)	292,618.77	77,623.30	20,369.78	17,143.62	17,821.67	24,702.60

Source: Quintero, J.A., M.I. Montoya, Ó.J. Sánchez, O.H. Giraldo, and C.A. Cardona. 2008. *Energy* 33 (3):385–399. Elsevier Ltd. With permission.

TABLE 11.3
Ethanol Yields and Total Capital and Operating Costs for Fuel Ethanol Production from Two Feedstocks

Item	Sugarcane	Corn
Ethanol yield (L/ton of feedstock)	77.19	446.51
Ethanol yield (L/[ha*year])	8,764.00	6,698.00
Total capital costs (thous. US$)	75,613.00[a]	36,447.50
Total operating costs (thous. US$/year)	36,255.20	70,670.30

Source: Quintero, J.A., M.I. Montoya, Ó.J. Sánchez, O.H. Giraldo, and C.A. Cardona. 2008. *Energy* 33 (3):385–399. Elsevier Ltd. With permission.

[a]   Includes the cost of the co-generation unit.

TABLE 11.4
Unit Costs of Fuel Ethanol (US$/L of Anhydrous Ethanol)

Item	Corn-Based Process	Share/%	Cane-Based Process	Share/%
Raw materials	0.2911	70.84	0.1611	66.45
Utilities	0.0604	14.70	0.0033	1.35
Operating labor	0.0017	0.41	0.0028	1.14
Maintenance and operating charges	0.0053	1.30	0.0117	4.83
Plant overhead and general and administrative costs	0.0322	7.84	0.0218	8.97
Depreciation of capital [a]	0.0202	4.91	0.0418	17.26
Co-products credit	−0.0728	—	0.0272	—
Total	0.3381	100.00	0.2153	100.00

Source: Quintero, J.A., M.I. Montoya, Ó.J. Sánchez, O.H. Giraldo, and C.A. Cardona. 2008. *Energy* 33 (3):385–399. Elsevier Ltd. With permission.

[a]   Calculated by straight line method.

corn or lignocellulosic biomass, and the low cost of labor force, among other factors, makes this feedstock the more viable option for new ethanol production facilities. The commercialization of the co-products (e.g., press mud and concentrated stillage) allows a substantial economic balance improvement. The data presented in Table 11.5 shows a confirmation of the economic viability of this process.

One of the features of the simulation presented above is the inclusion of the co-generation unit. The simulation of this unit allows performing a more complete environmental evaluation of the overall technological configuration. The co-generation step employs the combustion of cane bagasse to cover the needs of both thermal and electric energy required by the whole ethanol production facility. In the following case study, the specific aspects of the co-generation simulation are presented.

TABLE 11.5

Some Economic Indicators of Two Processes for Fuel Ethanol Production Using Different Feedstocks

Economic Indicator	Ethanol from Sugarcane	Ethanol from Corn
Payout period (years)	4.13	3.85
Net present value (thous. US$)	174,453.00	130,251.00
Internal rate of return (%)	59.48	66.75

Source: Quintero, J.A., M.I. Montoya, Ó.J. Sánchez, O.H. Giraldo, and C.A. Cardona. 2008. *Energy* 33 (3):385–399. Elsevier Ltd. With permission.

CASE STUDY 11.2 SIMULATION OF THE CO-GENERATION PROCESS AND ENVIRONMENTAL EVALUATION OF FUEL ETHANOL PRODUCTION FROM SUGARCANE

The co-generation process has a strong influence on the environmental indicators of the cane-to-ethanol process, as shown in Chapter 10, Case Study 10.3. In order to get a comprehensive procedure to take into account this process, the simulation of the co-generation unit should be appropriately accomplished. In the previous work cited in the last case study (Quintero et al., 2008), this simulation was performed.

The co-generation technology corresponded to a circulating fluidized bed combustor/turbogenerator (CFBC/TG) system. This system has been contemplated in the model process designed by the U.S. National Renewable Energy Laboratory (NREL) for co-generation using the lignin released during the conversion of lignocellulosic biomass into ethanol (Wooley et al., 1999). The CFBC/TG technology offers an increased efficiency in the generation of steam and power related to conventional co-generation units working with low pressure boilers, which usually generate steam of 280° to 300°C and 20 to 21.7 atm (Agüero et al., 2006; Macedo and Nogueira, 2005). Mass balance data of CFBC/TG systems reported in the work of Wooley et al. (1999) were utilized for conceptual design and simulation of the co-generation unit using cane bagasse. This unit was simulated through several process modules of Aspen Plus. The burner was described through a stoichiometric reactor considering the incomplete combustion of bagasse organic components (lignin, cellulose, hemicellulose, etc.), taking into account the formation not only of CO_2, but also of CO. In the same way, reactions for NO_x formation were included. The boiler was studied as a heat exchanger where the feed water enters at 121°C and 97.5 atm and the generated steam exits at 510°C and 84.9 atm. A pump elevating the pressure of the feed water up to 97.5 atm was included in this analysis. The simulation of the co-generation unit also took into account that the combustion gases leaving the boiler can be used not only for preheating the air required to burn the bagasse, but also for drying the wet bagasse generated in the mills. As studied by Maranhao (1982), a previous drying enables the reduction of bagasse moisture down to 40%, improving the combustion and increasing the amount of generated steam. The analysis of the co-generation system also included a cyclone for separating most particulate matter from the flue gases leaving the bagasse dryer. The electricity production using a turbogenerator was simulated through the compressor module of Aspen Plus considering a negative change of pressure and selecting the

TABLE 11.6

Main Atmospheric Emissions from the Co-Generation System Using Sugarcane Bagasse

Pollutant	Emission			Source
	kg/ton bagasse	kg/ton steam	mg/m³	
CO_2	840.6511	335.5544	159,388	This case study
	706.6800	390.0000	—	EPA (1993, 1996)
CO	8.1478	3.2523	1,544	This case study
	—	—	1,526	Gheewala (2005)
NO_x	0.7592	0.3031	144	This case study
	0.5436	0.3000	—	EPA (1993, 1996)
	—	—	92	Gheewala (2005)

Source: Quintero, J.A., M.I. Montoya, Ó.J. Sánchez, O.H. Giraldo, and C.A. Cardona. 2008. *Energy* 33 (3):385–399. Elsevier Ltd. With permission.

isentropic type of compressor. Thus, it is possible to simulate the power generation and calculate the properties of the exhausted steam. In this paper, a multistage turbine was taken into consideration for producing three types of steam: high pressure steam (used for preheating the water feeding the boiler), low pressure steam (used for the energy supply of most process units like heaters, sterilizers, and column reboilers), and very low pressure steam (employed for stillage evaporation).

The performed simulation of the co-generation system for the cane-based process showed a good agreement with the reported industrial data. The CO_2, CO, and NO_x emissions obtained in this work are close to the average emission factors reported by the U.S. Environmental Protection Agency (EPA) and other authors for combustion of bagasse in sugar mills (EPA, 1993, 1996; Gheewala, 2005; Table 11.6). Also, the presence of moisture in the bagasse has an important influence on the amount of thermal energy released during its combustion. This fact is confirmed by simulation. Thus, the use of nonpreheated wet bagasse (about 50% moisture) implies a 7.3% reduction in the amount of produced steam with a pressure of 84.9 atm compared to the case when the bagasse is dried down to 40% moisture. This very high-pressure steam from the boiler is used to power turbines. In contrast, the reduction of available steam reaches 13.18% when low-pressure boilers (29 atm) are used (Maranhao, 1982). The total amount of exhausted steam from the turbogenerator covers all the needs of thermal energy required by the ethanol production facility. This fact dramatically improves the economic performance of the overall process. A fraction of the energy released in the turbogenerator is also employed to cover the needs of mechanical energy required during cane milling. Moreover, the power produced (33 MW) meets plant requirements (13.90 kWh/ton cane, according to simulation results), remaining a significant surplus that can be sold to the electric network. Herein, the electricity surplus was not considered during the economic evaluation. Nevertheless, the co-generation system that was examined generates 99.06 kWh/ton cane of electric energy available for sale. The indicated amount of power is within the range of modern co-generation technologies based on extraction–condensation turbogenerators, which can reach 90 to 150 kWh/ton cane of electricity surplus (Macedo and Nogueira, 2005). This type of co-generation unit has been proposed for new Brazilian sugar mills and distilleries, being an excellent option for new ethanol production facilities in Colombia.

The environmental performance of the technological configuration for fuel ethanol production from sugarcane was carried out applying the WAR (WAste Reduction) algorithm by using the software WAR GUI (WAR graphical user interface). For this, the data on mass and energy balances obtained from the simulation with Aspen Plus were employed. As discussed in Chapter 10, Case Study 10.3, highest weightings (10.0) were assigned to the four local toxicological categories in comparison to the weightings of the four global environmental physical categories (2.5). These assignments were done to give higher importance to the local conditions taking into account that Colombia is an agricultural country with vast hydric resources and rich biodiversity that should be protected. Results using equally weighted categories can be found in a previous work (Montoya et al., 2006).

The total output rate of environmental impact for cane ethanol process is shown in Figure 11.3 and the potential environmental impact generated within the system is shown in Figure 11.4. The improved environmental performance (in the terms of potential environmental impact (PEI) leaving the system per mass of products) of the sugarcane-based process can be explained by the operation of the co-generation unit. In this unit, one of the process by-products, the bagasse, is employed as renewable fuel in order to generate all thermal and mechanical energy required by the process as well as the power needed. When bagasse is burned, the negative effect of CO_2 released into the atmosphere during its combustion is compensated positively by the CO_2 fixed from the atmosphere during the sugarcane growth. Therefore, the utilization of bagasse as a solid biofuel does not necessarily imply a net increase of

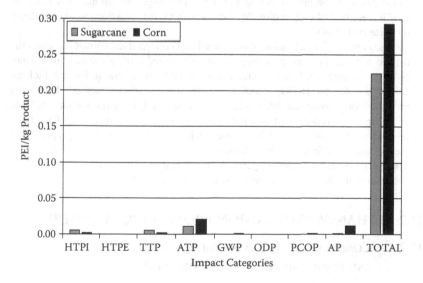

FIGURE 11.3 Total output rate of environmental impact for the studied processes. The impact is expressed as the PEI leaving the system per mass of product streams. (From Quintero, J.A., M.I. Montoya, Ó.J. Sánchez, O.H. Giraldo, and C.A. Cardona. 2008. *Energy* 33 (3):385–399. Elsevier Ltd. With permission.)

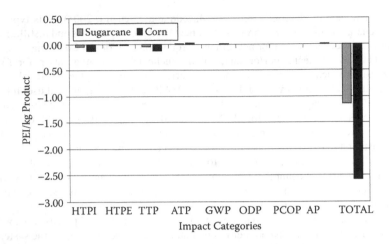

FIGURE 11.4 Potential environmental impact generated within the studied processes. The impact is expressed as the PEI generated within the process per mass of product streams. (From Quintero, J.A., M.I. Montoya, Ó.J. Sánchez, O.H. Giraldo, and C.A. Cardona. 2008. *Energy* 33 (3):385–399. Elsevier Ltd. With permission.)

CO_2 in the atmosphere. However, the environmental impact of the involved mass and energy flows in the whole life cycle is not assessed by the WAR GUI software. It only performs the evaluation of the potential environmental impact of the streams leaving or generated during the conversion process. To overcome the problem of CO_2 emissions, the flue gases stream leaving the co-generation unit was analyzed as if it were free of carbon dioxide. Thus, zero CO_2 net emissions were considered on a life cycle basis.

The commercial utilization of press mud and concentrated stillage indicates the drastic reduction of two polluting streams generated during the cane-to-ethanol process. Stillage is a highly contaminating liquid stream due to its high biological oxygen demand (BOD; 30,000 to 60,000 mg/L). Several methods have been proposed for its treatment, but few have been employed. The evaporation of stillage to obtain a co-product used as a fertilizer for cane plantations is one of the most popular applications, along with its use for irrigation. However, the properties of the soil may be affected by the utilization of huge volumes of concentrated stillage. These effects have not been properly studied within the framework of the life cycle assessment methodology and further studies are required.

11.2 ETHANOL PRODUCTION FROM STARCHY MATERIALS

11.2.1 Configuration Involving the Separate Hydrolysis and Fermentation (SHF) of Corn Starch

Starch is a high yield feedstock for ethanol production. Glucose is obtained by the hydrolysis of starch. Then, the glucose solution undergoes fermentation toward ethanol. From each 100 g of starch, 111 g of glucose theoretically can be obtained, which implies a stoichiometric ratio of 9:10. The output/input ratio of energy for

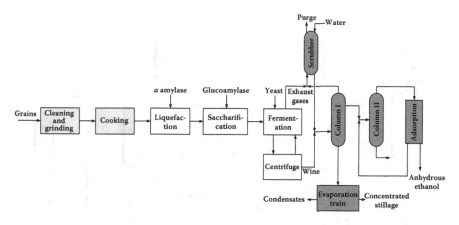

FIGURE 11.5 Simplified diagram for production of fuel ethanol from cereal grains by dry-milling technology.

corn ethanol is in the range of 1.1 to 1.2 (Prakash et al., 1998; Sánchez and Cardona, 2005). Fuel ethanol production from materials with a high content of starch needs some additional process steps compared to the process that employs sucrose-based materials as the feedstock. The process flowsheet comprises a pretreatment step of the starchy materials to make starch more susceptible to enzymatic hydrolysis. In this step, known as liquefaction, starch is partially hydrolyzed at a high temperature. In the following step of saccharification, liquefied starch undergoes a deeper hydrolysis where fermentable sugars (glucose) are obtained (Figure 11.5). After glucose fermentation, the process does not significantly differ from that one that employs materials with a high content of sucrose. Nevertheless, depending on the specific type of employed starchy feedstock, certain co-products used for animal feed can be produced during the evaporation of stillage.

When cereals are used as the feedstock for producing fuel ethanol, the raw material enters the process as grains, which should undergo cleaning and milling. Either wet-milling or dry-milling processes can carry out grain milling of such cereals as corn, wheat, and barley. The wet-milling process implies that only the starch enters fuel ethanol production process, as discussed in Chapter 4, Section 4.2. In this process, all the components of the kernel should be separated prior to the cooking step. These components represent value-added products that are used for food and feed. Moreover, part of the produced starch can be deviated to the production of sweeteners, such as the high fructose corn syrup (HFCS). During the dry milling of grains, the whole kernel enters the ethanol production line, which means that all its components are processed along with starch. Nonutilized components of the kernel are built up in the bottoms of the first distillation column and are concentrated to form a product utilized as animal feed. In general, the liquefaction, saccharification, and fermentation steps are the same for both types of technologies.

The overall production process of bioethanol from corn by the dry-milling technology includes the breakdown of this polysaccharide to obtain an appropriate

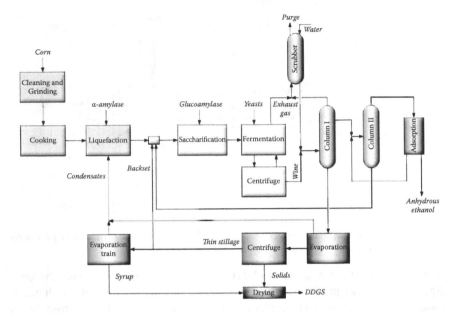

FIGURE 11.6 Technological configuration of the dry-milling process for production of fuel ethanol from corn grains by separate hydrolysis and fermentation (SHF). The scheme includes the production of DDGS.

concentration of fermentable sugars, which are transformed into ethanol by yeasts (Figure 11.6). After washing, crushing, and milling the corn grains, the starchy material is gelatinized in order to dissolve the amylose and amylopectin (cooking step). This process is accomplished with the help of a jet cooker working at 110°C. In dissolved form, starch is accessible for enzymatic attack in the following step, the liquefaction that is carried out at 88°C. This step is considered a pretreatment process because of the partial hydrolysis of the starch chains using thermostable bacterial α-amylase that yields a starch hydrolyzate with a hydrolysis degree of approximately 10%. The obtained hydrolyzate has a reduced viscosity and contains oligomers, such as dextrins.

Then, this liquefied starch enters the saccharification process where it is hydrolyzed by microbial glucoamylase to produce glucose. This process operates at 60°C. The saccharified starch is cooled and sent to the next step where it is fermented by the yeast *S. cerevisiae* and converted into ethanol at 30°C. Fermentation gases, mostly CO_2, are washed in an absorption column to recover more than the 98% of the volatilized ethanol from the fermenter and sent to the first distillation column. The culture broth containing 11% (by weight) ethanol is recovered in a separation step consisting of two distillation columns. In the first (concentration) column, aqueous solutions of ethanol are concentrated to 50%. In the second (rectification) column, the concentration of the ethanolic stream reaches a composition near the azeotrope (95.6%). The dehydration of the ethanol is achieved through adsorption in a vapor phase with molecular sieves. The stream obtained during the regeneration of molecular sieves containing 70% ethanol is recycled to the rectification column.

The stillage from the concentration column is evaporated and the obtained solids are separated by centrifugation. These solids are dried for producing the distiller's dried grains with solubles (DDGS), a co-product used for animal feed due to its high content of proteins and vitamins. The remaining liquid or thin stillage is evaporated in a double effect evaporator. The obtained syrup is combined with the DDGS and dried. The condensed water from the evaporators is recirculated to the liquefaction stage, while the bottoms of the rectification column and one fraction of the thin stillage (backset) are recycled back to the saccharification step.

11.2.2 CONFIGURATION INVOLVING THE SIMULTANEOUS SACCHARIFICATION AND FERMENTATION (SSF) OF CORN STARCH

New tendencies in the corn-to-ethanol industry are aimed at dry-milling processes. The increase in ethanol production capacity in the United States is mainly represented by dry-milled corn ethanol plants (Sánchez and Cardona, 2008; Tiffany and Eidman, 2003). In this regard, the simultaneous saccharification and fermentation (SSF), a reaction–reaction integrated process, has been proving its potential for improving the overall process, as was discussed in Chapter 9, Section 9.2.2.1. This process only differs from the SHF process in that the saccharification and fermentation are carried out simultaneously in the same unit at 33°C (Figure 11.7). The remaining steps are similar for both processes.

CASE STUDY 11.3 SIMULATION OF FUEL ETHANOL PRODUCTION FROM CORN BY SSF AND COMPARISON TO SUGARCANE-BASED PROCESS

As mentioned in Case Study 11.1, the aim of simulating the process to obtain bioethanol from dry-milled corn can provide valuable information on the suitability of different technological configurations, in this case, the configuration involving the SSF of the starch contained in the corn grain. This simulation was performed in previous works (Cardona et al. 2005b; Quintero et al. 2008) under Colombian conditions and corresponds to the scheme depicted in Figure 11.8. As in the case of cane-to-ethanol conversion, the downstream processes are practically the same (see Case Study 11.1). The differences lie in the biological transformation step. After washing, crushing, and milling the corn grains (dry-milling process), the starchy material is gelatinized in order to dissolve the amylose and amylopectin. In dissolved form, starch is accessible for enzymatic attack in the following liquefaction step. In this step, a partial hydrolysis (about 10%) of the starch chains using thermostable bacterial α-amylase is achieved. The hydrolyzate obtained has reduced viscosity and contains starch oligomers called dextrins. Then, the liquefied starch enters the SSF process where it is hydrolyzed by microbial glucoamylase to produce glucose. This sugar is immediately assimilated by the yeast S. cerevisiae in the same reactor and converted into ethanol. As mentioned above, the dry-milling technology allows the production of DDGS using the recovered solids from the bottoms of the concentration column. In this case, no co-generation is contemplated;

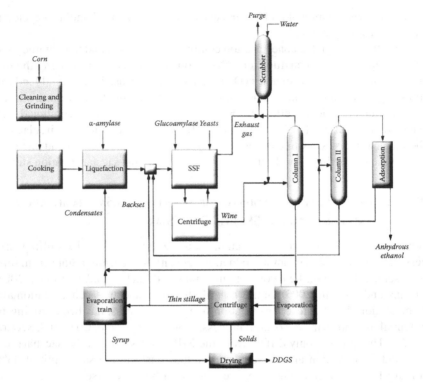

FIGURE 11.7 Technological configuration of dry-milling process for production of fuel ethanol from corn grains by simultaneous saccharification and fermentation (SSF).

thus, the acquisition and utilization of fossil fuels to supply the steam for the process are required.

The simulation of this process was carried out using Aspen Plus as well. Main input data employed for process simulation are shown in Table 11.7. As in the case of cane ethanol, the simulation used a production capacity of about 17,830 kg/h anhydrous ethanol. The simulation approach described in Chapter 8, Case Study 8.1 and others was also applied for this case study. Enzymatic hydrolysis and continuous fermentation processes were simulated based on a stoichiometric approach that considered the conversion of starch into glucose as well as the transformation of glucose into cell biomass, ethyl alcohol, and other fermentation by-products. The economic analysis was performed by using the Aspen Icarus Process Evaluator package and the same local conditions of Case Study 11.1.

Some simulation results of main streams for this process are shown in Table 11.8. The compositions of the streams calculated by simulation agree very well with those reported for commercial processes. The DDGS generally contains 9% moisture and 27 to 32% protein (McAloon et al., 2000). Results of many analyses done during a five-year period (1997 to 2001) to determine the composition of DDGS obtained in corn dry-milling ethanol production facilities in the United States (Belyea et al., 2004) revealed good agreement with the simulation data obtained (see Table 11.8). The results obtained for ethanol yield in the process analyzed, along with total operating and capital costs, are shown in Table 11.3. In this regard, the average

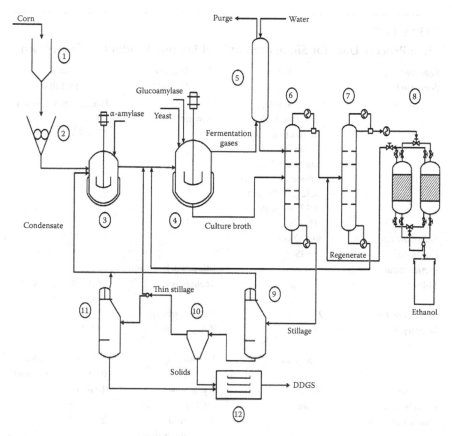

FIGURE 11.8 Simplified flowsheet of fuel ethanol production from corn: (1) washing tank, (2) crusher, (3) liquefaction reactor, (4) SSF reactor, (5) ethanol absorber, (6) concentration column, (7) rectification column, (8) molecular sieves, (9) first evaporator train, (10) centrifuge, (11) second evaporator train, (12) dryer. (From Quintero, J.A., M.I. Montoya, Ó.J. Sánchez, O.H. Giraldo, and C.A. Cardona. 2008. *Energy* 33 (3):385–399. Elsevier Ltd. With permission.)

yield of technified corn crop in Colombia is about 5 ton/ha for a harvesting time of four months (Quintero et al., 2004). This yield is lower than the corn yield in the United States, the major corn producer in the world. Note from Table 11.3 that the calculated ethanol yield from corn (in terms of produced ethanol per tonne of feedstock entering the plant) is greater than that from sugarcane because of the higher amount of fermentable sugars (glucose) that may be released from the original starchy material. However, the annual ethanol yield from each hectare of cultivated corn is 23.6% lower than that for sugarcane. This preliminary fact shows the comparative advantage of using sugarcane as feedstock for ethanol production under high-productivity conditions for cane cropping in Colombia.

Total operating costs are significantly higher for ethanol production from corn than from sugarcane under Colombian conditions (Table 11.3). This is mostly explained by the feedstock cost, as shown in Table 11.4, where operating costs were

TABLE 11.7

Main Process Data for Simulation of Fuel Ethanol Production from Corn

Feature	Value	Feature	Value
Feedstock	Corn	Product	Fuel Ethanol
Composition	Starch 60.6%[a], cellulose 3.46%, hemicellulose 4.6%, lignin 0.4%, glucose 8.7%, protein 2.2%, fatty acids 3.64%, ash 1.17, moisture 15.5%	Composition Flow rate	Ethanol 99.5%, water 0.5% 17,837 kg/h
Feed flow rate	50,630 kg/h		
Co-product	DDGS		
Pretreatment		Ethanol dehydration	
Milling		Technology	PSA with molecular sieves
Number of mills	2	Number of units	2
Hydrolysis (liquefaction)		Temperature	116 °C
Bioagent	α-amylase	Pressure	1.7 atm (adsorption) 0.14 atm (desorption)
Temperature	88°C	Cycle time	10 min
Residence time	5 min	DDGS processing	
Number of units	6	Number of evaporator trains	2
Starch conversion	99%	Number of evaporators (1st train)	2
Simultaneous saccharif. and fermentation		Number of evaporators (2nd train)	2
Bioagent	Glucoamylase and *Saccharomyces cerevisiae*	Average area of each evaporation unit (1st train)	1,186 m²
Temperature	33°C	Average area of each evaporation unit (2nd train)	42 m²
Residence time	48 h	Type of dryer	Indirect contact rotary dryer
Number of units	10		
Ethanol percentage	11%		

TABLE 11.7 (*Continued*)
Main Process Data for Simulation of Fuel Ethanol Production from Corn

Feature	Value	Feature	Value
Feedstock	Corn	Product	Fuel Ethanol
Conventional distillation			
Number of columns	2		
Pressure of columns	1 atm	Involved components	19
Ethanol content at distillate (1st column)	63%	Blocks	23
Ethanol content at distillate (2nd column)	90%	Streams	83
		Substreams in streams	3

Source: Quintero, J.A., M.I. Montoya, Ó.J. Sánchez, O.H. Giraldo, and C.A. Cardona. 2008. *Energy* 33 (3):385–399. Elsevier Ltd. With permission.

a All the percentages are expressed by weight.

TABLE 11.8
Flow Rates and Composition of Some Streams for Corn-Based Ethanol Process

	Streams			
Compounds	Corn (wt.%)	Purge (wt.%)	Ethanol (wt.%)	DDGS (wt.%)
Ethanol	—	0.05	99.50	—
Sugars	2.19	—	—	1.96
Starch	60.59	—	—	0.17
Fiber	8.21	—	—	33.21
CO_2	—	98.13	—	—
Fats	3.64	—	—	14.70
Protein	8.69	—	—	35.15
Water	15.50	1.81	0.50	9.82
Ash	1.18	—	—	4.76
Others	—	0.01	—	0.23
Total flow rate (kg/h)	50,629.99	17,247.83	17,836.83	12,483.97

Source: Quintero, J.A., M.I. Montoya, Ó.J. Sánchez, O.H. Giraldo, and C.A. Cardona. 2008. *Energy* 33 (3):385–399. Elsevier Ltd. With permission.

disaggregated. In comparison to corn, the greater cane demand for producing the same amount of ethanol (about six times the grain requirements) is compensated for by the lower cost of this raw material. In fact, the main part of fuel ethanol costs corresponds to the feedstock: 66.45% and 70.84% using sugarcane and corn, respectively. Usually the feedstock cost for Brazilian cane ethanol is about 60% of the production costs (Xavier, 2007), whereas for corn (mostly transgenic) the cost is about 63% in the United States (McAloon et al., 2000). These results confirm the validity of the data obtained by simulation, as well as the assumptions considered during the economic analysis. Steam and power generation through the combustion of cane bagasse reduces the utilities cost considerably. This makes a big difference in cane-to-ethanol processes and justifies the installation and operation of bagasse combustion systems. On the contrary, a corn-based process requires the consumption of fossil fuels that negatively affects its operating costs and environmental performance. Total capital costs are lower for the corn process (Table 11.3), even though it has a more complex configuration involving an additional enzymatic hydrolysis step. For the cane-based process and due to the higher amount of feedstock to be handled, a greater capacity of equipment is required. In addition, the co-generation system increases the required investment for such types of configurations. Nevertheless, the possibility of selling electricity contributes to offset these additional expenses.

The production costs structure of one liter of ethanol produced from corn (as shown in Table 11.4) is comparable to the costs structure estimated by the NREL for the mature process of ethanol production from corn via dry-milling technology in the United States. In the latter case, ethanol production costs reach US$0.232 per liter of anhydrous ethanol (McAloon et al., 2000). The main difference in the production costs are mostly explained by the higher corn prices in Colombia related to cheaper U.S. corn (0.076 US$/kg). In fact, the utilization of imported corn from the United States as feedstock for new ethanol production facilities located on the Colombian Caribbean coast has been proposed by different organizations including the Colombian government. In any case, the volatility of corn prices is a crucial factor to be accounted for. Production costs obtained in this case study are very close to those reported by Macedo and Nogueira (2005) for ethanol production from milo (a kind of sorghum) in the United States. It should be noted that co-product (DDGS) sales in corn ethanol production play a significant role in process sustainability.

The confirmation of the economic viability of the two analyzed processes is presented in Table 11.5. In relation to their profitability indicators, both processes are comparable although the production costs for sugarcane are clearly smaller. Moreover, ethanol from cane offers a higher NPV with a lower internal rate of return (IRR). Different evaluations simulating changes in the price of the main feedstock show that the process employing sugarcane is much more stable to these kinds of variations. Thus, a 71% increase in the price of corn (likely to occur under Colombian conditions) leads to negative NPV during the lifetime of the project. In contrast, this same increase in the price of sugarcane (whose price is much more stable in Colombia) only leads to a 38.7% reduction in NPV and 28.4% decrease in IRR. These results allow the conclusion that ethanol production process from sugarcane represents the best investment possibility under Colombian conditions.

CASE STUDY 11.4 COMPARISON OF THE ENVIRONMENTAL PERFORMANCE FOR SUGARCANE ETHANOL AND CORN ETHANOL

Using the simulation results from the last case study, the environmental assessment of bioethanol production from dry-milled corn employing the SSF process was carried out applying the WAR algorithm (Quintero et al., 2008). The results obtained were compared to the outcomes corresponding to the process for ethanol production from sugarcane.

The total output rate of environmental impact for corn ethanol process is shown in Figure 11.3 and the potential environmental impact generated within the system is shown in Figure 11.4. It is evident that ethanol production from sugarcane has also lower impact on the environment compared to the corn process: the latter process exhibits a higher PEI per mass of products. This environmental indicator is the index to be considered for the overall evaluation of both processes. On the other hand, the corn-based process has a more negative generated PEI meaning that the PEI of the substances entering the system is reduced by their transformation into other less dangerous compounds. For the sugarcane process, the higher value of generated PEI (although negative) indicates that the conversion of entering substances also occurs, but to a lesser degree. This feature can be related to the fact that this process requires the input of a greater amount of feedstock. The commercialization of DDGS implies the elimination of the polluting stillage stream.

The four case studies presented above demonstrate the power of the simulation approach employed. In order to obtain a general framework for comparing different technological configurations for ethanol production from diverse feedstocks according to technoeconomic and environmental criteria, the formulation of an overall comparison criterion is required. This is shown in the following case study.

CASE STUDY 11.5 COMBINED EVALUATION OF TECHNOECONOMIC AND ENVIRONMENTAL PERFORMANCE OF TWO BIOETHANOL PRODUCTION PROCESSES

The evaluation considering economic and environmental indexes of the two simulated processes for ethanol production from sugarcane and corn under Colombian conditions was carried out to obtain a combined index useful for the selection of the most appropriate technology as formulated in previous works (Quintero et al., 2008; Sánchez, 2008). This combination was done by following the procedure developed by Chen et al. (2002) where the economic and environmental objectives were aggregated into a single objective function using the analytic hierarchy process (AHP) approach. The AHP is one variant of multicriteria analysis that uses a number of pair-wise comparisons between quantitative or qualitative criteria to assess the relative importance of each criterion. These comparisons can be arranged in a hierarchical manner to form sets of attributes, and qualities (levels) within these attributes (Hussain et al., 2006).

The hierarchical structure for this case study is shown in Figure 11.9. Once mass and energy balances have been calculated by simulation, the economic and environmental evaluations are performed by using the corresponding tools (process

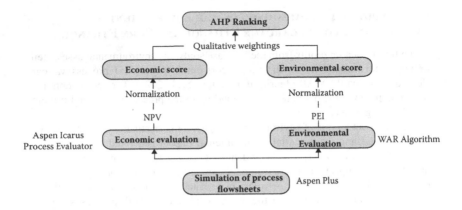

FIGURE 11.9 Analytical hierarchy structure (AHP) employed for the analysis of two processes for fuel ethanol production.

evaluation software and WAR algorithm, respectively). The indexes (NPV and PEI) for each process are determined from these evaluations. Alternatively, other economic indexes, such as IRR, can be used. The indexes are normalized, so that they do not exceed a normalization value, and converted to quantitative scores. The normalization value for each index was calculated as the sum of the index values in both processes. The economic score of a given process was determined as the ratio between the NPV of the process and the corresponding normalization value, i.e., the sum of the two calculated NPVs. The environmental score was calculated taking the difference between the corresponding normalization value and the PEI of the process and dividing the result obtained by the normalization value. The AHP score of a process design represents the sum of the products of the average process score for a given attribute and the weighting for that attribute, that is:

$$AHP = P_{Ecn} \cdot weight + P_{Env} \cdot weight \qquad (11.1)$$

where P_{Ecn} is the normalized economic score calculated from the NPV of the two analyzed processes, and P_{Env} is the normalized environmental score resulting from the PEI values of the two processes. The qualitative weightings of economic and environmental attributes were taken as 0.82 and 0.18, respectively. These values are suggested by Chen et al. (2002) who applied them to several chemical processes based on the survey carried out by Dechanpanya (1998) on the comparison of economic and environmental attributes for a chemical process from several faculty members and graduate students at Michigan Technological University using the AHP approach (Hussain et al. 2006).

The results obtained from the integration of the economic and environmental assessments following the proposed methodological approach are presented in Table 11.9. From these results, the sugarcane-based process exhibits a higher AHP score, indicating that it has better performance than the corn-based process when both economic and environmental criteria are analyzed in a combined manner. Changes in the evaluation of the AHP when different weightings are selected for both processes show that the sugarcane-based process will always have a better

TABLE 11.9

Results of Environmental and Economic Integration of the Fuel Ethanol Production Process Using Two Feedstocks

Feedstock	NPV/thous. US$	P_{Ecn}	PEI	P_{Env}	AHP
Sugarcane	174,453.00	0.573	0.224	0.567	0.571
Corn	130,251.00	0.427	0.293	0.433	0.429

Source: Quintero, J.A., M.I. Montoya, Ó.J. Sánchez, O.H. Giraldo, and C.A. Cardona. 2008. *Energy* 33 (3):385–399. Elsevier Ltd. With permission.

performance with any value of the economic weighting. When this weighting is increased (this is equivalent to the reduction of the environmental weighting), the AHP score will also increase. In contrast, the corn-based process shows slightly worse performance when the economic attribute has a higher weight. This means that the economic advantages for the sugarcane process are actually better than those displayed for the corn scenario. Therefore, assigning weightings for this case study does not affect the final qualitative result of the combined evaluation within the AHP framework. Thus, the process that employs sugarcane for producing fuel ethanol shows better performance.

The procedure proposed to analyze different flowsheet configurations proved to be a useful methodology for process synthesis and can support the decision making for further experimental studies at pilot scale and industrial levels. This is a key issue considering the limited resources for extensive and long-term research in such countries as Colombia.

11.2.3 CONFIGURATION FOR PRODUCTION OF CASSAVA ETHANOL

Cassava represents an important alternative source of starch not only for ethanol production, but also for production of glucose syrups. In fact, cassava is the tuber that has gained the most interest due to its availability in tropical countries, being one of the top 10 more important tropical crops. Ethanol production from cassava can be accomplished using either the whole cassava tuber or the starch extracted from it (Sánchez and Cardona, 2008). Starch extraction can be carried out through a high-yield, large-volume industrialized process as the Alfa Laval extraction method (FAO and IFAD, 2004), or by a traditional process for small- and midscale plants (see Chapter 3, Section 3.2.2.3). This process can be considered as the equivalent of the wet-milling process for ethanol production from corn. The production of cassava with high starch content (85 to 90% dry matter) and less protein and minerals content is relatively simple. Cassava starch has a lower gelatinization temperature and offers a higher solubility for amylases in comparison to corn starch. The hydrolysis of cassava flour has been proposed for glucose production in an enzymatic hollow-fiber reactor with 97.3%

conversion (López-Ulibarri and Hall, 1997), considering that cassava flour production is simpler and more economic than cassava starch production. However, it is considered that cassava ethanol would have better economic indicators if the whole tuber were used as feedstock, especially when small producers are involved (Sánchez and Cardona, 2008). Fuel ethanol production from whole cassava is equivalent to ethanol production from corn by dry-milling technology. For this, cassava should be transported as soon as possible from cropping areas because of its rapid deterioration due to its high moisture content (about 70%). Hence, this feedstock should be processed within three to four days after its harvest. One of the solutions to this problem consists in the use of sun-dried cassava chips (Sriroth et al., 2007). The farmers send the cassava roots to small chipping factories where they are peeled and chopped into small pieces. The chips are sun-dried for two to three days. The final moisture content is about 14% and the starch content reaches 65%.

The first step of the process in the distillery is the grinding of the dried cassava chips or fresh roots (if a permanent supply is ensured; Sánchez and Cardona, 2008). Milled cassava is mixed with water and undergoes cooking followed by the liquefaction process (Nguyen et al., 2008). Liquefied slurry is saccharified to obtain the glucose, which will be assimilated by the yeast during the next fermentation step. The process can be intensified through the SSF as in the case of corn (Figure 11.10). If fresh roots are employed, a fibrous material is obtained in the stillage after distillation. This material can be used as an animal feed similar to the DDGS produced in the corn-based process. The wastewater can be treated by anaerobic digestion to produce biogas, which can be used to produce steam and power for the process. Nevertheless, the amount of steam generated is not enough to cover the needs of the process. Hence, natural gas or other fossil fuel is required (Dai et al., 2006).

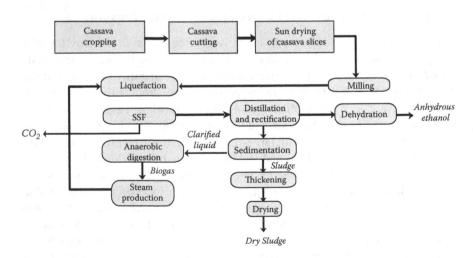

FIGURE 11.10   Simplified diagram for fuel ethanol production from cassava.

CASE STUDY 11.6   COMPARISON OF FUEL ETHANOL
PRODUCTION FROM CORN OR CASSAVA

Through process simulation, the yield of an ethanol production facility from cassava can be assessed. This acquires greater importance considering the lack of information published about this type of process. In a previous work (Cardona et al., 2005a), the performance of two processes for bioethanol production employing two different starchy feedstocks using a commercial simulation package (Aspen Plus) and especially developed models were compared. For this, a SSF was contemplated in both processes. For corn, dry-milling technology and an organization of streams, such as that shown in Figure 11.7, was considered. This flowsheet allows the production of a valuable co-product, the DDGS. In the case of cassava, the utilization of fresh roots was considered. This implies the generation of a fibrous residue.

The comparison results are summarized in Table 11.10. Presented data show that the higher yields correspond to the corn due to its higher starch content measured in wet basis. These data are close to ethanol yields from corn and cassava reported by the FAO (for instance, these data were published by Observatorio Agrocadenas Colombia, 2006). The production and protein content of the material to be used as a protein supplement in animal feed in the corn case (DDGS) are higher than in the case of cassava (fibrous residue) due to its low protein content. On the other hand, the high moisture content of the cassava implies that the condensates of the stillage evaporation step cannot be recycled to the cooking step. In fact, this explains the need for employing a higher amount of feedstock to achieve the same starting starch mass as in the case of corn. However, for some countries, cassava represents a better option considering the high agronomic yields compared to corn. For instance, the cassava yield in Colombia reaches 20 ton/ha while average corn yield only reaches 6 ton/ha (Observatorio Agrocadenas Colombia, 2006). This situation would represent ethanol yields per hectare of 3,336 L EtOH/ha cassava and 2,105 L EtOH/ha corn. These estimations clearly favor the utilization of the tuber especially if the agroecological conditions of this country are taken into account.

TABLE 11.10

Comparison of Two Starch-Containing Feedstocks for Fuel Ethanol Production in a Process Involving the SSF

Feedstock	Mass Flow of Feedstock/ kg/h	Produced EtOH/ kg/h	Ethanol Yield/L/ton feedstock	Produced DDGS/ kg/h	DDGS Yield/kg/ ton feedstock	Protein Content in DDGS/%
Corn	50,630	13,589.76	350.85	12,023.38	237.48	36.59
Cassava	115,755	14,771.96	166.80	4,084.91	35.29*	22.66[a]

Note: The mass flowrate of starch in each process is 30,675 kg/h.
[a] Solids equivalent to DDGS.

11.3 ETHANOL PRODUCTION FROM LIGNOCELLULOSIC MATERIALS

Numerous studies for developing large-scale production of ethanol from lignocellulosic biomass have been carried out in the world. One of the advantages of the use of lignocellulosic biomass is that this feedstock is not directly related to food production, which would implement the extra production of bioethanol without the need of employing vast extensions of cultivable land for cane or corn production. In addition, lignocellulosics is a resource that can be processed in different ways for the production of many other products, such as synthesis gas, methanol, hydrogen, and electricity (Chum and Overend 2001). However, the main limiting factor is the higher degree of complexity inherent to the processing of this feedstock. This complexity is related to the nature and composition of lignocellulosic biomass (see Chapter 3, Section 3.3.1). Two of the main biomass polymers need to be broken down into fermentable sugars in order to be converted into ethanol or other valuable products. But this degradation process is complicated, energy-consuming, and not completely developed. Consequently, the involved technologies are more complex leading to higher ethanol production costs compared to cane, beet, or corn. However, the fact that many lignocellulosic materials are by-products of agricultural activities, industrial residues, or domestic wastes offers huge possibilities for the production of fuel ethanol at a large scale as well as its global consumption as a renewable fuel. It is thought that lignocellulosic biomass will become the main feedstock for ethanol production in the near future (Cardona and Sánchez, 2007). According to Berg (2001), the output/input ratio of energy for the production of lignocellulosic ethanol reaches a value of 6, indicating a better energy efficiency than in the case of corn ethanol.

The classic configuration employed for converting lignocellulosic biomass into ethanol involves a sequential process in which the hydrolysis of cellulose and the fermentation are carried out in different units. As mentioned in Chapter 7, Section 7.1.4.1, the main feature of this configuration known as separate hydrolysis and fermentation (SHF) is that optimum conditions of pH and temperature for both processes can be ensured in an independent way. The general flowsheet of this technology is illustrated in Figure 7.6. Depending on the pretreatment method, the lignin can be recovered in this step or remain in the stillage from which it can be burned to generate steam. The solid fraction obtained in the pretreatment reactor is sent to the hydrolysis bioreactor where it comes in contact with microbial cellulases. This scheme involves the fermentation of the hemicellulose hydrolyzate contained in the liquid fraction exiting the pretreatment reactor using pentose-assimilating yeasts (like *Candida shehatae* or *Pichia stipitis*) in a parallel way to the glucose fermentation carried out with *S. cerevisiae*. In the alternative variant involving the simultaneous saccharification and fermentation (SSF), the hydrolysis and fermentation are performed in a single unit as discussed in Chapter 9, Section 9.2.2.3. The most employed microorganism for fermenting the hydrolyzates of the lignocellulosic biomass is *S. cerevisiae*, which ferments the hexoses contained in the hydrolyzate, but not the pentoses. This configuration is depicted in Figure 9.4.

11.3.1 Process Flowsheet Development for Production of Biomass Ethanol

Unlike the design of fuel ethanol production processes from sucrose- or starch-containing feedstocks, the use of lignocellulosic materials entails the analysis of multiple alternative variants for accomplishing the conversion of biomass into ethanol. This means that process synthesis procedures play a significant role during the design of processes based on biomass. In fact, there exists a significant variety of biomass conversion technologies proposed worldwide to produce fuel ethanol. This variety should be thoroughly assessed in order to select the most suitable technological configuration considering the local conditions and the process performance in terms of its technical, economic, and environmental effectiveness.

Many efforts have been put forth worldwide to improve the efficiency of biomass-to-ethanol conversion. In the United States, the production of ethanol from lignocellulosic biomass is being studied intensively. Ingram et al. (1999) have carried out significant research on the development of recombinant strains of enteric bacteria to use for cellulosic ethanol production. Current technology allows the use of a genetically engineered *Escherichia coli* strain with the natural ability of assimilating both pentoses and hexoses found in the liquid fraction resulting from the dilute-acid pretreatment of lignocellulosic biomass. The main *Zymomonas mobilis* genes encoding the ability for homofermentative production of ethanol have been integrated into the bacterial chromosome in this strain. The solid fraction from this pretreatment that contains cellulose and lignin undergoes SSF using a recombinant strain of *Klebsiella oxytoca* with the ability to ferment cellobiose and cellotriose, eliminating the need for supplementing *Trichoderma reesei* cellulases with β-glucosidase. This strain also has the genes to encode the production of ethanol. The proposed overall process can be observed in Figure 11.11a (Cardona and Sánchez, 2007). At present, research efforts are oriented toward the development of a single microorganism capable of efficiently fermenting both hemicellulosic and cellulosic substrates that will make possible the development of the direct conversion of biomass into ethanol.

The model process designed by the NREL comprises a previous hydrolysis of wood with dilute acid followed by a simultaneous saccharification co-fermentation (SSCF) process using cellulases produced *in situ* by genetically engineered *Z. mobilis* with the ability to transform both glucose and xylose into ethanol (Figure 11.11b; Cardona and Sánchez, 2007). The process is energetically integrated using the heat generated during the combustion of methane formed in the anaerobic treatment of wastewater from pretreatment and distillation steps (Wooley et al., 1999). In addition, the burning of lignin allows the production of energy for the process and a surplus in the form of electricity. The production of one liter of ethanol by this process is calculated at US$0.396, whereas the ethanol production cost from corn is US$0.232 (McAloon et al., 2000). A pilot plant designed for conversion of lignocellulosic biomass into ethanol was built and operated with the aim of supporting industrial partners for the research and development of biomass ethanol technology (Nguyen et al., 1996). In this plant, tests in continuous regime

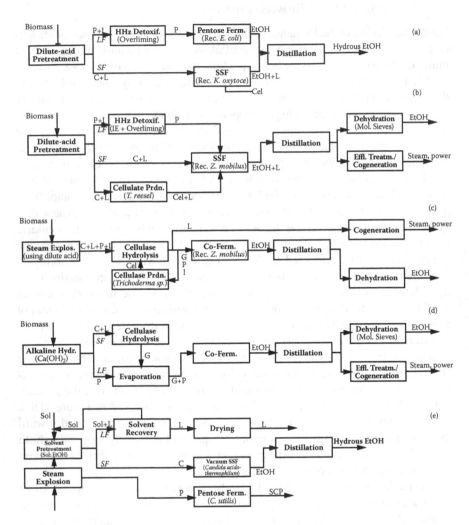

FIGURE 11.11 Some proposed flowsheet configurations for ethanol production from lignocellulosic biomass. (a) Process based on utilization of enteric bacteria (Ingram et al., 1999). (b) NREL model process (Wooley et al., 1999). (c) Iogen's process (Tolan, 2002). (d) Process proposed by Reith et al., (2002). (e) IIT Delhi process (Ghosh and Ghose, 2003). Main stream components: C = cellulose, L = lignin, G = glucose, P = pentoses, I = inhibitors, Cel = cellulases, EtOH = ethanol, Sol = solvent, SCP = single cell protein. LF = liquid fraction, SF - solid fraction, HHZ = hemicellulose hydrolyzate, Rec = recombinant, IE = ion exchange. (From Cardona, C.A., and Ó.J. Sánchez. 2007. *Bioresource Technology* 98:2415–2457. Elsevier Ltd. With permission.)

for the utilization of lignocellulosic residues of low cost and great availability, such as corn fiber, were carried out (Schell et al., 2004). The objective of these tests was the assessment of the operation of the integrated equipments and the generation of data concerning the process performance. This type of plant allows the acquisition of valuable experience considering the future implementation of the industrial process, the same as the feedback of the models utilized during the design step. In addition, feasibility studies carried out by NREL help industrial partners make decisions about the potential implementation of these technologies for fuel ethanol production (Kadam et al., 2000; Mielenz, 1997). Future trends for costs reduction in the case of the NREL process include more efficient pretreatment of biomass, improvement of specific activity and productivity of cellulases, the possibility of carrying out the SSCF process at higher temperatures, improvement of recombinant microorganisms for a greater assimilation of all the sugars released during the pretreatment and hydrolysis processes, and further development of co-generation system (Cardona and Sánchez, 2007). Nagle et al. (1999) proposed an alternative configuration involving a total hydrolysis of yellow poplar using a three-stage countercurrent dilute-acid process validated at experimental level. The obtained hydrolyzate is co-fermented by the recombinant strain of Z. mobilis. In this case, the lignin is recovered prior to the fermentation. Aspen Plus was utilized for generating the needed information to evaluate the economic performance of the whole flowsheet configuration through a spreadsheet model. Optimized values of the key process variables obtained from the simulation are utilized as target values for bench-scale research to design an advanced two-stage engineering-scale reactor for a dilute-acid hydrolysis process.

Some commercial firms have also invested funds in the development of an ethanol production process employing the lignocellulosic biomass. Iogen Corporation (Ottawa, Canada) developed an SHF process comprising a dilute-acid-catalyzed steam explosion and the removal of the major part of the acetic acid released during the pretreatment, the use of S. cerevisiae as a fermenting organism, distillation of broth, ethanol dehydration, and disposal of stillage in landfill (Tolan, 2002). Later modifications involve the co-fermentation of both hexoses and pentoses using genetically modified strains of microorganisms, such as yeasts or bacteria (Figure 11.11c). Using the recombinant Z. mobilis strain patented by NREL, Lawford and Rousseau (2003) tested two configurations for ethanol production using the conceptual design based on SHF developed by Iogen. These authors demonstrated that a configuration involving the continuous pentose fermentation using the recombinant Z. mobilis strain, and the separate enzymatic hydrolysis followed by continuous glucose fermentation using a wild-type strain of Z. mobilis is the most appropriate in comparison to the use of the co-fermentation process after the enzymatic hydrolysis or the use of an industrial yeast strain during the glucose fermentation (Cardona and Sánchez, 2007).

Reith et al. (2002) have reviewed different processes for production of biomass ethanol and concluded that verge grass, willow tops, and wheat milling residues could be potential feedstock for fuel ethanol production under the regulations of the Netherlands. These authors constructed a model using Microsoft™ Excel™

for the system description of generic biomass-to-ethanol process. This process involves the evaporation of the stream from the saccharification step in such a way that the sugar concentration allows a final ethanol concentration of at least 8.5 vol.% in the fermentation broth. In addition, pretreatment using $Ca(OH)_2$ was included in the analysis (Figure 11.11d). The advantage of using this type of pretreatment is that inhibitors are not formed making the detoxification step unnecessary. The evaluation showed that currently available industrial cellulases account for 36 to 45% of ethanol production costs, and therefore, a 10-fold reduction in the cellulase costs and a 30% reduction in capital costs are required in order to reach ethanol production costs competitive with starch ethanol. These evaluation approaches indicate the need for developing processes that contribute to improving one or all of the four critical areas related to cellulase research mentioned in Chapter 5, Section 5.2.3.1.

Ghosh and Ghose (2003) report on the model process for bioethanol production proposed by the Indian Institute of Technology (IIT) in Delhi (India). This process involves two pretreatment steps: steam explosion for xylose production followed by solvent pretreatment for delignification of biomass. The released pentoses are utilized for single cell protein production, while the cellulose undergoes simultaneous saccharification and fermentation. The SSF reactor is coupled with vacuum cycling and has a stepwise feeding of cellulose (Figure 11.11e). The process has been tested in a pilot plant using rice straw as a feedstock. However, the obtained product is hydrous ethanol (95% v/v) and the production costs (US$0.544/L) are higher than those expected for the production of dehydrated ethanol through the NREL model process (US$0.395). The consideration of adsorption separation stage (instead of distillation) increases the cost of ethanol by about 50%. The possibility of using alkali pretreatment was also assessed, but the costs increased due to lower by-product credits (low quality of obtained lignin as a fuel; Cardona and Sánchez, 2007).

Pan et al. (2005) report the preliminary evaluation of the so-called Lignol process for processing softwoods into ethanol and co-products. This configuration makes use of the organosolv process for obtaining high quality lignin allowing the fractionation of the biomass prior to the main fermentation. For this, the process utilizes a blend of ethanol and water at about 200°C and 400 psi. For ethanol production, SHF and SSF have been tested. Streams containing hemicellulose sugars, acetic acid, furfural, and low molecular weight lignin are also considered as a source of valuable co-products. Until now, the Lignol process has been operated only in a three-stage batch mode, but simulation studies indicate an improved process economics by operating the plant in continuous mode (Arato et al., 2005). Gong et al. (1999) report a fractionation process employing corncob and aspen wood chips as feedstocks and utilizing alkaline pretreatment with ammonia that favors the separation of lignin and extractives. After this step, the hemicellulose is hydrolyzed with dilute acid and released sugars are fermented by xylose-assimilating yeast. Finally the cellulose is converted into ethanol by batch SSF using a thermotolerant yeast strain. So and Brown (1999) performed the economic analysis of the Waterloo Fast Pyrolysis process comprising a 5% acid pretreatment, fast pyrolysis, levoglucosan hydrolysis, and the use of two cultures, *S. cerevisiae* and *P. stipitis*, to ferment hexoses and pentoses, respectively. These authors also analyzed the SSF process of dilute-acid pretreated feedstock that comprises the pentose fermentation by recombinant *E. coli* for xylose fermentation, and the SHF process of dilute-acid pretreated feedstock using a strain of *C. shehatae* for hexose and pentose fermentation. The evaluation indicates that the cost of the fast pyrolysis process is comparable to the other two processes in terms of capital costs, operating costs, and overall ethanol production costs (Cardona and Sánchez, 2007).

Due to the high costs of the feedstocks accounting for more than 20% in the case of the lignocellulosic biomass (Kaylen et al., 2000), the optimization of cellulose conversion is of great importance, especially if it is accompanied with the appropriate handling and utilization of all process streams. Although many related works can be found, the tendency is the optimization of separate process units. This implies that the integration of such separately studied units optimized at different scales does not always provide the correct information on the global process. This situation is particularly important in the case of the integration of the pretreatment step with the biological transformations. De Bari et al. (2002) undertook this problem emphasizing the scale-up features and the potential of produced by-products in the case of steam-exploded aspen wood chips. The pretreatment step was carried out in a continuous steam explosion pilot plant fed with 0.15 ton/h of dry matter that was coupled with the extraction step in order to separate the lignin and the hemicellulose and carry out the detoxification. The subsequent conversion to ethanol was made by SSF. The process was completed with a packed distillation column with a maximum reboiler capacity of 150 L working batch-wise. The experimentation allowed the definition of the best combinations of operation parameters, the selection of the best detoxification procedure, the determination of yields and operation conditions of SSF, the analysis of the distillation for its conversion to hydrogen or ethanol, and the determination of the chemical oxygen demand (COD) of the liquid stream from the distillation step. However, this work does not report if any analysis of the process was carried out from the viewpoint of thermodynamic and kinetics fundamentals of the studied system, or if process synthesis procedures were used for the definition of the selected configuration of the process. These tools may help predict the behavior of experimental systems. Similarly, pilot-plant data can provide feedback to the mathematical models used for the analysis of the system, the same as for the study of its stability and operability. For this point, the complementation with simulation tools is invaluable (Cardona and Sánchez, 2007).

In general, it is thought that reductions in processing or conversion costs of lignocellulosic biomass offer the greatest potential for making biobased products like ethanol competitive in the market place in comparison to oil-based products for which high raw material costs are characteristic (Dale, 1999). Therefore, the fundamental research on the development of cost effective processes for biomass processing represents the key for attaining the mentioned competitiveness. Lynd et al. (1999) argue that oil refineries are unlikely to have significant economies of scale advantages in comparison with the expected mature biomass refineries. In this way, the challenges associated with the biomass conversion are related to the recalcitrance of cellulosic biomass (conversion into reactive products like fermentable sugars) and to the product diversification (conversion of reactive intermediates into valuable products; Cardona and Sánchez, 2007).

In general, process synthesis procedures can be significantly enhanced using process simulation packages. These simulators have allowed the analyzing of several technological options and the gaining of insight about the process improvements (Cardona and Sánchez, 2007). Almost all the approaches for carrying out process synthesis can rely on simulation tools to evaluate different process alternatives. This is illustrated in the following case study.

CASE STUDY 11.7   ANALYSIS OF INTEGRATED FLOWSHEETS FOR ETHANOL PRODUCTION FROM BIOMASS

In previous works (Cardona and Sánchez, 2004, 2006), the analysis of several integrated process flowsheets for production of fuel ethanol from lignocellulosic biomass was performed. The flowsheets were compared with a base case representing a nonintegrated configuration. The comparison criterion was the energy consumption defined as the thermal and electric energy demanded during the production of ethanol from biomass.

The different flowsheet configurations were simulated using Aspen Plus. Shortcut methods based on the principles of the topological thermodynamics (analysis of the statics, see Chapter 2) were employed for the synthesis of the distillation train as highlighted in Chapter 8, Case Study 8.1. The amount of feedstock (lignocellulosic biomass) was the same for every combination of process configurations (160,950 kg/h). Wood chips were analyzed as feedstock during the simulations. The analysis was made taking into account the best variants of each configuration assuming that no technological limitations were present for the proposed technologies. For example, it was assumed that the cellulases used for hydrolysis were purchased from commercial suppliers, which ensured their availability and efficiency. It was also assumed that SSF and SSCF processes were fully developed.

The considered overall process included all the steps required for ethanol production from pretreatment until effluent treatment. The defined nonintegrated base case is shown in Figure 11.12. This configuration comprised

- The pretreatment step using dilute sulfuric acid
- Detoxification step for the liquid fraction of pretreated biomass (hemicellulose hydrolyzate) through ionic exchange followed by alkali neutralization (not shown in Figure 11.12)
- Pentose fermentation using the xylose-assimilating yeast *C. shehatae*
- Enzymatic hydrolysis of cellulose contained in the solid fraction of pretreated biomass
- Hexose (glucose) fermentation using *S. cerevisiae*
- Ethanol separation by distillation
- Ethanol dehydration by azeotropic distillation
- Effluent treatment step by evaporation of stillage with recovery of lignin

The alternative integrated configurations were synthesized through the combination improvements in some of the steps making up the overall process. Thus, two types of pretreatment and hydrolysis schemes (with deviation of the liquid fraction

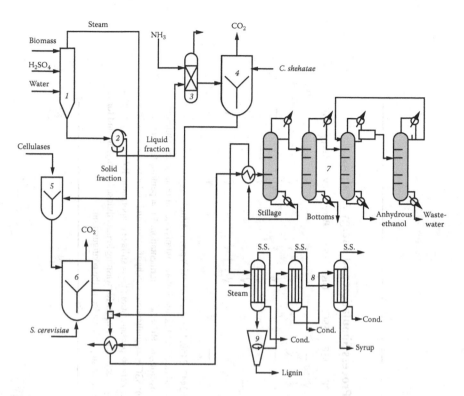

FIGURE 11.12 Simplified flowsheet for fuel ethanol production from lignocellulosic biomass (base case): (1) pretreatment reactor, (2) rotary filter, (3) ionic exchange, (4) pentose fermentation, (5) enzymatic hydrolysis, (6) hexose fermentation, (7) separation and dehydration of ethanol by azeotropic distillation, (8) evaporation train for effluent treatment, (9) centrifuge. S.S. = secondary steam, Cond. = condensate. (Adapted from Cardona, C.A., and Ó.J. Sánchez. 2006. *Energy* 31:2447–2459.)

of hemicellulose hydrolyzate or without it); three types of fermentation processes (separate hexose and pentose fermentation, SSF, or SSCF); two types of separation technologies (azeotropic distillation or pervaporation); and three types of effluent treatment schemes (without recycling of water or with two alternatives for recycling water) were selected for the subsequent simulation. The selection procedure included the technologies that are more perspective considering the use of qualitative improvements of the process and the viability of their implementation. For example, all the analyzed configurations included the use of dilute sulfuric acid for the pretreatment of biomass. In this way, six alternative configurations were synthesized and analyzed (Table 11.11).

Data for the comparison of energy consumption for each configuration were obtained from the simulation results (Table 11.12). Considering the results shown, it was evident that those alternatives involving a higher degree of process integration (SSCF, recirculation of water streams, coupling of distillation with pervaporation) presented lower energy costs. In particular, configuration 6 that included the SSCF process and ethanol dehydration by pervaporation had a 23% reduction in the energy consumption related to the nonintegrated base case. The second best

TABLE 11.11

Process Configurations Considered during Process Simulation and Energy Analysis

Flowsheet Variant	DA	DLF	Det	EH	HF	PF	SSF	SSCF	Dist	Az	Perv	Ev	RW₁	RW₂
Base case	√	√	√	√	√	√	—	—	√	√	—	√	—	—
Configuration 1	√	√	√	—	—	√	√	—	√	√	—	√	—	—
Configuration 2	√	—	√	—	—	—	—	√	√	√	—	√	·	—
Configuration 3	√	—	√	—	—	—	·	√	√	—	√	√	—	—
Configuration 4	√	—	√	—	—	—	·	√	√	√	—	√	√	—
Configuration 5	√	—	√	—	—	—	—	√	√	√	—	√	√	√
Configuration 6	√	—	√	—	—	·	—	√	√	—	√	√	√	√

Source: Cardona, C.A., and Ó.J. Sánchez. 2006. *Energy* 31:2447–2459. Elsevier Ltd. With permission.

Note: DA = dilute acid pretreatment; DLF = deviation of liquid fraction of hemicellulose hydrolyzate for pentose fermentation; Det = ion exchange detoxification; EH = enzymatic hydrolysis; HF = hexose fermentation; PF = pentose fermentation; SSF = simultaneous saccharification and fermentation; SSCF = simultaneous saccharification and co-fermentation; Dist = conventional distillation; Az = azeotropic distillation; Perv =- pervaporation; Ev = stillage evaporation; RW₁ = recycling of water for washing hemicellulose hydrolyzate; RW₂ = recycling of water for washing hemicellulose hydrolyzate and for pretreatment reactor. The symbol "√" indicates that a given step is included in the configuration.

TABLE 11.12

Comparison of Simulated Configurations according to Their Energy Consumption

Flowsheet Variant	Ethanol Yield (L/dry wood ton)	Unit Energy Costs (MJ/L EtOH)	Energy Costs (% of the base case)
Base Case	246.67	34.84	100.00
Configuration 1	262.68	33.12	95.06
Configuration 2	297.70	28.56	81.98
Configuration 3	300.18	27.83	79.87
Configuration 4	302.14	28.37	81.43
Configuration 5	305.34	27.84	79.92
Configuration 6	308.03	26.84	77.05

Source: Modified from Cardona, C.A., and Ó.J. Sánchez. 2006. *Energy* 31:2447–2459. Elsevier Ltd.

scheme corresponded to configuration 5, which had a higher degree of integration as well (Figure 11.13) and offered a 20% reduction in energy costs.

The effect of water recycling on the energy costs of the entire process should be noted. From the multiple recycling configurations, two basic schemes were selected. In the first case, the bottoms of the rectification column were mixed with a fraction of the liquid stream from centrifuge to utilize this combined stream for washing the hemicellulose hydrolyzate (configuration 4). This stream contains water and very small amounts of soluble compounds such as glucose, xylose, and acetic acid. The second case considered, besides the above-mentioned recycled water, the additional use of the evaporated water obtained in the effluent treatment step as process water for the pretreatment reactor (configurations 5 and 6), as suggested by Wooley et al. (1999).

The recycling of water has two main goals: (1) reduction of the amount of fresh water utilized in the process and (2) the increase of ethanol yield through more complete utilization of remaining fermentable sugars contained in the recycled wastewater. Increased yields lead to reduced energy consumption for producing the same amount of final product. In addition, the main effluent, the stillage from the first distillation column, resulted in more concentrates (10.1% solids) than the stillage corresponding to the base case (5.8% solids) as a consequence of the higher amounts of fresh water utilized throughout the latter process. Therefore, the recycling of water reduces the amount of water to be evaporated in the effluent and the cost of treatment of wastewater by subsequent treatment processes like anaerobic digestion. Thus, the simulation shows that the energy consumption during the partial evaporation of wastewater can be reduced from 11.60 MJ/L EtOH for the base case to 7.58 MJ/L EtOH for configuration 5 (34.72% reduction) (Cardona and Sánchez, 2006).

This case study illustrates the advantages and possibilities that process simulation offers during the synthesis of technological schemes with a high energy performance. In addition, the information obtained through simulation can be the base for life-cycle analysis of processes for bioethanol production as exemplified below.

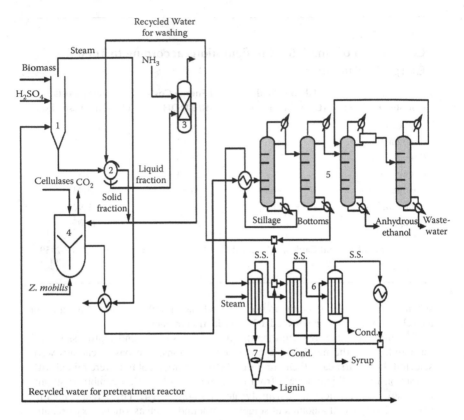

FIGURE 11.13 Simplified flowsheet for fuel ethanol production from lignocellulosic biomass (configuration 5): (1) pretreatment reactor, (2) rotary filter, (3) ionic exchange, (4) SSCF bioreactor, (5) separation and dehydration of ethanol by azeotropic distillation, (6) evaporation train for effluent treatment, (7) centrifuge. S.S. = secondary steam, Cond. = condensate. (Adapted from Cardona, C.A., and Ó.J. Sánchez. 2006. *Energy* 31:2447–2459.)

CASE STUDY 11.8 NET ENERGY VALUE OF ETHANOL FROM BIOMASS

In previous chapters, the controversial debate on the sustainability of corn ethanol was mentioned several times. Ethanol from lignocellulosic biomass may contain the key for improving the environmental performance of this liquid biofuel, especially if the energy employed for its production throughout the whole life cycle is considered. Besides the output/input ratio of energy discussed above, the net energy value (NEV) can be used for assessing the energy efficiency of produced ethanol. To this end, NEV of ethanol obtained by a particular process using a specific feedstock should be determined. The NEV is calculated by subtracting the energy required to produce a liter of ethanol during the whole life cycle from the energy contained in a liter of ethanol (Wang et al., 1999). In this way, several process configurations and feedstocks can be evaluated in order to elucidate the energy gains attained during the exploitation of the determined resource for the production and use of a biofuel like ethanol. In the previously mentioned work (Cardona and Sánchez, 2006), a preliminary estimation of the energy balance for the ethanol production process from lignocellulosic biomass was carried out. To this end, the information obtained during the simulation of the best flowsheet configurations was used (see previous case study), as well as literature data on biomass handling and transportation costs, in order to take into account the whole life cycle of the fuel.

The simulation results for the configurations with the best energy performance were utilized. In particular, configuration 5 of the previous case study (see Figure 11.13) was assessed from the point of view of its energy balance. The production of bioethanol according to the simulation of given configuration was about 615,000 L EtOH/d. For this analysis, the energy gain in the effluent treatment step was estimated. The nonfermentable component of biomass, the lignin that is recovered from centrifuge and sent to the boiler, has an average energy value of 25.4 MJ/kg. The different liquid effluents contain water, minerals, and residual materials. These liquid streams have a high biological oxygen demand and must be treated before discharge. Anaerobic digestion is generally carried out for reducing the organic matter content of the wastewater and releasing biogas. It is estimated that from 1 L of wastewater can be generated approximately 35 L of biogas. The biogas, containing about 60% methane and having an approximate calorific value of 20 to 24 MJ/m^3, is fed directly into the boilers for co-generation of both thermal and electric energy (Prakash et al., 1998). The collected wastewater corresponded to the effluent streams from the ion exchange used for inhibitor removal during the detoxification step, the aqueous stream from the bottoms of the fourth distillation column during the dehydration step (for the case of azeotropic distillation), the concentrated stream from the evaporators with a high solid content (syrup), and part of the condensates from evaporation step that was not recycled to the pretreatment reactor. The simulation provided the amount of produced lignin and wastewater mass flow rates (28,969 kg/h and 207.7 m^3/h, respectively). In this way, the thermal energy released during the burning of lignin and biogas covers all the requirements of steam for the process leading to a net thermal energy consumption equal to zero, as shown in Table 11.13. The amount of electricity co-generated by the process in the boiler can be estimated from published data (Wang et al., 1999; Wooley et al., 1999). According to Wang et al., a conservative value of electricity credit should be 50% of the data reported. Thus, these considerations indicate that 0.225 kWh/L EtOH of surplus power are produced in the process from biomass. This surplus can be sold to the grid for balancing the energy costs. The values of recovered energy in the effluent treatment step are given in Table 11.13.

The presented results demonstrated that the thermal energy required for the production of biomass ethanol could be offset by the energy carriers generated or released in the same process (lignin, biogas). Because the combustion of ethanol releases 21.2 MJ/L EtOH, a NEV for ethanol produced from lignocellulosic biomass of 19.83 to 21.11 MJ/L EtOH was calculated. For the estimation of energy input needed for ethanol production from lignocellulosic biomass, literature data on biomass handling and transportation costs were included for considering the entire life cycle of this biofuel (California Energy Commission, 2001; Wang et al., 1999). These costs are low compared with corn ethanol because of the nature of waste with biomass that does not require high energy inputs on fertilizer production, among other factors. The calculated NEV can be compared with those for corn and sugarcane ethanol (see Table 11.13), and for cellulosic ethanol from woody biomass estimated by Wang et al. (1999) at about 20.9 MJ/L EtOH. Consequently, the use

TABLE 11.13

Energy Allocation for the Production of Biomass Ethanol and Comparison with Other Feedstocks according to Their Net Energy Values (NEV)

Step	Energy Consumption (MJ/L EtOH)	Energy Recovery (MJ/L EtOH)	Reference
Conversion to ethanol in plant	24.30		
Pretreatment and SSCF	4.23		
Distillation and dehydration	9.77		
Evaporation	7.58		
Effluent treatment	2.72		
Released biogas		6.36	
Burned lignin		28.05	
Electricity credit		0.82	Wang et al. (1999)
Net thermal energy consumption in plant	0.00		
Feedstock handling	0.86– 1.06		California Energy Commission (2006) Wang et al. (1999)
Transportation	0.05– 1.13		California Energy Commission (2006) Wang et al. (1999)
Total	0.91– 2.19	0.82	
Energy use for ethanol production from biomass, MJ/L	0.09– 1.37		
Energy value of ethanol, MJ/L	21.20		Prakash et al. (1998)
NEV of biomass ethanol, MJ/L	19.83– 21.11		This case study
NEV of corn ethanol, MJ/L	5.57–6.99		Wang et al. (1999)
NEV of sugarcane ethanol, MJ/L	11.39		Prakash et al. (1998)

Source: Modified from Cardona, C.A., and Ó.J. Sánchez. 2006. *Energy* 31:2447–2459. Elsevier Ltd.

of biomass ethanol could improve the energy balance of the process and even have environmental benefits because of the reduction in the process requirements of a nonrenewable sources of energy, such as oil and natural gas.

11.3.2 OPTIMIZATION-BASED PROCESS SYNTHESIS FOR ETHANOL PRODUCTION FROM BIOMASS

Process synthesis procedures can be significantly enhanced using process simulation packages as shown in previous case studies. Simulation packages are essential for evaluating the amount of studied alternative process flowsheets. The described approach corresponds to knowledge-based process synthesis (see Chapter 2). In this regard, the task of process synthesis for the biotechnological production of fuel ethanol has mostly been undertaken using the approach based on the present knowledge. The other main approach is the optimization-based process synthesis that relies on the use of optimization for identifying the best configuration. For this, the definition of a superstructure, which considers a significant amount of variations in the topology of technological configurations of a given process, is required. The evaluation and definition of the best technological flowsheet are carried out through tools such as mixed-integer nonlinear programming (MINLP). The advantages and drawbacks of this strategy were disclosed in Chapter 2, Section 2.3.

Several cost-effective flowsheet configurations for the production of fuel ethanol from renewable resources like biomass have been reported in the literature. Most of the proposed flowsheets have been defined using different heuristic rules and knowledge-based rules, and have involved the use of such tools as commercial process simulators. However, a cost-effective flowsheet for bioethanol production utilizing an optimization-based approach has not been found in the literature. In the following case study, an optimization-based strategy was implemented in order to preliminarily synthesize several technological schemas for ethanol production from lignocellulosic biomass employing a net revenue function as a comparison criterion.

CASE STUDY 11.9 PRELIMINARY APPLICATION OF THE OPTIMIZATION-BASED STRATEGY TO PROCESS SYNTHESIS OF ETHANOL PRODUCTION FROM BIOMASS

In a previous work (Sánchez et al., 2006), a preliminary approximation to the process synthesis of ethanol production from lignocellulosic materials employing the optimization-based approach was presented. The analyzed system comprised the step of biological transformation of the pretreated feedstock through different technological options (separate hydrolysis and fermentation, simultaneous saccharification and fermentation, simultaneous saccharification and co-fermentation) and the step of ethanol dehydration using distillation. The initial stream entering the system contains the main components that are formed after the pretreatment using dilute acid, i.e., cellulose, pentoses (mainly xylose), glucose, lignin, and water. The system should process this stream in such a way that the final product stream has an ethanol content greater than 99.5% wt. For this preliminary study, the wastewater treatment step was not considered.

For tackling such a complex process as the bioethanol production, the Jacaranda synthesis package was employed, which has been described elsewhere (Fraga, 1998; Fraga et al., 2000). It has been successfully applied to the preliminary design of a hydrofluoric acid plant (Laing and Fraga, 1997), the generation of optimal downstream processing flowsheets of bioprocesses (Steffens et al., 1999b), and the process synthesis for the microbial production of penicillin (Steffens et al., 1999a). Any optimization-based strategy for process synthesis requires the implementation of models of process units. In this case, the biological transformations were described by kinetic models considering the cellulose hydrolysis, glucose formation and consumption, cell growth, and ethanol biosynthesis (South et al., 1995; see Chapter 7, Case Study 7.2). When co-fermentation using recombinant bacteria was

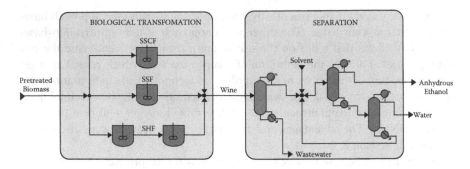

FIGURE 11.14 Superstructure of the biological transformation and separation sections for ethanol production from lignocellulosic biomass. Columns: 1 = concentration, 2 = extractive, 3 = solvent recovery.

taken into account, the model reported by Leksawasdi et al. (2001) was used (see Section 7.2.2). For the calculation of distillation units, the Fenske–Underwood–Gilliland (FUG) method was utilized considering the presence of binary azeotropes in the system ethanol–water. The separation section of the process, used to generate 99.5% pure ethanol, consisted of distillation units alone. As there was an azeotrope formed by water and ethanol, an extractive distillation step was used with ethylene glycol as the solvent.

The objective function used, in this case, is net revenue defined as the value of the ethanol produced minus the annualized cost of the process, which is a function of both capital and operating costs. For the continuous bioreactors, operating and capital costs are directly related to the residence time. The capital costs of distillation units are related to the vapor velocity inside the columns and to the number of stages, and the operating costs are linked to the energy consumption (mainly heat duty). The problem posed to Jacaranda consists of a superstructure, which is shown in Figure 11.14. The reaction section has a choice of three paths, the SSF reactor, the SSCF reactor, and the combination of cellulose hydrolysis followed by hexose fermentation (SHF). The separation section consists of three distillation steps: concentration column, extractive column, and recovery column with a recycle of the solvent to the extractive column. The superstructure is the basis for an MINLP model. This model has some characteristics that make it difficult to solve, such as the physical properties models used (NRTL [nonrandom two-liquid model] in this case) and the equations for the concentrations in the reactors as a function of residence time. These are present in the optimization problem as equality constraints and are difficult to satisfy. Furthermore, the use of different hot and cold utilities was allowed to meet the heating and cooling demands of any process alternative. Using discrete utilities means that the objective function is discontinuous even as a function of only the real valued variables. Furthermore, the capital cost function for the distillation units uses integer values for the number of stages determined by the FUG procedure, also leading to discontinuities in the objective function. The result is that the overall optimization problem is not solvable using standard mathematical programming approaches.

Jacaranda provides access to a number of optimization procedures including direct search methods (Kelley, 1999) and stochastic methods, such as genetic algorithms (Goldberg, 1989) and simulated annealing (van Laarhoven and Aarts, 1987).

In this case study, it was decided to use the genetic algorithm (GA) approach. The GA uses a replacement policy for the population at each generation, with an elite size of 1, a mutation rate of 10%, a crossover rate of 70%, and a roulette wheel selection procedure. The fitness function is based on the objective function value directly with infeasible solutions discarded if they arise (which they do with a frequency of approximately 5 to 6%).

For this first attempt at automated design for the production of ethanol from biomass, the number of degrees of freedom was limited. Specifically, the residence times of each reactor and the top and bottom key component recoveries in each distillation column were selected as the manipulating variables. Therefore, four residence time variables and six recovery variables were manipulated. The superstructure makes use of two binary variables for identifying the path taken through the reaction section of the process.

The results obtained identify the SSCF configuration as the best performing for the given process. This is reasonable given the high degree of integration achieved with this configuration, which makes possible the immediate consumption of the glucose formed during the cellulose hydrolysis. In this way, the inhibition of cellulose-degrading enzymes (cellulases) is avoided. In addition, the utilization of xylose allows an increase in the content of fermentation sugars and, therefore, in the overall amount of produced ethanol. This enhanced utilization of the feedstock is not characteristic for the SSF process. The SHF option implies the utilization of an additional bioreactor (the enzymatic hydrolysis and the fermentation are carried out in different units), which involves the increase in the capital costs for this configuration. Jacaranda allowed the determination of the values of the operating parameters corresponding to the separation section. In particular, the make-up of ethylene glycol and the recycle stream flow rate are determined automatically.

Early results demonstrate that the genetic algorithm used by Jacaranda handles the complexity of the problem design robustly with respect to the numerical difficulties that may arise. The solutions obtained show variability in the technological option. From 10 different runs, three of the solutions corresponded to SSCF configurations (two of them with the best values of the objective function), six solutions to the SSF process, and one solution to the SHF configuration.

Undoubtedly, the development of this approach will make possible the synthesis of technological flowsheets considering the structure of the system on a mathematical programming basis. The complementation with tools of a knowledge-based approach will allow gaining a deeper insight of the overall process needed for the synthesis of technological configurations with increased performance.

11.4 ROLE OF ENERGY INTEGRATION DURING PROCESS SYNTHESIS

Described types of integration in Chapter 9 are related to the integration of material flows either for their transformation or for the separation of their components. Similarly, the energy integration of the different steps for ethanol production is possible. Energy integration, particularly heat integration, looks for the best utilization of energy flows (heat, mechanical, and electrical) generated or consumed inside the process with the aim of reducing the consumption of external sources of energy such as electricity, fossil fuels (oil, natural gas) mainly used for steam generation, and cooling water. Pinch technology is one of the most widely applied approaches for heat integration in the process industry, especially in the petro-chemical industry. This technology provides the necessary tools for design of the heat exchanger networks including plant utilities. During preliminary design of the heat exchange network, pinch technology allows one to obtain the best values of many process parameters as the type of utilities and their specifications.

Pinch technology has been utilized for the design of heat exchanger networks (HEN) during ethanol production. For the case of ethyl alcohol production from molasses, a process flowsheet was simulated and optimized by heat integration emphasizing the separation step by distillation (Sobočan and Glavič, 2000). The simulation was made by shortcut and rigorous methods in order to perform the heat integration. For ethanol concentrations up to 95.7% by weight, the optimal configuration corresponded to one single column and not two, as had been pro-posed. This work demonstrates the usefulness of heat integration since the optimal design showed a 27% reduction in the total costs (Cardona and Sánchez, 2007).

CASE STUDY 11.10   HEAT INTEGRATION OF FERMENTATION, PRODUCT RECOVERY, AND STILLAGE EVAPORATION STEPS FOR FUEL ETHANOL PRODUCTION

In a previous work (Grisales et al., 2005), the heat integration approach was uti-lized for the analysis of the fermentation, distillation, and evaporation steps of the fuel ethanol production process from lignocellulosic biomass using azeotropic dis-tillation for ethanol dehydration. Low ethanol concentrations in the culture broth exiting the fermenter increase energy costs in the distillation train and, therefore, in the evaporation train utilized for obtaining concentrated stillage (the first opera-tion of the effluent treatment scheme). For this reason, it is of great importance that the application of energy integration be instituted in order to improve process performance and make it more environmentally friendly (via the reduction in the consumption of external nonrenewable sources of energy).

Process simulator Aspen Plus was employed for calculating mass and energy balances of the analyzed technological configuration. Through a graphical repre-sentation of the energy requirements of the process, the exchanged heat was identi-fied considering the external utilities (steam and cooling water). The process was represented by its hot and cold profiles, which were defined by the corresponding hot and cold composite curves (Figure 11.15). These curves show how much energy could be transferred from hot streams to cold streams within the process. To com-plete the global heat balance, hot utilities (vapor at 2 bar) and cold utilities (cooling water at 10°C) were utilized. For the definition of the required amount of hot and cold utilities, a grand composite curve was built. Consumed energy by hot and cold utilities was determined by simulation. For the design of HEN, a grid diagram was employed in which streams were represented with their respective supply and tar-get temperatures as well as the position of pinch. Through heuristic rules for pinch (Shenoy, 1995), different configurations of HEN were proposed and evaluated in terms of total recovered heat and operation costs. These HENs should ensure the target temperatures of the streams. Heat transfer areas were also calculated in order to define the capital costs. In particular, it was established that the hot stream exit-ing the top of the first distillation column (concentration column) should be split

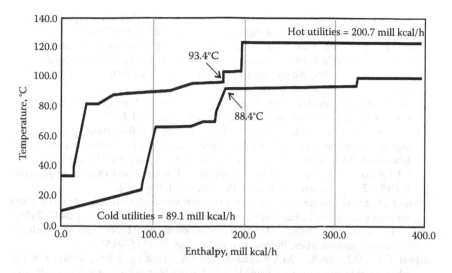

FIGURE 11.15 Representation of the heat balance of the process through hot and cold composite curves for a minimum temperature difference of 5°C. Minimum approximation of the curves corresponds to the pinch. Upper curve represents the hot streams; lower curve represents the cold streams.

into two substreams. These two substreams are organized in such a way that they transfer heat to the second effect of evaporation and to the heat exchanger utilized for preheating the evaporated liquid exiting this second effect which is sent to the third effect of evaporation. This configuration contrasts with the base case configuration where this stream is condensed and sent as a distillate to the second distillation column (rectification column) without taking advantage of its caloric energy.

Applying the described procedure, the energy saving of the new HEN was determined compared to the original network for the studied process steps. This information allowed quantifying the economic benefits that could be obtained if the defined HENs by means of pinch analysis were implemented. For instance, the external energy supplied to the process by the hot utilities was reduced by 22.8% for a minimum temperature difference of 5°C. The achieved energy recovery is 75.5% of the maximum possible energy recovery calculated in the targeting step (Grisales et al., 2005).

REFERENCES

Agüero, C.J., J.R. Pisa, and R.L. Andina. 2006. Considerations on the rational utilization of cane bagasse as a fuel (Consideraciones sobre el aprovechamiento racional del bagazo de caña como combustible, in Spanish). *Revista de Ciencias Exactas e Ingeniería de la Universidad Nacional de Tucumán* 27:1–8.

Arato, C., E.K. Pye, and G. Gjennestad. 2005. The Lignol approach to biorefining of woody biomass to produce ethanol and chemicals. *Applied Biochemistry and Biotechnology* 121–124:871–882.

Belyea, R.L, K.D Rausch, and M.E Tumbleson. 2004. Composition of corn and distillers dried grains with solubles from dry grind ethanol processing. *Bioresource Technology* 94:293–298.

Berg, C. 2001. *World Fuel Ethanol. Analysis and Outlook.* F.O. Licht. http://www.agra-europe.co.uk/FOLstudies /FOL-Spec04.html (accessed March 2004).

California Energy Commission. 2001. *Costs and benefits of a biomass-to-ethanol production industry in California.* Commission report no. P500-01-002, California Energy Commission. http://www.energy.ca.gov/reports/2001-04-03_500-01-002+002A. PDF.

California Energy Commision. 2006. *Energy Story.* California Energy Commision. http://www.energyquest.ca.gov/story/chapter08.html (accessed July 2006).

Cardona, C.A., and Ó.J. Sánchez. 2004. Analysis of integrated flow sheets for biotechnological production of fuel ethanol. Paper presented at the 7th Conference on Process Integration, Modelling and Optimisation for Energy Saving and Pollution Reduction (PRES 2004)—16th International Congress of Chemical and Process Engineering (CHISA 2004), August 22–26 2004, Prague, Czech Republic.

Cardona, C.A., and Ó.J. Sánchez. 2006. Energy consumption analysis of integrated flowsheets for production of fuel ethanol from lignocellulosic biomass. *Energy* 31:2447–2459.

Cardona, C.A., and Ó.J. Sánchez. 2007. Fuel ethanol production: Process design trends and integration opportunities. *Bioresource Technology* 98:2415–2457.

Cardona, C.A., Ó.J. Sánchez, M.I. Montoya, and J.A. Quintero. 2005a. Analysis of fuel ethanol production processes using lignocellulosic biomass and starch as feedstocks. Paper presented at the 7th World Congress of Chemical Engineering, July 10–14, 2005, Glasgow, Scotland, U.K.

Cardona, C.A., Ó.J. Sánchez, M.I. Montoya, and J.A. Quintero. 2005b. Simulación de los procesos de obtención de etanol a partir de caña de azúcar y maíz (Simulation of processes for ethanol production from sugarcane and corn, in Spanish). *Scientia et Technica* 28:187–192.

CENICAÑA. 2003. *Service of economic and statistical analysis.* Cali, Colombia: Centro Nacional de Investigación de la Caña de Azúcar (CENICAÑA).

Chen, H., Y. Wen, M.D. Waters, and D.R. Shonnard. 2002. Design guidance for chemical processes using environmental and economic assessments. *Industrial & Engineering Chemistry Research* 41:4503–4513.

Chum, H.L., and R.P. Overend. 2001. Biomass and renewable fuels. *Fuel Processing Technology* 171:187–195.

Dai, D., Z. Hu, G. Pu, H. Li, and C. Wang. 2006. Energy efficiency and potentials of cassava fuel ethanol in Guangxi region of China. *Energy Conversion and Management* 47:1686–1699.

Dale, B.E. 1999. Biobased industrial products: Bioprocess engineering. When cost really counts. *Biotechnology Progress* 15:775–776.

De Bari, I., E. Viola, D. Barisano, M. Cardinale, F. Nanna, F. Zimbardi, G. Cardinale, and G. Braccio. 2002. Ethanol production at flask and pilot scale from concentrated slurries of steam-exploded aspen. *Industrial & Engineering Chemistry Research* 41:1745–1753.

Dechanpanya, W. 1998. Chemical process analysis of economic and environmental performance: Creating the design enhancement to AHP ranking (DEAR) software design aid. Michigan Technological University, Houghton, MI, USA.

EPA. 1993. *Emission factor documentation for AP-42 Section 1.8 Bagasse combustion in sugar mills,* U.S. Environmental Protection Agency (EPA). http://www.epa.gov/ttn/chief/ap42/ch01/bgdocs/b01s08.pdf.

EPA. 1996. *Bagasse combustion in sugar mills.* AP 42 Supplement E Compilation of Air Pollutant Emission Factors Volume I: Stationary Point and Area Sources, U.S. Environmental Protection Agency (EPA). http://www.epa.gov/ttn/chief/ap42/ch01/final/c01s08.pdf.

ETPI. 2003. *The Sugar Sector Environmental Report*, Environmental Technology Program for Industry (ETPI). http://www.cpp.org.pk/etpirpt/SugarSectorReport.pdf.

FAO. 2007. *FAOSTAT*. Food and Agriculture Organization of the United Nations (FAO). http://faostat.fao.org (accessed February 2007).

FAO and IFAD. 2004. Cassava starch. Paper presented at the Proceedings of the Validation Forum on the Global Cassava Development Strategy, Rome, Italy.

Fraga, E.S. 1998. The generation and use of partial solutions in process synthesis. *Transactions of the Institution of Chemical Engineers* 76 (A1):45–54.

Fraga, E.S., M.A. Steffens, I.D.L. Bogle, and A.K. Hind. 2000. An object oriented framework for process synthesis and simulation. In *Foundations of computer-aided process design*, ed. M.F. Malone, J.A. Trainham, and B. Carnahan, Vol. 323 of AICHE Symposium Series.

Gheewala, S.H. 2005. Environmental assessment of power production from bagasse in Thailand: A life cycle evaluation. Paper presented at the International Workshop–Capacity Building on Life Cycle Assessment in APEC Economies, Bangkok, Thailand.

Ghosh, P., and T.K. Ghose. 2003. Bioethanol in India: Recent past and emerging future. *Advanced Biochemical Engineering and Biotechnology* 85:1–27.

Goldberg, D.E. 1989. *Genetic algorithms in search, optimization and machine learning*. Reading, MA, USA: Addison-Wesley.

Gong, C.S., N.J. Cao, J. Du, and G.T. Tsao. 1999. Ethanol production from renewable resources. *Advances in Biochemical Engineering/Biotechnology* 65 (207–241).

Grisales, R., C.A. Cardona, Ó.J. Sánchez, and L.F. Gutiérrez. 2005. Heat integration of fermentation and recovery steps for fuel ethanol production from lignocellulosic biomass. Paper presented at the 2nd Mercosur Congress on Chemical Engineering and 4th Mercosur Congress on Process Systems Engineering, ENPROMER 2005, Agosto de 2005, Rio de Janeiro, Brazil.

Hussain, S., L. Halpin, and A. McVittie. 2006. The economic viability of Environmental Management Systems: An application of Analytical Hierarchy Process as a methodological tool to rank trade-offs. Paper presented at the Corporate Responsibility Research Conference CRRC 2006, Dublin, Ireland.

Ingram, L.O., H.C. Aldrich, A.C.C. Borges, T.B. Causey, A. Martinez, F. Morales, A. Saleh, S.A. Underwood, L.P. Yomano, S.W. York, J. Zaldivar, and S. Zhou. 1999. Enteric bacterial catalysts for fuel ethanol production. *Biotechnology Progress* 15:855–866.

Kadam, K.L., L.H. Forrest, and W.A. Jacobson. 2000. Rice straw as a lignocellulosic resource: Collection, processing, transportation, and environmental aspects. *Biomass and Bioenergy* 18:369–389.

Kaylen, M., D.L. Van Dyne, Y.-S. Choi, and M. Blase. 2000. Economic feasibility of producing ethanol from lignocellulosic feedstocks. *Bioresource Technology* 72:19–32.

Kelley, C.T. 1999. *Iterative methods for optimization*. Philadelphia, PA, SIAM.

Laing, M.D., and E.S. Fraga. 1997. A case study on synthesis in preliminary design. *Computers and Chemical Engineering* 21 (Suppl.):S53–S58.

Lawford, H.G., and J.D. Rousseau. 2003. Cellulosic fuel ethanol. Alternative fermentation process designs with wild-type and recombinant *Zymomonas mobilis*. *Applied Biochemistry and Biotechnology* 105–108:457–469.

Leksawasdi, N., E.L. Joachimsthal, and P.L. Rogers. 2001. Mathematical modeling of ethanol production from glucose/xylose mixtures by recombinant *Zymomonas mobilis*. *Biotechnology Letters* 23:1087–1093.

López-Ulibarri, R., and G.M. Hall. 1997. Saccharification of cassava flour starch in a hollow-fiber membrane reactor. *Enzyme and Microbial Technology* 21:398–404.

Lynd, L.R., C.E. Wyman, and T.U. Gerngross. 1999. Biocommodity engineering. *Biotechnology Progress* 15:777–793.

Macedo, I.C., and L.A.H. Nogueira. 2005. Evaluation of ethanol production expansion in Brazil. In *Biofuels (Biocombustíveis)*, ed. Cadernos NAE/Núcleo de Assuntos Estratégicos da Presidência da República. Brasilia, Brazil: Núcleo de Assuntos Estratégicos da Presidência da República.

Maranhao, L.E.C. 1982. Individual bagasse drier. U.S. Patent 4326470.

McAloon, A., F. Taylor, W. Yee, K. Ibsen, and R. Wooley. 2000. *Determining the cost of producing ethanol from corn starch and lignocellulosic feedstocks*. Technical Report NREL/TP-580-28893, National Renewable Energy Laboratory, Washington, D.C.

Mielenz, J.R. 1997. Feasibility studies for biomass-to-ethanol production facilities in Florida and Hawaii. *Renewable Energy* 10 (2–3):279–284.

Montoya, M.I., J.A. Quintero, Ó.J. Sánchez, and C.A. Cardona. 2006. Evaluación del impacto ambiental del proceso de obtención de alcohol carburante utilizando el algoritmo de reducción de residuos (Environmental impact assessment of fuel ethanol production process using the waste reduction algorithm, in Spanish). *Revista Facultad de Ingeniería* 36:85–95.

Moreira, J.S. 2000. Sugarcane for energy—Recent results and progress in Brazil. *Energy for Sustainable Development* 4 (3):43–54.

Nagle, N., K. Ibsen, and E. Jennings. 1999. A process economic approach to develop a dilute-acid cellulose hydrolysis process to produce ethanol from biomass. *Applied Biochemistry and Biotechnology* 77–79:595–607.

Nguyen, Q.A., J.H. Dickow, B.W. Duff, J.D. Farmer, D.A. Glassner, K.N. Ibsen, M.F. Ruth, D.J. Schell, I.B. Thompson, and M.P. Tucker. 1996. NREL/DOE ethanol pilot-plant: Current status and capabilities. *Bioresource Technology* 58:189–196.

Nguyen, T.L.T., S.H. Gheewala, and S. Bonnet. 2008. Life cycle cost analysis of fuel ethanol produced from cassava in Thailand. *International Journal of Life Cycle Assessment* 13:564–573.

Observatorio Agrocadenas Colombia. 2006. *Segundo informe de coyuntura maíz 2006*. Ministerio de Agricultura y Desarrollo Rural. http://www.agrocadenas.gov.co/home. htm (accessed February 2007).

Pan, X., C. Arato, N. Gilkes, D. Gregg, W. Mabee, K. Pye, Z. Xiao, X. Zhang, and J. Saddler. 2005. Biorefining of softwoods using ethanol organosolv pulping: Preliminary evaluation of process streams for manufacture of fuel-grade ethanol and co-products. *Biotechnology and Bioengineering* 90 (4):473–481.

Prakash, R., A. Henham, and I.K. Bhat. 1998. Net energy and gross pollution from bioethanol production in India. *Fuel* 77 (14):1629–1633.

Quintero, J.A., M.I. Montoya, Ó.J. Sánchez, O.H. Giraldo, and C.A. Cardona. 2008. Fuel ethanol production from sugarcane and corn: Comparative analysis for a Colombian case. *Energy* 33 (3):385–399.

Quintero, L.E., X. Acevedo, and R. Rodríguez. 2004. *Production costs of technified yellow corn in Colombia*. Bogotá, Colombia: Observatorio Agrocadenas Colombia, Ministerio de Agricultura y Desarrollo Rural.

Reith, J.H., H. den Uil, H. van Veen, W.T.A.M. de Laat, J.J. Niessen, E. de Jong, H.W. Elbersen, R. Weusthuis, J.P. van Dijken, and L. Raamsdonk. 2002. Co-production of bioethanol, electricity and heat from biomass residues. Paper presented at the 12th European Conference and Technology Exhibition on Biomass for Energy, Industry and Climate Protection, Amsterdam, The Netherlands.

Sánchez, Ó.J. 2008. Síntesis de esquemas tecnológicos integrados para la producción biotecnológica de alcohol carburante a partir de tres materias primas Colombianas (Synthesis of integrated flowsheets for biotechnological production of fuel ethanol from three Colombian feedstocks, in Spanish), Departamento de Ingeniería Química, Universidad Nacional de Colombia sede Manizales, Manizales.

Sánchez, Ó.J., and C.A. Cardona. 2005. Producción biotecnológica de alcohol carburante I: Obtención a partir de diferentes materias primas (Biotechnological production of fuel ethanol I: Production from different feedstocks, in Spanish). *Interciencia* 30 (11):671–678.

Sánchez, Ó.J., and C.A. Cardona. 2008. Trends in biotechnological production of fuel ethanol from different feedstocks. *Bioresource Technology* 99:5270–5295.

Sánchez, Ó.J., E.S. Fraga, and C.A. Cardona. 2006. Process synthesis for fuel ethanol production from lignocellulosic biomass using an optimization-based strategy. Paper presented at the World Renewable Energy Congress IX and Exhibition, Florence, Italy.

Schell, D.J., C.J. Riley, N. Dowe, J. Farmer, K.N. Ibsen, M.F Ruth, S.T. Toon, and R.E. Lumpkin. 2004. A bioethanol process development unit: Initial operating experiences and results with a corn fiber feedstock. *Bioresource Technology* 91:179–188.

Sheehan, G.J., and P.F. Greenfield. 1980. Utilisation, treatment and disposal of distillery wastewater. *Water Research* 14:257–277.

Shenoy, U.V. 1995. *Heat exchanger network synthesis: The pinch technology-based approach.* Houston, TX: Gulf Publishing Company.

So, K.S., and R.C. Brown. 1999. Economic analysis of selected lignocellulose-to-ethanol conversion technologies. *Applied Biochemistry and Biotechnology* 77-79:633–640.

Sobočan, G., and P. Glavič. 2000. Optimization of ethanol fermentation process design. *Applied Thermal Engineering* 20:529–543.

South, C.R., D.A.L. Hogsett, and L.R. Lynd. 1995. Modeling simultaneous saccharification and fermentation of lignocellulose to ethanol in batch and continuous reactors. *Enzyme and Microbial Technology* 17:797–803.

Sriroth, K., B. Lamchaiyaphum, and K. Piyachomkwan. 2007. *Present situation and future potential of cassava in Thailand.* http://www.cassava.org/doc/presentsituation2.pdf (accessed February 2007).

Steffens, M.A., E.S. Fraga, and I.D.L. Bogle. 1999a. Multicriteria process synthesis for generating sustainable and economic bioprocesses. *Computers and Chemical Engineering* 23 (1455–1467).

Steffens, M.A., E.S. Fraga, and I.D.L. Bogle. 1999b. Synthesis of bioprocesses using physical properties data. *Biotechnology and Bioengineering* 68 (2):218–230.

Tiffany, D.G., and V.R. Eidman. 2003. *Factors associated with success of fuel ethanol producers.* Staff Paper Series P03-7, University of Minnesota, Minneapolis, USA.

Tolan, J.S. 2002. Iogen's process for producing ethanol from cellulosic biomass. *Clean Technologies and Environmental Policy* 3:339–345.

van Laarhoven, P.J., and E.H. Aarts. 1987. *Simulated annealing: Theory and applications.* Dordrecht, The Netherlands: Kluwer Academic Publishers.

Wang, M., C. Saricks, and D. Santini. 1999. *Effects of fuel ethanol use on fuel-cycle energy and greenhouse gas emissions.* Report no. ANL/ESD-38, Argonne National Laboratory, Center (Argonne, IL, USA) for Transportation Research. www.transportation.anl.gov/pdfs/TA/58.pdf.

Wilkie, A.C., K.J. Riedesel, and J.M. Owens. 2000. Stillage characterization and anaerobic treatment of ethanol stillage from conventional and cellulosic feedstocks. *Biomass and Bioenergy* 19:63–102.

Wooley, R., M. Ruth, J. Sheehan, K. Ibsen, H. Majdeski, and A. Galvez. 1999. *Lignocellulosic biomass to ethanol process design and economics utilizing co-current dilute acid prehydrolysis and enzymatic hydrolysis. Current and futuristic scenarios.* Technical Report NREL/TP-580-26157, National Renewable Energy Laboratory, Washington, D.C.

Xavier, M.R. 2007. *The Brazilian sugarcane ethanol experience,* Competitive Enterprise Institute. http://www.cei.org/pdf/5774.pdf

12 Food Security versus Fuel Ethanol Production

The Food and Agriculture Organization of United Nations clearly defines the basis for food security existence: *Food security exists when all people, at all times, have access to sufficient, safe, and nutritious food to meet their dietary needs and food preferences for an active and healthy life.* The Rome declaration on world food security in 1996 established the strategies of global dimensions to solve the problems of hunger and food insecurity (United Nations, 1996). However, at that time, the bioenergy or biofuels were not mentioned or considered as a threat.

Today, the opposition considers that the use of biofuels exacerbates world hunger based on the land competition for bioenergy and food crops. On the other hand, defenders believe that far from creating food shortages, biofuels represent the best opportunity for sustainable economic prospects in many developing countries. To discuss these positions objectively, different factors must be analyzed in order to explore the global panorama of the problem (or the opportunity). First, potentials for growing edible crops for biofuels or food should be identified. Second, the residues obtained in food crops production and processing are a source for cellulosic ethanol without using new lands for biofuels. The genetic modification of crops looking for high productivity is also an issue of high importance. Another important factor is the complex relationships between bioethanol and fossil oil dependence not only for energy but also for agrochemicals and fertilizers. But one of the most significant factors to be considered in this discussion is the economical issue (including agricultural market impacts) of producing crops for bioethanol instead of foods. Hereinafter, these factors are studied and discussed.

For this type of discussion, technical tools are needed. An integral strategy was proposed by the Bioenergy and Food Security (BEFS) Project, whereby the risks and opportunities of bioenergy and how it can affect food security in developing countries are analyzed. An overview of the project, its strategy and objectives, as well as some modifications given by the authors, is presented in this chapter.

12.1 CROP POTENTIALS FOR FOOD AND ENERGY

Among the bioenergy crops used for fuel ethanol production, sugarcane is the main feedstock utilized in tropical countries like Brazil and India. In North America and Europe, fuel ethanol is mainly obtained from starchy materials, especially corn. Different countries such as the United States and Sweden have defined strategic policies for the development of this technology in order to produce large amounts of renewable biofuels and diminish their dependence on imported fossil fuels. However, the possible land competition between food and biofuels is not often regulated by the government, being considered the market law as the natural judge for this kind of competition. Different scenarios must be analyzed depending on the type of feedstock and participation of the country in the supply chain (producer or consumer)

12.1.1 Corn in the United States

The United States is a developed country, which will have a future dependence on bioethanol imports (the U.S. is going to become more of a consumer than a producer). This is explained by the fact that this country actually plants about 32 million hectares of corn, but this acreage is not going to be strongly increased in the coming years. The energy products in the United States are expected to be covered by imports from its neighbor, Latin America.

The production of bioethanol in the United States has increased from 4.16 billion liters in 1996 to 24.6 billion liters in 2007 (Renewable Fuels Association, 2008). But the target consumption for 2022 is 56.7 billion liters annually of ethanol (U.S. Dept. of Agriculture, 2007). The country uses corn instead of sugarcane as a raw material (sugarcane is produced with agroecological limitations only in Florida). The energy balance concerning corn conversion into ethanol is negative (for each unit of energy supplied by ethanol, more energy is used to produce it). The bioethanol productivity per hectare of corn is three times less than in the case of sugarcane.

Consequently the bioethanol production costs for corn are 80% higher than for sugarcane. The U.S. ethanol industry is viable only because there are major subsidies for bioethanol production.

High-fructose corn syrup (HFCS) is an important food product from corn. It is more economical because the U.S. price of sugar is twice the global price and the price of corn is substantially low due to government subsidies. Overall analysis shows that, while the United States is a consumer of corn, it also is an exporter of this grain (Table 12.1).

The most important foods in the United States from corn are flakes and HFCS. Table 12.1 shows that these products are not greatly affected by biofuels production. However, the real concerns about food security in the United States are outside the country. In past years, Mexico has become a net importer of genetically modified corn, absorbing the massive American surplus. As maize cultivation in Mexico becomes an economically impractical proposition, the farmers abandon the land to migrate to Mexico City or to the United States. In this case, the consumers suffer the consequences. As an example, between 1994 and 2003, the price of tortillas (Mexican national food) quadrupled. The problem could increase for Mexico if American corn is used to produce fuel ethanol instead of tortillas. This, however, depends on the market and negotiations between the United States and producers in Latin America.

If this occurs, the price of corn will increase substantially. But an import complement to the food security discussion in corn–ethanol production is the existence of the by-product called dried distillers grains (DDGs), which is used

TABLE 12.1

Use of Corn Produced in the United States from September 2007 to August 2008

Description	Tonnes of Corn *1000	Share/%
High fructose corn syrup (HFCS)	17.278	3.76
Glucose syrup and dextrose	8.302	1.81
Corn starch	9.225	2.01
Fuel ethanol	106.639	23.21
Beverage ethanol	4.771	1.04
Total other food uses and export	313.242	68.17
Total corn production	459.459	100

Source: National Agricultural Statistics Service. 2007. National statistics. U.S. Dept. of Agriculture, Washington, D.C. http://www.nass.usda. gov/QuickStats/

as livestock feed. DDGs are the high protein feed produced through distilling, vaporizing, and drying after fermentation in alcohol production from corn. DDGs contain abundant amino acid, citrine, and diversiform minerals, which are beneficial to the growth of the animals. Countries in Europe (especially Ireland), Mexico, Taiwan, Indonesia, Venezuela, Malaysia, and Israel are today importers of DDGs for animal feeding. Therefore, any change in the actual protectionist policies of the United States for corn and ethanol production could drastically affect the price of different food products inside and outside this country.

12.1.2 SUGARCANE IN COLOMBIA

Colombia presents a scenario of a developing country producer and consumer with possibilities of exports. Most of Colombia's sugarcane is grown in the Cauca Valley, a rich agricultural valley with agroecological characteristics allowing the highest yield of sugarcane in the world (up to 130 tonnes/hectare).

Sugarcane cultivation requires a tropical or subtropical climate, with a minimum of 600 mm of annual rainfall. It is one of the most efficient photosynthesizers in the plant kingdom, able to convert up to 2% of incident solar energy into biomass. Sugarcane is cultivated in almost all parts of the world, but only for a few months of the year, a period called *safra*. The only place in the world where there is no safra and, therefore, sugarcane is cultivated and produced year-round is in Colombia.

Colombian sugar production in 2008 was down 11% to 2,036,134 tons, from total sugar production of 2,277,120 tons in 2007. However, for 2009, an increase of about 15% is expected. Meanwhile, total Colombian sugar exports in 2008 fell 33% from the export volume in 2007 of 720,000 tons (Colombian Ministry of Agriculture and Rural Development, 2008).

TABLE 12.2

Use of Sugarcane Produced in Colombia 2006

Cropped land with cane for sugar-cakes or loaves/ha	340,000
Cropped land with cane for sugar or ethanol/ha	200,218
For sugar production/ha	170,218
For ethanol production/ha	30,000
Sugar production/ton	2,356,617
Sugar consumption/ton	1,503,561
Sugar exports/ton	841,237
Fuel ethanol production/L	268,456,000

Source: Colombian Ministry of Agriculture and Rural Development, 2008.

Fuel ethanol production in Colombia in 2008 rose 400 million liters, but the country expects to produce 1 billion liters of ethanol per year by 2010, more than doubling the current output, and the country plans to have enough production by the end of the year for export. It is believed that the U.S. Congress will approve the proposed U.S.–Colombia Free Trade Agreement, which will allow Colombia to permanently ship its ethanol surplus to the United States duty free and not be subjected to any quotas. The Colombian government believes that agricultural exports and food security would not be affected by expanding ethanol output because the cane and other ethanol feedstock for future production will come from new crops and unused land.

In Colombia today, the food and bioethanol competition based on cane use is not an important issue; however, in the near future, it could be an important topic of discussion. The country uses 340,000 hectares (Table 12.2) with low productivity (30 to 80 ton/ha). The sugar is handcrafted and prepared as brown blocks rather than as a crystalline powder, by pouring sugar and molasses together into molds and allowing the mixture to dry. This results in sugar cakes or loaves called in Colombia *panela* (*jaggery* or *gur* in India). This product is not important on the international market, but has a very important internal market, as high nutritional raw material for national beverages. The possibility of using part of the existing cane hectares for ethanol instead of *panela* could be a catastrophic scenario for the food security in Colombia.

12.1.3 SUGARCANE IN BRAZIL

Brazil is a developing country that is a high producer and high consumer of bioethanol, and one that today exports biofuel to more than 10 countries in the world. Additionally, although more than 90% of all electrical energy in Brazil comes from hydroelectric sources (Colombia), sugar mills and distilleries sell the excess electrical energy they produce back to the grid. This electricity comes

TABLE 12.3

Use of Sugar Cane Produced in Brazil 2007

Cropped land with cane for sugar and other products/ha	7,680,000
Cropped land with cane for ethanol/ha	3,240,000
Sugar production/ton	33,400,000
Sugar consumption/ton	11,950,000
Sugar exports/ton	21,450,000
Fuel ethanol production/L	26,236,000,000
Fuel ethanol export/L	5,320,000,000

Source: Based on Foreign Agricultural Service. 2008. GAIN report BR8013. U.S. Dept. of Agriculture, Washington, D.C.; Food Outlook. 2007. Global market analysis 2007. FAO.

from burning the sugarcane trash and bagasse. Today, in Latin America, Brazil is a major producer (Table 12.3) extending its expertise to other countries, including technology and investments. Sugarcane in Brazil has a life cycle of 12 to 18 months and yields a range of 50 to 130 ton/ha.

Brazil is also the world's largest consumer of sugar, with per capita consumption around 55 kg/year, just beating out Mexico and the United States. But, the reality today in Brazil is that sugarcane bioethanol production, as in other countries with available land, has little effect on food production. This is explained by the fact that there is enough capacity for supporting new requirements in tool or expanded agricultural activities during the coming years. Another important fact to be considered is that today Brazil is the second largest producer of fuel ethanol in the world and simultaneously one of the largest food suppliers in the international market.

12.1.4 SUGARCANE IN TANZANIA

Tanzania is representative of an underdeveloped country that does not produce or consume bioethanol, but which has a high interest in having other countries using its fertile, virgin lands for bioenergy purposes. Tanzania is a sugarcane producer, but not a fuel ethanol producer as of yet. A Swedish company is among seven foreign firms that want to buy or rent large chunks of fertile land along the Rufiji Delta. The firm is looking for about 400,000 hectares for the production of ethanol from sugarcane. It has already put 20,000 hectares into the crop and a further 50,000 hectares are yet to be developed (Cardona et al., 2009).

Sugarcane is an important commercial crop in Tanzania. It is the main source of sugar produced for both export and domestic consumption. Tanzania is well situated for the sugarcane production in East Africa. The country has a wide variety of climate and weather and an area of 945,087 km². Rainfall may be considered the limiting factor for most crops, sugarcane inclusive. About 21% of the country can expect 90% probability of receiving slightly higher than 750 mm of rainfall and only about 3% can expect more than 1,250 mm.

The yields and technology for sugarcane growing are very low. Consequently, the feedstock prices are very high. The production of ethanol in Tanzania from sugarcane juice or molasses can be economically competitive with global production once the costs of feedstock, namely, sugarcane, are reduced by 70 to 80% of the current cost (Cardona et al., 2009).

Actually, this country is under very serious analysis by FAO (a BEFS project with the participation of the authors of this book) regarding the biofuels potential and food security risks. Some considerations about this country will be discussed in detail later in this chapter.

12.1.5 LIGNOCELLULOSICS: NONFOOD ALTERNATIVE

Most of the fuel ethanol produced in the world is currently sourced from starchy biomass or sucrose (molasses or cane juice), but the technology for ethanol production from nonfood plant sources is being developed rapidly so that large-scale production will be a reality in the coming years (Cardona Alzate, 2008; Lin, 2006). Moreover, when using nonfood raw materials, food security is not affected by this industry and improves ethanol's social and environmental impacts. The negatives of lignocellulosic biomass are the access (including transport costs), pretreatment cost for breaking its complex structure, and the production of nondesired products that can inhibit the enzymes and microorganism activities during hydrolysis and fermentation steps.

The importance of a particular type of biomass depends on the chemical and physical properties of the large molecules from which it is made. The chemical structure and major organic components in biomass are important in the development of processes for producing fuel and chemicals derived from it. Biomass contains varying amounts of cellulose, hemicellulose, and lignin, and a small amount of extractive (Bridgewater, 1999).

Worldwide generation of lignocellulosic residues is estimated to be more than about 4 billion tons each year. However, there are concerns about the importance of harvest residues in maintaining soil quality. Adding harvest residues to soils is very important in the provision of plant nutrients and for the water binding capacity of soils. Low levels of soil organic carbon contribute much to poor agricultural yields in large parts of sub-Saharan Africa and the tropical and subtropical areas of Asia. If this practice is not controlled, use for bioethanol of crop residues will exacerbate soil degradation and aggravate food insecurity (Lal, 2008; Reijnders, 2008).

Kim and Dale (2004) analyzed the size of the bioethanol feedstock resource at global and regional levels taking into consideration wasted crops (crops lost in distribution) and lignocellulosic biomass (crop residues and sugar cane bagasse). These authors estimate that the global potential ethanol production from these feedstocks accounts for 491 gallons/year, which is 16 times higher than current ethanol production and which could replace 32% of the global gasoline consumption. Rice straw is the feedstock that potentially could produce the largest amounts of ethanol, followed by wheat straw. In general, the total volume of energy from agricultural residues is estimated at 12 EJ (ExaJoule = 10^{18} J), as shown in Hall et al. (1993).

12.2 BIOETHANOL AND FOSSIL OIL DEPENDENCE

The extent to which bioethanol can displace petroleum-based fuels depends on the efficiency with which it can be produced. Sustainable and less energy-dependent agricultural biomass production for bioethanol is a challenge. Serious studies have been undertaken to analyze this problem in detail (Koga, 2008). In this work, different crops in Japan, such as wheat, potato, bean, and sugar beet, were assessed on energy output/inputs. Tractor operations, truck transportations, grain drying, and the use of materials necessary for agricultural production (chemical fertilizers, biocides, and agricultural machines) were considered. Chemical fertilizer consumption, for example, contributed significantly to the energy use, representing 20 to 43% of the total energy inputs. In Koga's study, comparisons demonstrated that the sugar beet is the most promising biomass-derived energy feedstock crop in this region, due to its high energy output/input ratio and net energy gain (energy output/input). The author analyzed the requirements for transformation technologies less dependent on fossil fuels to reach sustainable bioethanol production systems in this country.

Similar results were also found by scientists from the United Kingdom who emphasized the fact that the identification of crop production methods that maximize energy efficiency and minimize greenhouse gas emissions is vital for food or bioenergy crops (Deike et al., 2008; Tzilivakis et al., 2005). Results obtained by these authors for this region technically and quantitatively established how the extra distance travelled by the organic beet from the farm to the factory increased the energy input per ton above that of the conventional situations.

The discussion must include three different concepts. First, in the case of corn, ethanol soil and water are used and the disposal of wastewater polluted by nitrogen and phosphate fertilizers, as well as by pesticides and herbicides, should be noted as possibly damaging to the land for other food crops. Second, bioethanol inputs include the fuel to power machinery needed to grow and harvest the feedstock, such as corn; the petroleum used in manufacturing the required fertilizers, pesticides, and herbicides for the feedstock; the energy expended to transport the feedstock to the processor; and the energy used by the ethanol processing plant. And, third, the high consumption of fertilizers by the bioethanol crops could increase their price in the market and affect food crops' viability. The last is usually minimized by bioethanol defenders advocating the potential use of by-products, like vinasses (stillages), as fertilizer for the bioenergy crops, but many industrial processes have demonstrated noncompatibility between these residues and the soil.

Another important discussion related to second-generation ethanol is the possible competition between biomass for livestock food and fuel ethanol. A large part of lignocellulosics is used for livestock food, converting "unsuitable for human consumption" materials into highly valued food protein contained in meat, milk, and eggs. From this point of view, agricultural residues do not represent a waste stream, but an important source for the food production chain. In the case of using lignocellulosic residues (actually used as livestock food) for non-feed purposes, adaptations must be developed in the food system to compensate for protein losses, e.g., by growing beans or supplementary livestock feed crops (Nonhebel, 2007). The author of this work concluded that land requirements for such adaptations are substantial and larger than the area needed for energy crops that produce equivalent amounts of energy. Thus, from a land-use perspective, using the suitable residues for livestock feed while generating bioenergy from dedicated energy crops is the most preferable option. This result, however, is a contradiction for those who consider total biomass availability as the way of solving competition between food and biofuels.

12.3 BIOENERGY AND TRANSGENICS

The productivity and other characteristics of biofuel crops are factors to be considered as a potential response to food security concerns. Plants have not been domesticated for high-scale biofuel production, although the biotechnology advance is the quickest and most efficient way to rationally transform plants to biofuel feedstock. Here the differences between highly developed and developing countries are so big that influence of biofuels on food security is not a priority. For example, the United States has up to 60 million hectares of transgenic (genetic modified) crops mainly for livestock and biofuels; at the same time, Colombia uses 50,000 hectares for textile and other uses. So, for the transgenic soy and corn, where productivity per hectare could be four times more than in the case of the same native crops, one hectare in the United States replaces four hectares in Colombia. The challenges or opportunities in biotechnology for improving biofuel crops have been studied in detail by Gressel (2008). The author of this work, as many other scientists in the world, considers that transgenic crops are imperative for the production of biofuel. Based on this work and other reviews, the expected contribution from biotechnology to bioethanol production, from the point of view of food security, is presented in Table 12.4.

Global population may possibly exceed 6 billion by 2050. Approximately 90% of the global population will reside in Asia, Africa, and Latin America. Today, the population of these countries suffers from malnutrition problems and energy insecurities. Transgenic crops represent promising technologies that can make a vital contribution to global food and biofuels security. However, harmony and high level research (to avoid the possible dangerous aspects of using transgenics directly for human food) are the key words to reach these purposes.

12.4 BIOENERGY AND FOOD MARKET

The International Food Policy Research Institute (IFPRI) estimates that rising bioenergy demand accounts for 30% of the increase in weighted average grain

TABLE 12.4

Some Biotechnology Improvements in Bioethanol Production Contributing to Food Security

Case	Improvement	Effect on Food Security
Bioethanol production from cereals, lignocellulosics, and sugarcane is limited by the maximal weight yield per hectare.	Genetic modified plants could reach higher productivities per hectare including the raising of starch, celluloses or sugars content.	Land is used more efficiently for bioethanol crops; food crops are then less affected.
Second generation bioethanol from cultivated lignocellulosic crops or residues requires pretreatment that uses heat and acid to remove lignin.	Lignin could be partially replaced by transgenically reducing or modifying lignin content and upregulating cellulose biosynthesis.	More economic technology could be increased use of residues instead of food crops.

prices between 2000 and 2007 (Rosegrant, 2008). The impact was 39% of the real increase in maize prices.

However, these estimations can not be very accurate if the complexity of the markets and the different interactions between parameters affecting the price of crops are not scientifically studied in detail. The key question to answer is: Why small or large farmers can choose to grow crops for biofuels instead of food? Many challengers would say that it is not a business for farmers, and only government subsidies are the reason these projects can exist.

The choice now faced by farmers is between two alternatives. They can choose either to produce energy crops or to do nothing at all if the actual crop they have is profitable. However, profits in biofuels depend on how many hectares will be used for bioethanol feedstock. Usually sugarcane in South America and wheat and sugar beet in Europe require more than 10 hectares to be profitable depending on the prices in their annual contracts. Small-scale producers below 10 hectares can make a profit only when they belong to strong cooperatives or associations. On the set-aside land, farmers can grow bioethanol crops for a supplementary income in small-size plantations.

In the case of Africa where food security plays a key role, increased demand for biofuels is a positive development for African farmers. They have been getting paid less and less for their products, but now prices for farm products are on the rise, and farmers' incomes are rising. After years of suffering from falling sugar prices, African farmers are finally seeing prices on the upswing due to increased demand for sugarcane ethanol, which in turn makes it more profitable to grow the crop. Countries, such as India, Mexico, and the Sudan, are world producers of sweet sorghum and they see the opportunity for farmers in biofuels. This crop provides food, livestock feed, and biofuel. It grows in dry conditions; tolerates heat, salt, and waterlogging; and provides a steady income for poor farmers. To produce ethanol, the sorghum stalks are crushed to yield sweet juice that is fermented. But that grain can also be used for food or for chicken and cattle feed. The grain, just in case of market fluctuations, is a source of starch that can be used for bioethanol. This integral use of the crop gives stability to the farmers, which makes the business more attractive for combined production of raw materials for food and ethanol.

But one of the more important driving forces in farmer incomes from bioethanol crops is the price of oil. Farmers will rejoice when seeing fuel ethanol prices rise. Lastly, fuel ethanol prices simultaneously increase as oil prices increase. Most of the countries producing fuel ethanol define the prices of this product based on subsidies, raw material price in the market, and oil prices (Bernardi, 2001; Cohen et al., 2008).

12.5 BIOENERGY AND FOOD SECURITY PROJECT

Even as world hunger increases and government policy responses have limited effect on high food prices, there is an opportunity for agricultural (including small holders) in developing countries. In this context, the Food and Agriculture Organization (FAO) defines food security based on four dimensions:

1. Availability (production)
2. Access (income and prices)
3. Utilization (nutrition)
4. Stability (price volatility)

But in the context of the bioenergy and food security program (BEFS), the focus consists on availability and access (FAO, 2008). The BEFS project is designed in the framework of FAO and the German Gesellschaft für Technische Zusammenarbeit (GTZ) collaboration to mainstream food security concerns into national and subnational assessments of bioenergy potential. As reported by the FAO, the key questions the project will address are:

- What are the best types of bioenergy systems to help diversify agricultural output (energy feedstock), contribute to rural development, and increase rural employment and incomes?
- How could bioenergy production benefit the environment and increase energy and food security for producers farming biomass as a source of energy for themselves, on-farm use, local communities, or commercial markets?
- How could diversification of domestic energy supply provide increased energy access to rural enterprises and reduce the household energy burdens of rural women?
- Is there anything different about bioenergy that could mitigate or overcome factors of exclusion that contribute to food insecurity and rural poverty?

- How can low-income, food deficit countries ensure that food security concerns are addressed, given the complex linkages between agriculture, energy, environment, and trade?
- What are the implications on available food supplies and food prices in terms of competition for natural or human resources?
- How will bioenergy affect agricultural systems, particularly for poorer households dependent on their own food production?
- Who (public, private, or civil society) is best placed to deal with potential conflicts arising from competition for food, feed, or fuel use of biomass?

FAO considers that increased competition for land and water resources may result, and higher and less stable food prices may be one of many possible consequences. Bioenergy may also provide ways to support rural development and raise farm incomes.

Resuming, the BEFS project will analyze the risks and opportunities of bioenergy and how it could affect food security in partner countries.

12.5.1 The BEFS Project Is Developed in Basic Phases

Phase 1: Develop analytical framework and guidance to assess the bioenergy and food security nexus.

Phase 2: Assess bioenergy potential and food security implications.

Phase 3: Strengthen institutional capacities, exchange knowledge, pilot sustainable and food-secure bioenergy practices, and recommend standards and policies

BEFS partners include Cambodia, Peru, Tanzania, and Thailand. The project has already begun in Tanzania. The project provides to these countries a science-based quantitative methodology to minimize food security risks. This approach helps to build their own capacity and management, at the same time, appreciating food security concerns. The project itself is not just an assessment. BEFS produces a permanent economic forecast and food security monitoring, which emphasizes deepening insights for developing countries' bioenergy potentials

Figure 12.1 shows the analytical framework of the BEFS project. Every module is linearly connected, but entirely independent in relation to other modules. One axis is the basis of all the modules: consideration of food security as primary. The purposes and activities of the modules are discussed below.

12.5.2 Purposes and Activities of Modules

12.5.2.1 Module 1

Biomass potential helps stakeholders to understand:

- The extent and location of areas suitable for the relevant bioenergy crops.

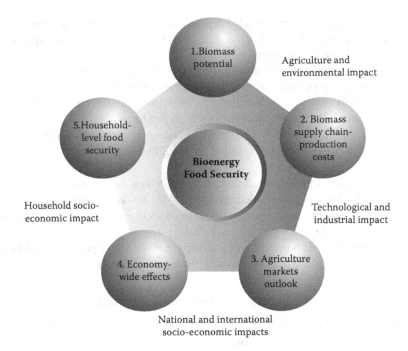

FIGURE 12.1 The bioenergy and food security (BEFS) analytical framework. (Based on FAO, 2008; Cardona Alzate et al., 2009)

- Assist farmers in bioenergy developments in their land-use planning.
- Highlight the advantages and disadvantages of different agricultural production systems and level of inputs.
- Detail land requirements for current and future food to safeguard food production.

12.5.2.2 Module 2

Biomass supply chain and production costs assess the agro-industry development and biofuel production chains by looking at

- Accessibility to technology infrastructure and availability of necessary human skills.
- Opportunities for rural development through production systems, e.g., feedstock supplier inclusive of smallholder and combined commercial–smallholder schemes.
- Processing waste by-products into valuable co-products focusing on use in local settings.
- Costs of production of the biofuel at the factory gate and distribution to domestic and international markets.
- Implications for economic viability of production chains.

12.5.2.3 Module 3

Agricultural market outlook projects the impacts of biofuel production and bio-fuel policies on agricultural markets in the context of the analyzed country over a 10-year outlook period solving the following questions:

- How much land will be required to satisfy *future food* demand?
- What is the *outlook for main food crops* in the analyzed country under different conditions, e.g., biofuel production, lower oil prices? What is the impact for major agricultural commodities in the country?
- What are the implications of biofuel policies, domestic and global, for biofuel development in the analyzed country?

12.5.2.4 Module 4

Economy-wide effects uses innovative tools including:

- Tools that cover 87 countries and country groupings.
- Use of GLOMAB FAO's multicountry–CGE model for agriculture and bioenergy developed in FAO (Aziz, 2009).
- Diverse set of biomass coverage: maize, cassava, sugarcane (ethanol), oilseeds, palm oil (biodiesel), agricultural residues, woody biomass (cellulosic ethanol, biopower).
- Separate treatment for temperate biodiesel (using soybeans) from tropical biodiesel (using palm oil).
- Inclusion of first generation biofuels (starch- or sugar-ethanol, biodiesel) and second generation biofuels (cellulosic ethanol).
- Explicit treatment of biopower, not just biofuels.

12.5.2.5 Module 5

Household-level food security assesses how price increases will affect different groups:

- Select food security crop list.
- Trade position of the country according to the single crop.
- Based on household income and expenditure data by crop, assesses the household welfare impacts by population group, focusing on the *poorer quintile*.
  - Net consumers: Those who buy more food than they sell will be hurt by higher prices.
  - Net producers: Those who sell more food than they buy benefit from higher prices.

12.5.2.6 Module 6

Household welfare impacts are based on the net welfare impact calculated from the difference between production and consumption.

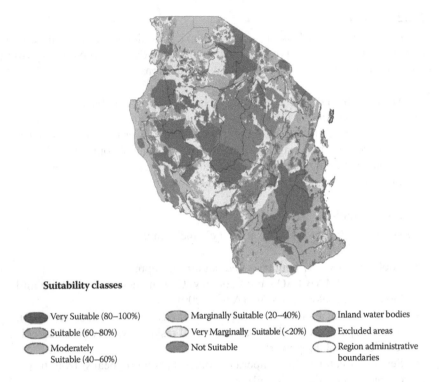

Suitability classes

⬤ Very Suitable (80–100%)   ⬤ Marginally Suitable (20–40%)   ⬤ Inland water bodies

⬤ Suitable (60–80%)   ⬤ Very Marginally Suitable (<20%)   ⬤ Excluded areas

⬤ Moderately   ⬤ Not Suitable   ⬤ Region administrative
Suitable (40–60%)   boundaries

FIGURE 12.2   Suitability index for cassava under tillage-based production system at low level of input (area available) in Tanzania.

12.5.3 PRELIMINARY BEFS RESULTS FOR TANZANIA

Potential bioenergy feedstocks were selected after in-country discussions and government's indications: sugarcane and cassava for bioethanol and palm oil for biodiesel. The suitability index (resulting from a geographic information system) for the cassava case in the framework of tillage-based production is shown in Figure 12.2. It is seem that very suitable or suitable areas practically do not exist in Tanzania. Potential for cassava growing is in the range of moderately suitable.

To delineate the industrial configurations that would be more adaptable to the Tanzania context, globally available bioethanol technologies were reviewed and analyzed against the technology and human capacity assessment (Sánchez and Cardona, 2008). Three technological configurations were designed as shown in Table 12.5. The configurations are differentiated based on the level of complexity of the technologies involved in each of the main processing steps. Two types of raw materials were considered as shown in Table 12.6.

In the production cost of cassava ethanol, the feedstock was also the highest factor contributing between 65 to 71% of the total cost. The production cost from fresh cassava feedstock is slightly lower than processed dried chips (see Table 12.7). Taking into consideration the infrastructural limitations in Tanzania

TABLE 12.5

Technology Levels for Bioethanol Production from Cassava in Tanzania

Level	Hydrolysis, Fermentation	Separation	Purification	Effluent Treatment (Vinases)	Co-Generation
Low	Enzymatic hydrolysis plus batch fermentation	Sedimentation (gravity)	Distillation at atm pressure	Evaporation plus incineration	None
Medium	Batch simultaneous sacharification and yeast fermentation	Centrifuge	Molecular sieves	Anaerobic digestion	None
High	Continuous simultaneous sacharification and bacterial fermentation	Centrifuge	Molecular sieves	Anaerobic digestion and use of solids animal feed	Yes

TABLE 12.6

Bioethanol Production Scenarios for Tanzania

Raw Material	Scenario	Parameter	Description
Cassava	7	Stand alone	Fresh cassava (single plant), increased cassava yield
Cassava	8	Stand alone	Dried cassava (single plant), increased cassava yield

TABLE 12.7

Cassava Ethanol Production Cost Depending on Scenario and Level of Technology

	Level of Technology		
Scenario	Low US$/L	Medium US$/L	Advanced US$/L
Scenario 7 Fresh	0.67	0.616	0.559
Scenario 8 Dry	0.695	0.6832	0.592

and the fact that cassava roots perish quickly after harvesting, this production route may not be the most appropriate. Thus, scenario 8 provides a more viable alternative whereby fresh cassava roots are first dried to extend the shelf life, then are transported to an ethanol processing plant. Scenario 8 facilitates greater opportunity for small farmers in isolated rural areas to participate. The costs presented in Table 12.7 do not include co-generation or use of by-products.

Ruvuma is a representative region in Tanzania located at the southern boundary with Mozambique. The welfare effects in this region were analyzed based on a 10% of producer price increase for different crops including cassava. They found positive welfares in specific cases for cassava. For bean and sugarcane, the welfare effects were always negative.

Concluding partial results showed that in order to be able to reap the benefits of bioenergy investments, Tanzania has to consider strengthening and developing local markets, local production capacity, and its infrastructure. Analysis so far does not include all crops and full integration across modules. The results are under discussion in the country.

The preliminary conclusions drawn by the BEFS project indicate that bioenergy development, which safeguards food security, is only sustainable in Tanzania when a bioenergy project:

- Does not hinder the natural resource base
- Involves smallholders, increases employment, and takes into account the specific risks for subsistence farmers

- Increases access to markets and infrastructure
- Builds domestic skills and expertise
- Ensures local benefits and sustainability of the industry
- Monitors welfare impacts at the household level
- Respects and protects the livelihoods of women
- Strengthen farmers' negotiating power
- Further enhances institutional capacity

12.6 CONCLUDING REMARKS

Farmers, refiners, and consumers are the same actors in the food and biofuels market. Any discussion regarding the possible contradictions between the use of crops and residues for food and fuels should not be globalized. The negative energy balance in some countries (consumers) is a problematic issue and fuel ethanol for automotive transport, for example, represents the unique stable alternative. Additionally, environmental improvements are reached when fuel ethanol is blended with gasoline. On the other hand, other countries can find in biofuels like ethanol the beginning of economic development for rural areas. Here critical positions are related to the possibility of using rural areas for food projects instead of fuel ethanol programs.

However, the real regulator of land use for food or biofuels is the market. In order to avoid nonequilibrium development of the fuel ethanol market based on energy crops, drastic, but fair, regulations from the governments must exist. Every country has to create appropriate rules and laws for developing fuel ethanol programs based on its specific supply and demand characteristics. One way to diminish possible impacts of fuel ethanol production on food security is the use of compensation strategies. For example, if residual biomass increases value as a raw material for producing biofuels instead of livestock food, then alternative food crops must be allocated together and simultaneously with the biofuel project. In the case of commercial sugar or starchy feedstocks for bioethanol, accurate calculations and predictions of food and energy consumption should be the basis of any discussion. If sugar, for example, is going to be exported but the prices are not stable and profitable, fuel ethanol production is a sustainable alternative not competing with food. Fuel ethanol producers in some countries having enough land for food have demonstrated that internal and export prices of food are not affected. However, countries like the United States alarmed the world when subsidized corn ethanol production affected neighbors' food security, especially Mexico.

Finally, the authors of this book consider that the main problem we have today in this important discussion is the existence of much speculative information about biofuels and food security. Most of this information is used incorrectly for political and economical purposes. In this context, technological platforms based on scientific and real information are the only way to consider exactly the influence of biofuels like ethanol on the food security.

REFERENCES

Aziz, Elbehri. 2009. Global Model for Agriculture and Bioenergy (GLOMAB): Application to food security in Peru and Tanzania. Paper presented at the Kiel Institute for World Economy. FAO–UN. http://www.narola.ifw-kiel.de/das-narola-projekt/veranstaltungen/2-workshop-febr.-2009/prasentationen/round2_elbehri.pd

Bernardi, M. 2001. Linkages between FAO agroclimatic data resources and the development of GIS models for control of vector-borne diseases. *Acta Tropica* 79:13.

Bridgewater, A.V. 1999. Principles and practice of biomass fast pyrolysis processes for liquids. *Journal Analysis Applied Pyrolysis* 51:22.

Cardona Alzate, C.A, et al. 2008. Fuel ethanol production from sugarcane and corn: Comparative analysis for a Colombian case. *Energy Journal* 33:14.

Cardona Alzate, C.A., E. Felix, and A. Von Brandt. 2009. Bio-energy and food security: An environmental, technical and economic analysis for developing countries. *Biofuels in Latin America: Ongoing Research, Experiences and Potential for the Region.*

Cohen, M.J. et al. 2008. Impact of climate change and bioenergy on nutrition. FAO and IFPRI.

Deike, S., B. Pallutt, and O. Christen. 2008. Investigations on the energy efficiency of organic and integrated farming with specific emphasis on pesticide use intensity. *European Journal of Agronomy* 28 (3):9.

FAO. 2008. BEFS project to analyze linkages between bioenergy and food security. http://www.fao.org/nr/ben/befs/index.html

Food Outlook. 2007. Global market nnalysis 2007. Ed. E. a. S. D. D. FAO.

Foreign Agricultural Service. 2008. GAIN report BR8013. U.S. Dept. of Agriculture, Washington, D.C.

Gressel, J. 2008. Transgenics are imperative for biofuel crops. *Plant Science* 17.

Hall, D.O. et al. 1993. Biomass for energy: Supply prospects. *Renewable Energy: Sources for Fuels and Electricity* 58.

Historic US fuel ethanol production. 2008. Renewable Fuels Association, Washington, D.C. http://www.ethanolrfa.org/industry/statistics

Kim, S., and B.E. Dale. 2004. Global potential bioethanol production from wasted crops and crop residues. *Biomass and Bioenergy* 14.

Koga, N. 2008. An energy balance under a conventional crop rotation system in northern Japan: Perspectives on fuel ethanol production from sugar beet. *Agriculture, Ecosystems & Environment.* 125 (1–4):9.

Lal, R. 2008. Crop residues as soil amendments and feedstck for bioethanol production. *Waste Management* 28 (4):11.

Lin, Y., and S. Tanaka. 2006. Ethanol fermentation from biomass resources: Current state and prospects. *Applied Microbiology Biotechnology* 69:627–642.

National Agricultural Statistics Service. 2007. National statistics. U.S. Dept. of Agriculture, Washington, D.C. http://www.nass.usda.gov/QuickStats/

Nonhebel, S. 2007. Energy from agricultural residues and consequences for land requirements for food production. *Agricultural Systems* 94:6.

Reijnders, L. 2008. Ethanol production from crop residues and soil organic carbon. *Resources, Conservation and Recycling* 52 (4):5.

Rosegrant, M.W. 2008. Biofuels and grain prices: Impacts and policy responses. U.S. Senate Committee on Homeland Security and Governmental Affairs, Washington, D.C.

Sánchez, Ó.J., and C.A. Cardona. 2008. Trends in biotechnological production of fuel ethanol from different feedstocks. 99 (13):25.

Tzilivakis, J. et al. 2005. An assessment of the energy inputs and greenhouse gas emissions in sugar beet (*Beta vulgaris*) production in the UK. *Agricultural Systems* 85 (2):18.

U.N. Food and Agriculture Organization. 1996. Rome declaration on world food security. http://www.fao.org/docrep/003/w3548e/w3548e00.htm

13 Perspectives and Challenges in Fuel Ethanol Production

Demand for energy, enough food, and a good environment is the most important concern in the entire world today. Then, the possibility of obtaining a renewable, available, safe, and effective source of energy is one of the challenges that humanity should address. The biofuels, particularly bioethanol, are an environmentally clean source of energy. However, production costs of fuel ethanol are higher than production costs of gasoline. Nevertheless, many groups and research centers in different countries are continuously carrying out studies aimed at reducing ethanol production costs for a profitable industrial operation. Diverse research trends and process improvements could be successful when trying to lower ethanol costs. These research tendencies are related to the nature of utilized feedstocks (looking for the most productive and cheap raw materials), tools of process engineering (mainly process synthesis, integration, and optimization), food security, and environmental impacts.

13.1 FEEDSTOCKS

The three kinds of feedstocks used for fuel ethanol production correspond to resources that are present in almost all the countries. In particular, all populated regions in the world account for vast amounts of lignocellulosic waste materials that eventually can be converted into ethanol. Tropical countries exhibit comparative advantages in the availability of sucrose- and starch-containing feedstocks for ethanol production in comparison to European or North American countries. In fact, the dynamics of the global ethanol market could require these countries to supply the growing demand of those countries that have implemented or will implement ambitious programs for the partial substitution of fossil fuels with renewable liquid fuels. These programs may have dissimilar motivations other than environmental concerns, but humankind and global climate will be benefited in any case.

For the three main types of feedstocks, the development of effective, continuous fermentation technologies with near 100% yields and elevated volumetric productivities is one of the main research subjects in the ethanol industry. To this end, many of newly proposed technologies for reducing the product inhibition effect on the cell growth rate should be scaled up at the industrial level. Additionally, past research tendencies for cell-free ethanol production, using only the enzymes involved in the conversion of glucose to ethanol, may offer a practical and beneficial alternative. This progress should complement the intense efforts oriented to the selection and development of microbial strains with particular traits, such as specific flocculating properties or increased tolerance to ethanol, inhibitors, and salts.

Consequently, an important part of the research trends on fuel ethanol production is geared to the reduction of feedstock costs, especially through the utilization of less expensive lignocellulosic biomass. In general, most of the research efforts are oriented to the conversion of lignocellulosics into fermentable sugars and then to ethanol. One of the key factors for enhancing the competitiveness of the biomass-to-ethanol process is the economic and concentrated access to large quantities of biomass distributed in rural areas of the world. Another important issue in transformation technologies for lignocellulosic feedstocks is the design of low-cost methods for its delignification. After that, the increase in the specific activity of cellulases and the decrease in their production costs play a crucial role in the process costs. The technology of recombinant DNA will provide important advances for the development of the fuel ethanol industry. The development of genetically modified microorganisms capable of converting starch or biomass directly into ethanol and with a proven stability under industrial conditions will allow the implementation of the consolidated bioprocessing of the feedstocks.

The massive utilization of fuel ethanol in the world requires that its production technology must be cost effective and environmentally sustainable. In particular, ethanol production costs should be lowered. For current technologies employed at the commercial level, the main share in the cost structure corresponds to the feedstocks (above 60%) followed by the processing expenditures. In general, the use of sucrose-containing materials such as cane molasses allows producing ethanol with the lowest costs compared to using the starchy materials (mostly grains).

Although the ethanol yield from corn is higher than that from sugarcane, the lower annual yield of corn per cultivated hectare makes it necessary to use larger cropping areas. On the other hand, the lignocellulosic biomass represents the most promising feedstock for ethanol production. The availability and low cost of a wide range of lignocellulosic materials offer many possibilities for the development of bioindustries that could support the growth of the international biofuel market and contribute to the reduction of greenhouse gas emissions worldwide. A summary about research perspectives of feedstocks for ethanol production is presented in Table 13.1.

13.2 PROCESS ENGINEERING

Process engineering could provide the means to develop economically viable and environmentally friendly technologies for the production of fuel ethanol. Process synthesis will play a very important role in the evaluation of different technological proposals, especially those related to the integration of reaction–separation processes, which could have major effects on the economy globally. Similarly, the integration of different chemical and biological processes for the complete

TABLE 13.1

Research Trends and Priorities for Improving Fuel Ethanol Production from Different Feedstocks

Issue	All Feedstocks [a]	Sucrose-Containing Materials	Starchy Materials	Lignocellulosic Biomass
Feedstock	Reduction in costs of feedstocks by improving crop yields, pest resistance, and cropping systems	Increase in crop productivity	Utilization of native starchy material other than cereal grains (cassava, indigenous roots, etc.)	Evaluation of the use of dedicated energy crops
			Development of corn hybrids with higher extractable starch or with higher fermentable starch content	Genetic modification of herbaceous plants for changing their carbohydrate content
			Genetic improvement of corn (e.g., "self-processing grains")	Economic utilization of different and alternative wastes, such as municipal solid waste (MSW)
Pretreatment		Removal of impurities and toxic substances from molasses	Reduction of energy costs of liquefaction	Reduction of milling power
				Optimization of steam explosion and dilute-acid pretreatment
				Development of LHW, AFEX, and alkaline hydrolysis
				Reduced formation of inhibitors
				Recycling of concentrated acids

Continued

TABLE 13.1 (Continued)
Research Trends and Priorities for Improving Fuel Ethanol Production from Different Feedstocks

Issue	All Feedstocks[a]	Sucrose-Containing Materials	Starchy Materials	Lignocellulosic Biomass
Hydrolysis			Low temperature digestion of starch	Increase in specific activity, thermal stability and cellulose-specific binding of cellulases (e.g., by protein engineering) Reduction of costs of cellulases production (10-fold reduction) Cellulases production by solid-state fermentation Recycling of cellulases Improvement of acid hydrolysis of MSW
Fermentation	Continuous fermentation with high cell density and increased yields and productivity	Reduction of inhibition by ethanol Microorganisms with increased osmotolerance or flocculating properties	Recombinant strains of yeasts with increased productivity and ethanol tolerance High cell-density fermentation (e.g., immobilized cells, flocculating yeasts, membrane reactors) Very high gravity fermentations	Increase in conversion of glucose and pentoses to ethanol Recombinant strains with increased stability and efficiency for assimilating hexoses and pentoses, and for working at higher temperatures Development of strains more tolerant to the inhibitors Increase of ethanol tolerance in pentose-fermenting microorganisms

[a] Refers to the three analyzed groups of feedstock: sucrose-containing materials, starchy materials, and lignocellulosic biomass.

utilization of the feedstocks should lead to the development of large *biorefineries* that allow the production of large amounts of fuel ethanol and many other valuable co-products at smaller volumes, improving the overall economical effectiveness of the conversion of a given raw material. Integration opportunities may provide the way for qualitative and quantitative improvement of the process so that not only technoeconomical, but also environmental criteria can be met.

The increasing energy requirements of the world's population will augment the pressure on R&D centers, both public and private, for finding new renewable sources of energy and for optimizing their production and utilization. The use of bioethanol as an energy source requires that the technology for its production from lignocellulosic biomass be fully developed by the middle of this century. This need is much more urgent for those countries that do not have the agroecological conditions for the cultivation of energy-rich crops like sugarcane, as is true with North American and European countries. Even from governmental biofuel programs in the United States, the retrofitting of the ethanol industry from corn starch to lignocellulosic residues (corn stover, woody materials, and municipal solid waste) has been recommended. Some countries, such as Brazil and Colombia, are in an excellent situation in this field considering the great availability of the three types of analyzed feedstocks. Although the more logical option is sugarcane, social benefits for rural communities when other alternative feedstocks, such as cassava or typical agricultural and tropical residues, are taken into account. Here, the process engineering strategy for assessing the real possibilities of tropical countries to develop fuel ethanol production is a real and rapid approach to be used by governments, investors, and decision makers.

Current development of the ethanol industry shows that complex technical problems affecting the indicators of global process have not been properly solved. The growing cost of energy, the design of more intensive and compact processes, and the concern of the populace about the environment have forced the necessity of employing totally new and integrated approaches for the design and operation of bioethanol production processes, quite different from those utilized for the operation of the old refineries. The spectrum of objectives and constraints that should be taken into account for the development of technologies for biofuels production grows wider and more diverse. The socioeconomic component involved during the production of biofuels in the global context should be noted. Practically every country can produce its own biofuel. In this way, the feedstock supply for ethanol production is "decentralized" and does not coincide with the supply centers of fossil fuels. This would make it possible for countries that are now dependent on oil to use biofuels at a high scale and, thus decrease their dependency on fossil fuels. In addition, human development indexes could be improved in two ways: the creation of new rural jobs and the reduction of gas emissions that produce a greenhouse effect. However, ethanol production costs are higher than those of the fossil fuels, especially in the case of biomass ethanol. Nevertheless, during the past two years, oil prices have persistently increased. There is no doubt that the price of gasoline and other oil-derived fuels have a subsidy paid by all taxpayers of the world and that is not necessarily made effective in gas stations. This "subsidy" is intended to compensate the inversions made for maintaining the status quo of international relationships. Logically, we also pay the consequences of the measures taken to "offset" this state of affairs: social instability and, unfortunately, to a certain degree, terrorism.

Therefore, the relatively higher production cost of ethanol is the main obstacle to be overcome. To undertake this, process engineering plays a central role for the generation, design, analysis, and implementation of technologies improving the indexes of the global process, or for the retrofitting of employed bioprocesses. Undoubtedly, process intensification through integration of different phenomena and unit operations, as well as the implementation of consolidated bioprocessing of different feedstocks into ethanol (that requires the development of tailored recombinant microorganisms), will offer the most significant outcomes during the search for efficiency in fuel ethanol production. Great efforts should be focused on the development of consolidated bioprocessing (CBP) of biomass, as lignocellulosics is the most promising feedstock for ethanol production. Additionally, the intensification of biological processes indicates a better utilization of the feedstocks and the reduction of process effluents improving the environmental performance of the proposed configurations. Attaining this set of goals is a colossal challenge to be faced through the fruitful interaction between biotechnology and chemical engineering. The most important and promising research priorities linked to process engineering for improving the global process are briefly summarized in Table 13.2.

Finally, regarding process engineering, this approach has more importance today when oil prices can change drastically depending on not-easy-to-predict factors. Then bioethanol "fashion" should be supported by numbers and projections based on serious studies.

13.3 FOOD SECURITY IMPACTS

Possible competition between bioenergy and food is so complex that many studies, depending on the context (or purposes of the study), will be contradictory. But, many countries (consumers) are dependent on fossil fuels and have to contend with their unstable prices. The negative energy balance has become a problematic issue, and fuel ethanol for automotive transport, for example, represents the unique stable alternative. Additionally, environmental improvements are reached when fuel ethanol is blended with gasoline. Other countries can find in biofuels like ethanol the beginning of an economic development for rural areas.

The main problem we have today in this important discussion is the existence of a large quantity of speculative information about biofuels and food security. Most of this information is incorrectly used for political and economical purposes. The real regulator of land use for food or biofuels is the market. Displacing gasoline demand in the coming years will require the combined development of second-generation technologies and large-scale international trade in ethanol fuel. Without second-generation technologies, large-scale production of ethanol, especially from sugarcane, in developing countries will increase and along with

TABLE 13.2

Research Trends and Priorities for Improving Fuel Ethanol Production by Means of Process Engineering

Issue	All Feedstocks	Sugarcane	Starchy Materials	Lignocellulosic Biomass
Process synthesis	Process synthesis by different approaches (e.g., optimization-based process synthesis)	Integration of ethanol production from sugarcane with facilities for ethanol production from cane bagasse	Integration of corn dry-milling plants with facilities for ethanol production from corn fiber	Precise assessment of available biomass based on access and composition Process flowsheet development considering different pretreatment methods
Process analysis	Improvement of process control and operation (e.g., modeling, nonlinear analysis) Improvement of simulation and optimization tools (e.g., optimization under uncertainty, metabolic-flux models)			Full pilot-plant process analysis, especially for continuous processes

Continued

TABLE 13.2 (*Continued*)
Research Trends and Priorities for Improving Fuel Ethanol Production by Means of Process Engineering

Issue	All Feedstocks	Sugarcane	Starchy Materials	Lignocellulosic Biomass
Process integration	Integration of fermentation and separation processes for reduction of product inhibition Application of membrane technology (e.g., for ethanol removal or dehydration) Energy integration (e.g., by pinch technology)	Development of efficient co-generation technologies using cane bagasse	Development of consolidated bioprocessing (CBP) Recombinant microorganisms for conversion of starch into ethanol	Increase of effectiveness of SSF and SSCF processes (e.g., by improvement of cellulase activity) Development of CBP Increase of ethanol tolerance in microorganisms converting cellulose into ethanol Development of recombinant microorganisms for CBP Develop. of efficient co-generation technology using solid residues of the process (as BIG/CC)
Other process engineering issues	Improvement of environmental performance considering the whole life cycle of bioethanol Production of valuable co-products (retrofit to biorefineries) Integration with petrochemical industry (e.g., ETBE prdn.) and biofuels production (e.g., biodiesel)			

this, new problems related to food security will appear. In order to avoid a non-equilibrium development of a fuel ethanol market based on energy crops, drastic but fair regulations from governments must be instituted.

Each country has to create appropriate rules and laws for developing fuel ethanol programs based on its specific supply and demand characteristics. Additionally, the overall energy policies that include bioenergy should be carefully designed to include all possible energy sources in harmonized and balanced form. It is not logical, for example, that renewable energies coming from the sun and wind are more developed in countries having limited access to these sources than in tropical countries.

With purposes of an open discussion about the future of food security concerns on fuel ethanol production, Table 13.3 presents an overview of the perspectives, challenges, and risks of different topics involved in this issue.

13.4 ENVIRONMENTAL IMPACTS

It is expected that the newer forms of biofuels, including ethanol, could really be cleaner and more efficient than traditional forms of biofuels. Favorable CO_2 balance and other emissions reduction when bioethanol is burned in engines could help mitigate global climate change. Even if aldehyde emissions slowly increase, the use of catalysts normally destroys this contaminant easily.

Biofuel production may introduce new environmental risks and new challenges for the producing countries. This will particularly be the case when natural resource constraints cause greater trade-offs between food production and biofuel production. Respect for the rainforests and all protected areas should be the core of any regulations and policies. There are serious concerns in Latin America and Africa for the deforestation activities.

Fuel ethanol production includes the generation of a large quantity of residues. Effluent treatment will always be an important topic of research. Reduction of air pollution must not cause the increase of soil or water contamination. There are cases of noncompatibility of stillages in Colombia with the soil quality in the sugarcane plantations.

Using ethanol as a vehicle fuel has the potential to reduce nonrenewable energy consumption and greenhouse gas (GHG) emissions. However, acidification, eutrophication, and photochemical smog could increase when compared to using gasoline as liquid fuel. This information should be analyzed for every feedstock, location, and blending to avoid confusion about the environmental performance of fuel ethanol projects.

Life cycle analysis (LCA) enables one to investigate environmental performance of fuel ethanol used in different concentrations in gasoline. This analysis can include a serious waste reduction (WAR) algorithm for numerical calculation of potential environmental impacts. This type of strategy is an important component for identifying practices that will help to ensure that a renewable fuel, such as ethanol, may be produced in a sustainable manner.

TABLE 13.3

Perspectives, Challenges, and Risks of Food Security on Different Topics Involved in Fuel Ethanol Production

Topic	Perspectives of Research and Development	Challenges	Risks	Type of Solution
Feedstocks Amylaceous	New crops for rural development	Practical use of integration technology	Use of cassava for ethanol instead of food in African countries	Policies Technology
Sugar based	Limited increase due to land use competition	Increase of productivity per ha in African and central American countries	Sugar prices could increase dramatically	Policies
Lignocellulosics	Tools for assessing availability and access High quality delignification technologies Overall low cost processing	Quantitative and qualitative analysis	Reductions in soil productivity and possible increases in prices for livestock food	Technology, policies
Overall food security in the world	Market price structure and international trade agreements in fuel ethanol import and export	Definition of fair prices for rural producers Lignocelluloses biomass as a profitable alternative	Best prices of crops for energy than for food use	Policies, technology
Energy security	Integral development of bioenergy policies Other renewable energies (solar, geothermal, wind, etc.) Other fuels development (natural gas, carbon derivatives, etc.)	Strategically integrated use of all energy resources in every country	Countries can find economically more attractive to produce bioenergy instead of food	Policies

The dominant factor determining most environmental impacts, such as greenhouse gas emissions, acidification, eutrophication, and photochemical smog formation, is soil-related nitrogen losses. Usually the source of soil nitrogen can include fertilizers.

However, the most important discussion about environmental advantages of fuel ethanol use is the fact that fluxes of avoided GHG emissions from this biofuel production system are found to be much less than from afforestation or reforestation. In countries like Brazil, Colombia, and Peru, where deforestation is advancing significantly, this issue is a real concern and should be watched closely by the entire world (Table 13.4).

TABLE 13.4

Perspectives, Challenges, and Risks Related to the Environment on Different Topics Involved in Fuel Ethanol Production

Topic	Perspectives	Challenges	Risks	Type of Solution
Environmental concerns	Precise CO_2 balance calculations Emissions analysis for different blending strategies Stillages disposal	New tools for environmental impacts assessment	Negative overall impacts to air, soil, and water	Policies, technology
Land use	Environmental policies for protecting rainforest and water Deforestation analysis Afforestation and reforestation strategies	New *in situ* technologies for using residues Use of nonproductive lands Strategies for stopping deforestation	Use of rainforest by poor countries Negative balance in water cycle use	Policies

Index

A

Additive, 8
Adsortion, 210
Anaerobic digestion, 290
Atmospheric emissions, 315, 334

B

Bacteria, 138
Beet, 48–50, 52
  molasses, 52
  sugar, 48
BEFS. See Bioenergy and Food Security
  Project
Bioenergy and Food Security Project, 359,
  368–375
Biofuels, 1–22
  biodiesel, 4
  bioethanol, 5
  gaseous, 3
  liquid, 4
  solid, 3
Biomass, 98, 148, 168

C

CBP. See Consolidated bioprocessing
Cellulose, 118, 122
Co-fermentation, 181, 224, 270
Composting, 291
Concentration, 199
Consolidated bioprocessing, 149, 241, 384
Culture broth, 199

D

Dehydration, 201, 213
Detoxification, 98–107, 125
Distillation
  azeotropic, 202–206
  extractive, 206–208
  hetero-azeotropic, 233
  saline extractive, 209

E

Effluent treatment, 285, 293, 296,
Emergy, 306
Energy integration, 351

Environmental impact, 315, 331
Enzymes, 160–163, 165, 168,170
Exergy, 307
Extractive fermentation, 272

F

Feedstocks, 43–70, 77–105, 379
  conditioning, 77, 78
  lignocellulosic. See Lignocellulosic
    materials
  pretreatment, 80
  starchy materials. See Starch
  sugars See Fermentable sugars
Fermentable sugars, 43–53
  beet sugar, 48
  cane sugar, 45–48
  sucrose See Sucrose-containing
  sugar beet, 48–50
  sugarcane. See Sugarcane
Fermentation, 133–161
  batch, 157, 236
  continuous, 161, 236
  fed-batch, 181
  inmobilized cells, 163, 165
  lignocellulosic-base media; 168
  modeling, 176–181
  pentose, 170
  saccharomyces cerevisiae, 79, 132, 155
  semicontinuous, 158
  separate hydrolysis and, 168–170, 224, 322
  sucrose-base media, 156
  starch-base media, 166
  very high gravity, 167
Figure of merit, 306
Food security, 359, 384
Fossil Oil, 365

G

GAMS. See Generic Algebraic Modeling
    System, 37
Gasoline oxygenation, 5, 7, 12
  di-isopropyl ether, 7, 11
  ethanol, 12–22
  ethers, 8
  ethyl tert-butyl ether 7, 10
  methanol, 7, 12
  methyl tert-butyl ether, 7, 9
  tert-amyl ethyl ether, 7, 11

tert-amyl methyl ether, 7, 11
tetraethyl lead, 8
Gas stripping, 246
Generic Algebraic Modelling System, 37, 269
Glucolysis, 134

H

Hexoses, 179
Hydroelectric energy, 2
Hydrolysis, 117–126

I

Inmobilized cells, 163–166

K

Knowledge-Based Methods, 30–34

L

LCA. See Life cycle assessment.
Life cycle assessment, 27, 299, 304–306
Lignocellulosic materials, 77–95, 148, 336, 364
    corn stover, 69
    cereal straws, 70
    co-fermentation hydrolyzates, 171
    pretreatment. See Pretreatment methods
    structure, 62
    waste, 70
Liquid extraction, 263

M

Membranes, 249–263
Microorganisms, 131, 134, 140, 148
Multiple steady-states, 182
Mutagenesis, 140

P

Pentoses, 179
Pervaporation, 211, 261
Pretreatment methods, 83
    biological, 95
    chemical, 90
    physical, 84
    physical-chemical, 85
Pressure-swing, 202, 211
Process
    analysis, 30
    design, 27–29, 173
    integration, 221

reaction-reaction, 224
reaction-separation, 244
Process synthesis, 29–40
    knowledge-based, 30–35
        conflict-based approach, 33
        evolutionary modification, 31
        hierarchical decomposition, 32
        phenomena-driven design, 33
        thermodynamics-based, 33
        statics analysis, 34
    optimization-based, 35–39, 348
        hybrid methods, 39
        MINLP, 36, 37
        NLP, 38
        superstructures, 38

R

Rectification, 199
Renewable energy, 1
Residues, 285

S

SHF. See separate hydrolysis and; Fermentation
Simultaneous
    saccharification
        co-fermentation, 240
        fermentation, 225–240, 325
    saccharification, yeast propagation, and
        fermentation, 230
Solid wastes, 285
SPI. See Sustainable process index
SSCF. See Simultaneous saccharification and
    co-fermentation
SSF. See saccharification – fermentation;
    Simultaneous
SSYF. See saccharification, yeast propagation,
    and fermentation; Simultaneous
Starch, 53–61, 143, 322
    cassava, 60, 333
    corn, 56–59, 166, 360
    grains, 59
    pretreatment, 80
    saccharification, 115–117, 166
    VHG. See Very High Gravity; Fermentation
Steam explosion, 87
Stillage, 286–290, 292, 314
Sucrose-containing, 50, 311
    beet juices, 51
    cane, 51
    sugarcane molasses. See Sugarcane
Sugarcane, 43–45, 51, 313, 361–363
    molasses, 51
    bagasse, 68
Sustainable process index, 307

T

Thermodynamic topological analysis, 228, 245
Transgenics, 366

V

Vacuum, 246
Vinasses, 286. *See also* Stillage

W

WAR. See Waste reduction algorithm.
Waste reduction algorithm, 299–304
Wastewater treatment, 293

Y

Yeast, 78, 136

Printed in the United States
by Baker & Taylor Publisher Services

Printed in the United States
by Baker & Taylor Publisher Services